AF764365

Advances in Spatial Science

Editorial Board

Manfred M. Fischer
Geoffrey J.D. Hewings
Peter Nijkamp
Folke Snickars (Coordinating Editor)

Springer

Berlin
Heidelberg
New York
Hong Kong
London
Milan
Paris
Tokyo

Titles in the Series

C. S. Bertuglia, M. M. Fischer and *G. Preto* (Eds.)
Technological Change, Economic Development and Space
XVI, 354 pages. 1995. ISBN 3-540-59288-1
(out of print)

H. Coccossis and *P. Nijkamp* (Eds.)
Overcoming Isolation
VII, 272 pages. 1995. ISBN 3-540-59423-X

L. Anselin and *R. J. G. M. Florax* (Eds.)
New Directions in Spatial Econometrics
XVX, 420 pages. 1995. ISBN 3-540-60020-5
(out of print)

H. Eskelinen and *F. Snickars* (Eds.)
Competitive European Peripheries
VIII, 271 pages. 1995. ISBN 3-540-60211-9

J. C. J. M. van den Bergh, P. Nijkamp and *P. Rietveld* (Eds.)
Recent Advances in Spatial Equilibrium Modelling
VIII, 392 pages. 1996. ISBN 3-540-60708-0

P. Nijkamp, G. Pepping and *D. Banister* (Eds.)
Telematics and Transport Behaviour
XII, 227 pages. 1996. ISBN 3-540-60919-9

D. F. Batten and *C. Karlsson* (Eds.)
Infrastructure and the Complexity of Economic Development
VIII, 298 pages. 1996. ISBN 3-540-61333-1

T. Puu
Mathematical Location and Land Use Theory
IX, 294 pages. 1997. ISBN 3-540-61819-8

Y. Leung
Intelligent Spatial Decision Support Systems
XV, 470 pages. 1997. ISBN 3-540-62518-6

C. S. Bertuglia, S. Lombardo and *P. Nijkamp* (Eds.)
Innovative Behaviour in Space and Time
X, 437 pages. 1997. ISBN 3-540-62542-9

A. Nagurney and *S. Siokos*
Financial Networks
XVI, 492 pages. 1997. ISBN 3-540-63116-X

M. M. Fischer and *A. Getis* (Eds.)
Recent Developments in Spatial Analysis
X, 434 pages. 1997. ISBN 3-540-63180-1

R. H. M. Emmerink
Information and Pricing in Road Transportation
XVI, 294 pages. 1998. ISBN 3-540-64088-6

P. Rietveld and *F. Bruinsma*
Is Transport Infrastructure Effective?
XVI, 384 pages. 1998. ISBN 3-540-64542-X

P. McCann
The Economics of Industrial Location
XII, 228 pages. 1998. ISBN 3-540-64586-1

L. Lundqvist, L.-G. Mattsson and *T. J. Kim* (Eds.)
Network Infrastructure and the Urban Environment
IX, 414 pages. 1998. ISBN 3-540-64585-3

R. Capello, P. Nijkamp and *G. Pepping* (Eds.)
Sustainable Cities and Energy Policies
XI, 282 pages. 1999. ISBN 3-540-64805-4

M. M. Fischer and *P. Nijkamp* (Eds.)
Spatial Dynamics of European integration
XII, 367 pages. 1999. ISBN 3-540-65817-3

M. M. Fischer, L. Suarez-Villa and *M. Steiner* (Eds.)
Innovation, Networks and Localities
XI, 336 pages. 1999. ISBN 3-540-65853-X

J. Stillwell, S. Geertman and *S. Openshaw* (Eds.)
Geographical Information and Planning
X, 454 pages. 1999. ISBN 3-540-65902-1

G. J. D. Hewings, M. Sonis, M. Madden and *Y. Kimura* (Eds.)
Understanding and Interpreting Economic Structure
X, 365 pages. 1999. ISBN 3-540-66045-3

A. Reggiani (Ed.)
Spatial Economic Science
XII, 457 pages. 2000. ISBN 3-540-67493-4

D. G. Janelle and *D. C. Hodge* (Eds.)
Information, Place, and Cyberspace
XII, 381 pages. 2000. ISBN 3-540-67492-6

P. W. J. Batey and *P. Friedrich* (Eds.)
Regional Competition
VIII, 290 pages. 2000. ISBN 3-540-67548-5

B. Johansson, Ch. Karlsson and *R. R. Stough* (Eds.)
Theories of Endogenous Regional Growth
IX, 428 pages. 2001. ISBN 3-540-67988-X

G. Clarke and *M. Madden* (Eds.)
Regional Science in Business
VIII, 363 pages. 2001. ISBN 3-540-41780-X

M. M. Fischer and *Y. Leung* (Eds.)
GeoComputational Modelling
XII, 279 pages. 2001. ISBN 3-540-41968-3

For information about further volumes please see end of the book

Bernard Fingleton
Editor

European Regional Growth

With 110 Figures
and 66 Tables

Springer

Dr. Bernard Fingleton
University of Cambridge
Department of Land Economy
19 Silver Street
Cambridge CB3 9EP
UK

ISBN 3-540-00366-5 Springer-Verlag Berlin Heidelberg New York

Bibliographic information published by Die Deutsche Bibliothek
Die Deutsche Bibliothek lists this publication in the Deutsche Nationalbibliografie;
detailed bibliographic data available in the internet at *http.//dnb.ddb.de*

This work is subject to copyright. All rights are reserved, whether the whole or part of the material is concerned, specifically the rights of translation, reprinting, reuse of illustrations, recitation, broadcasting, reproduction on microfilm or in any other way, and storage in data banks. Duplication of this publication or parts thereof is permitted only under the provisions of the German Copyright Law of September 9, 1965, in its current version, and permission for use must always be obtained from Springer-Verlag. Violations are liable for prosecution under the German Copyright Law.

Springer-Verlag Berlin Heidelberg New York
a member of BertelsmannSpringer Science + Business Media GmbH

http://www.springer.de
© Springer-Verlag Berlin Heidelberg 2003
Printed in Germany

The use of general descriptive names, registered names, trademarks, etc. in this publication does not imply, even in the absence of a specific statement, that such names are exempt from the relevant protective laws and regulations and therefore free for general use.

Cover design: Erich Kirchner, Heidelberg

SPIN 10908938 42/3130 – 5 4 3 2 1 0 – Printed on acid-free paper

Preface

Looking into the future and trying to visualize long-run regional disparities is difficult. There is no agreed single theory to guide our model building, although some insight can be gained from analysis of the past and by looking at the different theoretical positions to see if there is any agreement at least in terms of outcome if not in terms of mechanism. This is what is attempted in this book, which brings together specialists with a common interest in European regional growth and in applying quantitative analytical and simulation techniques in order to provide different perspectives on the topic. Many of the Chapters in the book employ the methods of spatial econometrics. Spatial econometrics is a suite of statistical and econometric tools dedicated to the analysis of spatial data, in other words data that is set within the context of socio-economic theories and which can be represented most informatively in mapped form. These methods have emerged over the last half-century from a series of influential papers and books in the fields of regional science, spatial statistics and quantitative geography, and are ripe for serious application in the context of the EU's development. In particular they provide the means by which to analyze spatially indexed data where the time dimension is essentially suppressed because, for instance the variables of interest are not available or are of questionable accuracy, which is often the case when one is interested in regional, rather than national, variations across a number of countries. In a typical geo-economic analysis the spatial interdependence of regional economies is of paramount importance, due for instance to simultaneous spillovers and externalities extending across regional boundaries, and correct estimation and inference demands that models reflect this kind of interaction. Moreover, unlike time, which is usually of necessity quite attenuated, space is highly differentiated so that there are numerous locationally specific observations. In other words we commonly have a panel in which the number of series greatly outnumbers the number of time periods, which may in the extreme case be just one or two. Panel data analysis is also part of spatial econometrics and panels with cross-sectional correlation are becoming a focus of interest among mainstream econometricians, although typically here the number of time periods would be large and the number of regions quite small, and so given the current state of the art are less appropriate to the type of study reported in this volume. In contrast, cross-sectional methods are more easily adapted to a spatially exhaustive analysis of the EU regions. Together the various methods of spatial econometrics and spatial data analysis provide useful tools for the growing number of economists who are interested in confronting the emerging theories of geographical economics or quantitative economic geography with data as a way of improving theoretical constructs by means of added realism.

Realizing this project would have been almost impossible without the help and stimulation provided by numerous friends and colleagues, and also without the support of my wife and family and my University. In particular I would like to acknowledge the part played by the European Commission which provides financial support for COST action A17, of which I am a member of the Management Committee, and which to date has funded meetings in Brussels, Vanersborg, Ankara, Oslo, Barcelona and Cambridge. I would like to offer a big thanks to all friends and colleagues at these COST meetings who have been influential in an informal way, and also to those who have contributed chapters to this volume. In addition, I would like to thank many colleagues and friends involved with the British and Irish section of the Regional Science Association, the Western Regional Science Association, the Regional Science Association International, and students and colleagues at the Universities of Trento and Cagliari and the European University Institute in Florence, Italy, the Federal University of Minas Gerais in Brazil, the University of Cambridge, UK, the University of California at Santa Barbara and MIT in Cambridge, USA, who have, in one way or another, allowed me to present ideas that appear in my own contributions to this volume. Finally, I would like to most warmly thank the Economics editor Marianne Bopp for her help, assistance and very constructive advice and guidance throughout the production of this book. I would also like to thank the referees who gave expert and useful advice. Last but not least, I would like to thank all the contributors including my co-authors Barry Moore, Danilo Igliori and Enrique López-Bazo for their patience, tolerance and enthusiasm. Their skill in producing manuscripts that were of a high quality and their willingness to keep to strict deadlines and compatible styles has ensured the book's smooth passage from initial conceptualization to final product.

Bernard Fingleton
Cambridge University

Contents

Introduction

Bernard Fingleton

University of Cambridge, UK

From the perspective of the Editor, I see this book as a product of several distinct themes which have motivated and influenced the contributing authors, the outcome being some interesting and useful additions to our understanding of European regional growth. The first theme is the ongoing economic integration of the European continent, which was initially conceived in 1950 as the European coal and steel community under the guiding influence of Jean Monet and which today manifests itself most completely in the form of the Euro-zone and by the progression of the 'candidate' countries towards full membership of the EU. The movement towards European cooperation and integration over the past half-Century has attracted the interest of academics, from diverse disciplines, who have been eager to analyze the assorted causes and consequences of this economic and social phenomenon on a truly continental scale. The contributions in this volume represent some of the latest, but certainly not the last, words on this important topic.

From the perspective of regional economics, economic geography and regional science, the issues of substance relate strongly to the regional implications of the removal of trade barriers and of common interest rates[1]. In particular, when one looks from a more micro-level or regional perspective, the one-size-fits-all interest rate policy evidently has implications that call for compensating instruments, since regions are highly differentiated perhaps in a way that national economies are not, and the exposure of some regions to asymmetric shocks, often in the form of job-destroying impulses, with no recourse to different interest and therefore exchange rates, may be a cost too great to bear irrespective of the benefits. Even if we were to smooth away regional specificities so that symmetric shocks became more prevalent, unemployment would remain a threat, the argument being that with a common interest rate and more-synchronised regional business cycles, Europe as a whole would be more volatile and subject to job threatening interest and exchange rate fluctuations via-á-vis other currencies. These considerations mean that unemployment, whether manifest in official statistics or concealed in the form of withdrawal or exclusion from the labour market[2], which remains the EU's major economic problem, will almost undoubtedly continue to be a serious concern.

---

[1] See King(1997).

[2] In the UK, in the former coalfields, job losses were not matched by rises in unemployment, partly because many men of working age became sickness benefit claimants.

Apart from unemployment, the other big issue is the eve-present threat of inflation. Enhanced competition may keep inflation in check, but monetary union does not mean that inflation will be eliminated or equalized across regions, and this is therefore of major relevance for regional analysis. With the existence of fixed (actual or nominal) exchange rates between regions of the common currency zone, but with differences in real exchange rates reflecting real underlying factors causing differential competitiveness between regions, the inevitable outcome is differential inflation. In the international context the real exchange rate measures relative prices in the same currency of the same good in different countries, as is the case at the regional level. In order to maintain international competitiveness, a country may adjust its nominal exchange rate to offset the impact of domestic inflation and the effects of globalization, and this may be used to support particular regions specializing in manufacturing production. This is not an option within a common currency zone. As King (1987) points out, 'the inflation rate in Bologna may well differ from that of Bari. But we do not notice this because statistics are not published. Even in the monetary union between Belgium and Luxembourg there have been periods in which inflation differed significantly (that is, by several percentage points) between the two countries'. Interest rates will be set by the European Central Bank to satisfy pan-European inflation targets and as a discretionary response to economic shocks. They will not necessarily equate to the needs of specific regions, with great scope for disharmony between what is for the global good and what best suits the region when viewed from the local level. Some regions made be better suited to a more relaxed target as a stimulus to demand in a lagging economy, whereas others may be require higher interest rates to hold down inflation. In either case, some kind of average is unlikely to perfectly satisfy all needs.

Of course, there are many benefits to set against these costs; otherwise the EU would not have gone down this road. More open trade and bigger markets is probably the main benefit, since this may itself promote growth. It is possible that the stimulus to trade as a result of the lowering of barriers will enhance the potential for scale economies and establish better conditions for the creation and diffusion of innovations. The enhanced competition resulting from openness should help put a brake on inflation and boost productivity. There is a prominent[3] body of research (Grossman and Helpman, 1991) supporting such ideas and there are the more obvious benefits to the man-in-the-street associated with reduced transaction costs. The verdict has been returned by most EU members on the economic costs and benefits of monetary union in their decision to join the Euro-zone, but my personal feeling is that the regional perspective is valuable in providing its own emphasis, accentuating the need for supplementary policies targeted at inevitable and perhaps not yet fully appreciated downside effects. The UK, past and present, is a good example of polarized economic development within a single currency area, and of the policies that have and could be used to create a greater alignment of regional economies. As within the Euro zone, common exchange rates and interest

[3] See the Appendix to Chapter 1 where reference is briefly made to endogenous growth theory.

rates are undoubtedly inappropriate to specific UK regions, and this has invited stronger regional transfers and policy instruments to support lagging regions. The experience of the UK in the post-war period up to the mid 1970s, when Keynesian-led ideas of demand management were much in vogue, shows that instruments can be used to bring about major reallocations of economic activity. One such instrument was the use of labour cost subsidies in the assisted areas (the so-called Regional Employment Premium) which effectively amounted to a regional devaluation, though without the disadvantages that, say, a devalued regional currency exposed to fluctuating exchange rates would have incurred. This spread the extra import costs normally associated with devaluation nationally and so avoided the disadvantages associated with national devaluation. Even today, in a low taxation, non-interventionist free-market climate, significant transfers still occur between the regions of the UK, so that spending per capita exceeds taxation per capita in lagging regions, and vice versa in the more prosperous regions. This spending, for instance on social security benefits, and various public services such as education and training and health services, is delivered to regions on a more or less equitable basis even though the tax base might not support them. However, current fiscal policy instruments are quite ineffective in countering the increasing regional polarization of the UK that we see today. Similar concerns that at the pan-European level without stronger regional policies there will be growing inequality and that social and economic cohesion will be jeopardized become apparent from various contributing chapters in this book, which highlight the diverse nature of the EU's regional economies and the need for sensitive and specific management and monitoring at a localized level.

A second theme which runs through the various contributions relates to recent developments in theory and conceptual analysis which have reinvigorated economic geography and regional economics. The most notable of these is the advent of new economic geography theory (Fujita, Krugman and Venables, 1999) and the developments springing from neoclassical growth theory (Solow, 1956, 1994, Swan, 1956) involving aspects of endogenous growth. Growth theory remains strongly influenced by a neoclassical view of how the economy operates, with an emphasis on diminishing returns to capital, constant returns to scale[4] and exogenously determined technical progress. This has remained popular partly because it produces some familiar results for mainstream economists; the output/capital ratio is endogenous so that as labour becomes scarce compared with capital, firms substitute capital for labour until diminishing returns lead to equilibrium[5]. Also, while sophisticated and rigorous, it is also extremely flexible and easily translated into a simple estimable reduced form.

This means that while, in its most basic form, neoclassical growth theory produces some rather unreal predictions when faced with a diverse set of economies, with modification to account for regional diversity, it is of considerable interest to the regional economist, as seen in the work of Barro and Sala-i-Martin (1995).

---

[4] Solow(1994) argues that this simplification is not essential and that the main results are preserved without it.

[5] More detail is provided in Chapter 1.

However, in the particular context of Europe, we see limited factor mobility, particular with regard to labour as a result of significant economic, linguistic and cultural barriers, and there remains an important role for regionally differentiated social networks, institutional and Governmental structures. These have had a tendency to interfere with any inclination there might be for the smooth unalloyed operation of market forces and progression towards a common equilibrium, the key prediction of simple neoclassical growth theory. Rather, empirical evidence given in various Chapters of this book and elsewhere suggests that the European regions have tended to follow diverse paths to their own equilibrium positions.

New economic geography or geographical economics combines, in a timely fashion, the impact of changing barriers to trade and the presence of increasing returns, building on the cumulative causation school but adopting a much more mainstream approach by incorporating assumptions about profit and utility maximisation at the micro-level. The new theory embodies in a formal way the concept of increasing returns, but this in itself is not new and many regional economists have long acknowledged increasing returns to be fundamental to an understanding of persistent regional economic differences. In a space-economy governed by the mechanisms of new economic geography, large agglomerations can emerge as a result of the hidden hand of market forces rather than due to any particular natural advantages, due to the interaction of increasing returns to scale and transportation costs. Once conditions are such that agglomeration processes commence, they tend to be self-reinforcing and the dynamics lead to concentration of activity in the core, to the detriment of the periphery (unless transport costs become so low that distance is irrelevant), as one of several possible stable equilibria. This possibility is of considerable interest in the European context, where increased integration is tantamount to a reduction in trade (or simply transport) costs, and of considerable interest to European policy makers, for whom increasing social cohesion across the diverse regions of Europe is a stated policy goal. This theory is suggesting that precisely the opposite is a logical outcome of the interplay of pure economic forces.

However there are some theoretical and practical limitations to new economic geography. Neary (2001) for instance emphasises the lack of identification within the model, a single parameter controls internal equilibrium increasing returns, the price elasticity of demand and the elasticity of substitution, it becomes impossible to disentangle the separate effects. Also, there is almost exclusive reliance on particular mathematical artifices such as embodied within the Dixit-Stiglitz model of monopolistic competition and iceberg transport costs. There is no strategic interaction between firms, which is at odds with research in industrial economics and with the reality of spatial competition. Also exogenous parameters vary across time and space and this variation is very likely to be in response to other variables excluded from the model (see Fingleton, 2001). In other words endogenous variables are assumed to be exogenous. Furthermore, the theory is in essence a system of simultaneous nonlinear equations, and there are important excluded effects, most notably technological externalities, making it somewhat intractable from the econometric point of view. This means that while some contributions in the book acknowledge a debt to the new theory, it is too restrictive in its current form to provide the precise platform for econometric analysis.

The third context for the contributors to this book is the development of analytical tools enabling the sophisticated visualization of empirical reality and the rigorous confrontation of theory by data. One important aspect of this is the role played by the development of geographical information systems (GIS), which enables the imaging of patterns and simulation outcomes in a way that was not possible until a few years ago. We see the manifestation of this technology at various points in this book, and although it remains in the background it is an important aspect of the work of many regional economists, regional scientists and economic geographers. The second suite of analytical tools comprises econometric methods dedicated to the analysis of spatial data (Anselin, 1988, Anselin and Florax 2003). One of the most important developments in recent years has been the growing recognition that space matters and that spatial effects are more readily recognised not simply as nuisance phenomena to be statistically eliminated in order to produce optimal estimates of parameters of relevance, but as being important in their own right as a manifestation of real empirical phenomena with a basis in theory. This heightened relevance of spatial effects partly derives from the importance attributed to externalities in the contemporary growth literature, notably from technological externalities deriving from knowledge spillovers within regions and across regional boundaries, with knowledge acknowledged as the important driving force of the new information economy of Europe. It also has received a stimulus from the development of new, more precise and informative tools of exploratory spatial data analysis. Several Chapters in the book demonstrate in diverse ways how spatial econometric tools can be applied to enhance our understanding the dynamics and persistent diversity among the regional economies of Europe, and this, plus the reality of Europe's economic integration and the developments of theory, provides a significant part of the conceptual framework used by the contributors to this volume.

Chapter 1, entitled *Models and simulations of GDP per inhabitant across Europe's regions: a preliminary view* (Fingleton), examines the empirical evidence for GDP per capita convergence of the EU's over the period 1987-97. The Chapter provides an initial descriptive account and suggests some possible geo-economic futures, inspired by empirical models based on neoclassical growth theory. The Appendix briefly summarises neoclassical and endogenous growth theory which provide a background for this and some subsequent chapters.

Chapter 2, *An Exploratory Spatial Data Analysis of European Regional Disparities, 1980-1995* (Ertur and Le Gallo) investigates the space-time dynamics of regional per capita GDP and average growth rates over the 1980-1995 period, paying particular attention to the role of spatial interaction. They use recently developed methods of Exploratory Spatial Data Analysis as an initial descriptive step to identify global and local spatial autocorrelation, and thus characterize the way economic activities are located in the European Union and the dynamics of the pattern of location.

Chapter 3, *A Spatial Econometric Analysis of Convergence across European Regions, 1980-1995* (Le Gallo, Ertur and Baumont) emphasises and illustrates the efficacy of spatial econometric models tools used to estimate regional growth models. While the empirical research presented in the chapter does not aim at formally testing economic geography models, the authors also relate their results to Krugman's formalization of inter-regional equilibrium with increasing returns and trade costs (Krugman 1991) and recent theoretical results for the analysis of regional growth. In addition, they carry out simulations using a spatial econometric model to illustrate the diffusion of random shocks across Europe.

Chapter 4 *Spatial Convergence Clubs and the European Regional Growth Process, 1980-1995* (Baumont, Ertur and Le Gallo) emphasises spatial heterogeneity, meaning that economic behaviour is not stable over space. They characterize patterns of economic development by spatial regimes and/or groupwise heteroskedasticity, identifying clusters of rich and poor regions. The spatial regimes are interpreted as spatial convergence clubs, capturing a North-South polarization pattern for the European regions. They show that the convergence process is different within each of the regimes, with no convergence process in the north, and a weak one in the south. The use of spatial econometric tools introduces a spatial spillover effect, with a region's average GDP per capita growth rate positively affected by that of neighbouring regions. They conclude that the geographic environment plays then an important role in the study of growth processes.

Chapter 5 *Spatial distance in a technology gap model* (Caniëls and Verspagen) incorporates space into a technology gap model. The model assumes that regions interact with each other in terms of knowledge diffusion, with regions differing according to their R&D activity and their ability to assimilate knowledge diffusing from other regions. Ceteris paribus, knowledge from regions close by diffuses more easily than knowledge from more remote regions, and using simulations, they illustrate the effect of this on the convergence process.

Chapter 6 *The Spatial Distribution of Labour Productivity in the European Regions. A study in Spatial Econometrics* (Vayá and Suriñach) demonstrate how the findings of traditional, non-spatial, location or inequality analyses can be improved by the use of Exploratory Spatial Data Analysis techniques. The analysis includes a simulation exercise generating different level of activity, with simulated distributions linked to New Economic Geography theory. Their analytical framework is applied to the distribution of labour productivity in the European Union (EU) regions for the period 1975-1996.

Chapter 7 *Trade, Human Capital and Innovation: The Engines of European Regional Growth in the 1990s* (Badinger and Tondl) investigates the growth factors affecting the EU regions in the 1990s. They test the hypothesis that regional growth is determined by endogenous growth factors, trade and technological catching-up, in a growth accounting framework, using a spatial autoregressive econometric model. The estimates suggest that growth of EU regions is positively related to the accumulation of physical and human capital, with innovation activity as well as international technology transfer important for growth. Further, they observe that technological catching-up is promoted by intensive foreign trade, a result which underlines the importance of trade openness for EU regions. They show that a region's growth is closely related to the growth of surrounding regions so that high growth regions tend to cluster together. They also conclude that higher education standards and research activity are appropriate goals for EU regional, technology and social policy.

Chapter 8 *The technology gap and the European regional growth dynamics* (Greunz) contrasts with the most empirical research in this field, which is based on single equation methods, by adopting an explicit multi-equation causal structure. This provides insights not available through the estimation of reduced forms. The model is estimated on the basis of a total of 150 European regions over the period 1989-1996, and using Brussels, Hamburg and Île-de-France as references, it suggests that about 7% of European regions will converge to the level of development achieved by these three best performing European regions. Approximately 82% of European regions move towards their own steady states, which are relatively high for 41% of them. For 26% of European regions, relative backwardness seems to be the consequence of their inability to raise sufficient 'social capability'. This fact could have implications for European socio-economic cohesion and the economic sustainability of European enlargement.

Chapter 9 *Employment Growth of Small Computing Services Firms and the Role of Horizontal Clusters: Evidence from Great Britain 1991-2000* (Fingleton, Igliori and Moore) is one of two that focus on regions within a specific country, examining the impact of a new measure horizontal clustering on 1991-2000 Small and Medium Enterprise (SME) employment growth in the Computing Services sector in 408 local authority areas of Great Britain. The Chapter is set in the context of two overlapping topics that have received particular attention by governments and researchers throughout the world, particularly in Europe. The emergence of local economies based on high-technology clusters and the role of SMEs in the generation of employment. Spatial econometric models indicate that there is a significant association between the intensity of horizontal clustering and employment change, and that local and neighbouring demand and supply conditions and human capital are also important. It is also apparent that external effects spill-over across area boundaries, pointing to the importance of local spin-offs and knowledge flows creating technological externalities that transgress area boundaries.

Chapter 10 *Externalities and local economic growth in manufacturing industries* (Usai and Paci) emphasises the technological milieu of an area as the crucial mechanism explaining industrial concentration, with technology being spatially bounded rather than completely exogenous and freely available in the whole economy. The central issue addressed in the Chapter is the identification of factors determining technology, and they distinguish between several types of externalities and also consider other elements such as the availability of human capital, the characteristics of the social environment, and the accessibility to public infrastructures. Their empirical analysis makes use of a large database of socio-economic indicators for the Italian Local Labour Systems (LLS), comprising 784 self-contained labour market areas covering the whole national territory, which is used to illustrate the spatial heterogeneity of local economic growth. One interesting result from their econometric modelling is the negative endogenous spatial lag, implying that at the local industrial level there are polarisation phenomena at work and that positive externalities are self-contained within the boundaries of local labour systems once a large set of local determinants has been controlled for.

Chapter 11 *Regional Disparities in Income and Unemployment in Europe* (Cappelen, Castellacci; Fagerberg and Verspagen) presents new evidence indicating that the regional convergence process that characterized the European Union from the 1950s onwards came to an end around 1980 and that since then there has generally been little change. They argue that any convergence that has occurred has tended to be at the country level, involving catching up by the relatively poor Southern countries that joined the EU during the 1980s. Moreover, within countries, there has at best been standstill if not divergence. Their econometric analysis finds that the reform of regional policies in the EU has increased growth in the poor regions and reduced dispersion of regional incomes per capita, but with regard to unemployment, the picture is more mixed, and at best regional support plays a small positive role in reducing unemployment taking into account the indirect effects operating via the effect of growth on unemployment. Thus although the change in EU-support policy may have promoted growth, they conclude that if the purpose of regional policies is to promote social cohesion, there is still a job to be done.

Chapter 12 *Regional growth in Western Europe: an empirical exploration of interactions with agriculture and agricultural policy* (Bivand and Brunstad) throws light on apparent variability in regional convergence in relation to agriculture, which is a sector subject to powerful political measures. Agricultural policy interventions have tended to be neglected, a topic they explore in the context of geographically weighted regression (GWR) which is used to assess coefficient variability, missing variables, and other specification problems. Their empirical results give support to the role of agricultural subsidies in accounting for variations in regional growth, arguing that subsidies either prevent or slow down transfer of factors of production to more productive sectors or regions, or distort farmers' incentives to choose the most productive options. With regard to methodology, they argue that GWR estimation shows that the usual strategy of assuming the absence of non-stationarity may hide interesting differences in regional growth patterns, and that GWR should be considered to be part of the full spatial econometrics toolbox required to unravel European regional growth stories.

Chapter 13 *Explaining the distribution of manufacturing productivity in the EU regions* (Fingleton and López-Bazo) simulates manufacturing productivity levels across the EU. The simulations are based on an econometric model influenced by recent urban economic and geographical economics theory, which includes both internal and external increasing returns and spatial externality (spillover) effects. The model implies market interdependence involving a competitive manufacturing sector and producer services under monopolistic competition. The assumption of technological externalities and the presence of cross-region spillovers in the model lead to a specification that is typical of recent approaches in spatial econometrics, which seeks to avoid bias by a specification involving spatial interdependence. The Chapter estimates density functions and stochastic kernels in order to visualize the long-run stochastic distributions under various assumptions. It shows that if the impact and levels of output growth, the technology gap and human capital differences across EU regions remains the same as evident in the recent past, then the long-run distribution is going to become more concentrated, although level of productivity differences will remain in equilibrium. There is evidently no absolute converge on the distant horizon.

References

Anselin L. (1988) Spatial Econometrics: Methods and Models Dordrecht: Kluwer

Anselin L. and R. Florax (eds) (1995) *New Directions in Spatial Econometrics*, Berlin: Springer-Verlag

Anselin L. and R. Florax (eds), (2003) *New Advances in Spatial Econometrics*, Heidelberg: Springer-Verlag

Barro R. and Sala-i-Martin X. (1995), *Economic Growth*, New York: McGraw Hill Inc

Fingleton B. (2001) Theoretical economic geography and spatial econometrics: dynamic perspectives, *Journal of Economic Geography*, 1, 201-225

Fujita M., Krugman P. and Venables A.J. (1999) *The Spatial Economy: Cities, Regions, and International Trade*, Cambridge: MIT Press

Grossman G.M. and Helpman E. (1991) *Innovation and Growth in the Global Economy*, Cambridge: MIT Press

King M. (1998) *The political economy of European Monetary Union*, European Investment Bank Lecture Series, Department of Economics, European University Institute, Florence

Krugman P. (1991) *Geography and Trade*, Leuven: Leuven University Press

Neary J.P. (2001) Of hype and hyperbolas: introducing the new economic geography, *Journal of Economic Literature*, XXXIX, 536-561

Solow R.M. (1956) A contribution to the theory of economic growth, *Quarterly Journal of Economics*, 70, 65-94

Solow R.M. (1994) Perspectives on growth theory, *Journal of Economic Perspectives,* 8, 45-54

Swan T. W. (1956) Economic Growth and capital accumulation, *Economic Record*, 32, 334-361

1 Models and Simulations of GDP per Inhabitant Across Europe's Regions: A Preliminary View

Bernard Fingleton

University of Cambridge, UK

1.1 Introduction

The intention in this chapter is to illustrate the geographical distribution across the EU regions of a selection of important socio-economic indicators and to describe some recent economic growth trends. Some attention is also given to convergence, which might be explained by neoclassical growth theory and a process of diminishing returns (see Barro and Sala-i-Martin 1991), but which might also be observed for regions in the sway of some other convergence-inducing mechanism, for instance the impact of structural funds, the effects of policy instruments on the migration of labour or on capital mobility, or the diffusion of innovations. Bearing in mind these alternatives, the chapter estimates cross-sectional models based on neoclassical growth theory as a basis for exploring via simulation alternative scenarios.

1.2 European Regional Disparities

The economic geography of Europe is characterised by wide disparities in the levels of a number of socio-economic variables that are both a cause of and a response to differences in the growth and levels of income per capita. In this section we illustrate disparities in some of the variables that are evidently linked to per capita income variations.

Consider first unemployment rate variation across the EU as illustrated in Figure 1.1. Long-term unemployment differences are probably one of the most revealing indicators of regional economic welfare variations and are a good if somewhat imperfect[1] measure of the social exclusion that accompanies long term

---

[1] The discouraged worker effect results in workers being excluded from the labour market and becoming economically inactive, as in the ex-coalfields of the UK (see my remark in the Introduction).

absence from the labour market. To show how this varies by region, we use an indicator of long-term unemployment (MUNLTR) defined as the number of long-term unemployed divided by each region's working population (the sum of employed and unemployed). In order to stabilize the year-to-year noise we take the mean of this ratio over the period 1987-1995. This also has the advantage of eliminating missing values and thus providing comprehensive coverage across all 200 regions. Regions with high levels of persistent unemployment are likely to be ones in which part of the workforce has become deskilled and this negative impact on the availability of human capital could be a cause of poor growth, as well as also being a signal that such regions have had a poor economic performance in the past. An important feature of Figure 1.1 is the significant spatial variation in long term unemployment rates, with a high concentration in peripheral and Southern regions.

A second dimension of human welfare is health. It is revealing therefore to look at an indicator of public health, which we define as the mean mortality (in thousands, averaged over the period 1985-95) of a region divided by the total population aged 60 and above. Since we are endeavoring to control for the effects of age, this reflects regional differences in infant mortality and other types of premature death, which may be associated with qualitative differences in lifestyle, diet, climate, and employment conditions, and may reflect differences in wealth and hence spending on healthcare. In comparing Figures 1.2 and 1.1, it appears that Ireland, part of Western Scotland, Eastern Germany and Northern France in particular have endured both high levels of long term unemployment and low levels of public health.

Another indicator of interest is the relative size of the dependant population given by the share of the total population less than 20 years old or 60 and above as an average of each year in the period 1985-95. Areas with high shares, such as some of the more rural regions highlighted by Figure 1.3, certainly have different social and economic conditions to those in which a larger share of the population is of an age that can potentially be self-sufficient and require less in terms of fiscal transfers for education and health services. These regions typically have larger proportions of children, students or people who have retired, and so it is likely that the economic burden on individuals in work will be higher, both as a result of family circumstances and possibly also because of higher taxation, although there will also be a net transfer of resources from national governments or from the EU to support local health and education services.

Apart from providing insights regarding the different social and economic circumstances across the EU, these variables also provide a more specific indication of the levels of human capital available for production in each region. However, probably the most important indicator of the available human capital is the skill of the workforce as given by the level of educational attainment of the population. We show in Figure 1.4 the reported share of persons aged 25-59 with a high level of educational attainment in 1999 (EDUC_H). Although there are undoubtedly variations due to varying national standards, it is none the less revealing that regions with high levels of attainment are those urbanised, non-peripheral regions which one would consider to be the productive core of Europe. Consider also the economic structure of each region. Regions specialised in high value added manufacturing, research and development and service activities will also have workforces with commensurate skills. We illustrate this dimension of human capital in a negative way by mapping the share of total employment working in agriculture in 1987 (see Figure 1.5). While this should not axiomatically result in lower human capital in these other sectors, there is a broad correlation that is worthy of note. In addition, there are also evidently regional variations in creative skill, effort and opportunity that should be considered when analysing economic performance. New inventions of products and processes is often associated with the development of new productive capacity with obvious consequences for the economic vitality of a region, although inventiveness is also likely to be a consequence of the capacity of firms and institutions to invest in research and development, and therefore cannot be simply seen as a causal factor. Figure 1.6 shows the number of European patent applications per million inhabitants, averaged over the years 1996-97-98 (PAVE[2]) which is taken from the REGIO database measuring research and development activity. It is apparent that it is the already well-endowed regions that are making the greatest effort to ensure that they remain at the forefront by investing most in research and development.

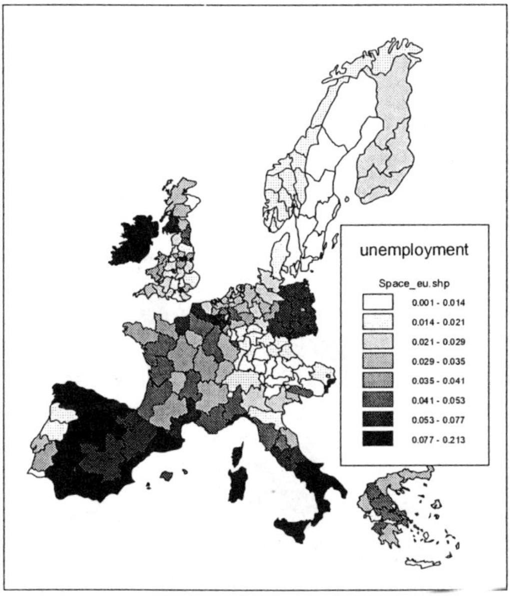

Fig. 1.1 Unemployment (long term)

[2] See Appendix for the method of construction of this variable.

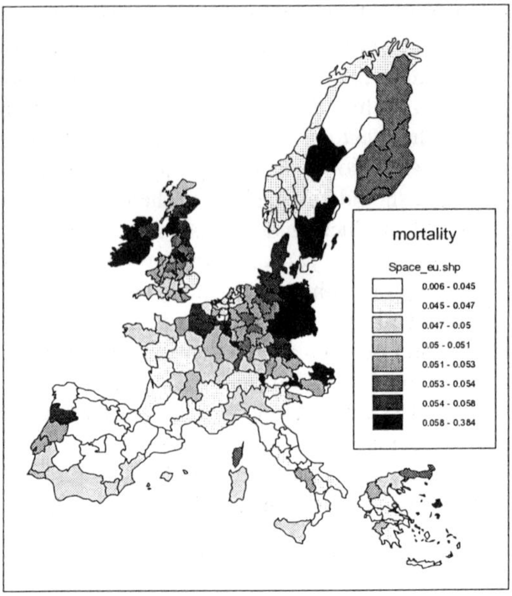

Fig. 1.2 Mean mortality

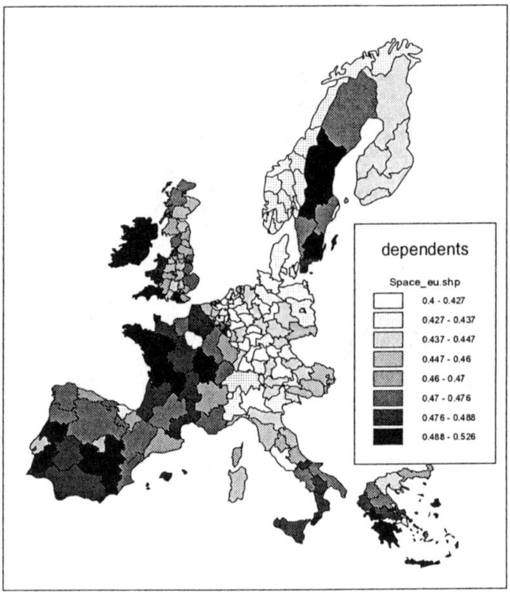

Fig. 1.3 Share of the total population less than 20 years old or 60 and above

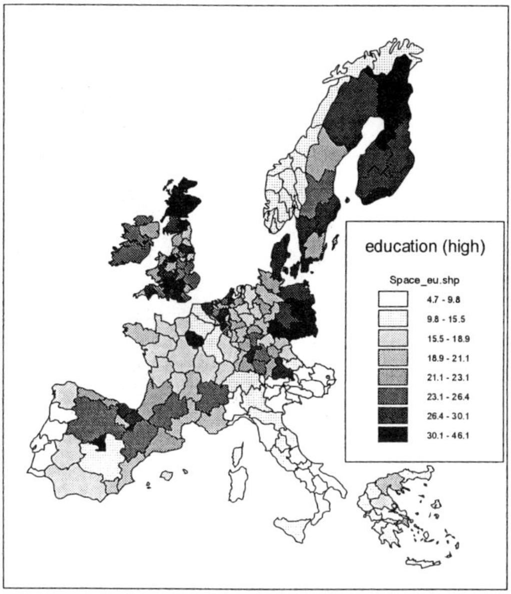

Fig. 1.4 Share of persons aged 25-59 : high level of educational attainment

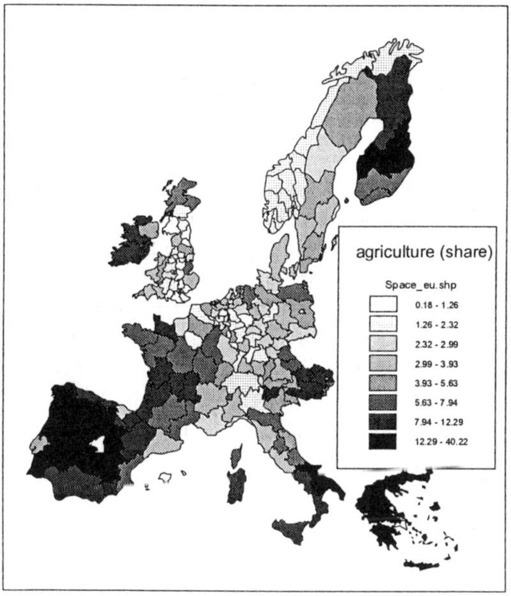

Fig. 1.5 Share of total employment in agriculture

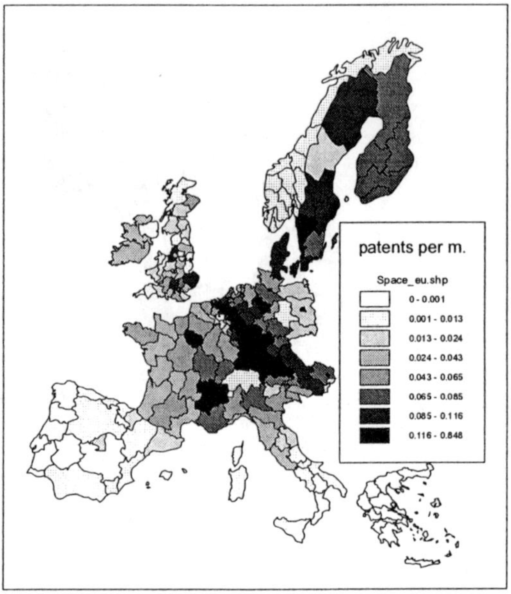

Fig. 1.6 Patent applications per million inhabitants

1.3 Recent Trends in GDP per Capita

Recent trends in the evolution of regional disparities across the EU can be succinctly captured by distribution maps and by simple statistical measures of dispersion. This is the intention of the preliminary descriptive analysis presented in this section, which is a precursor to subsequent model-based analyses where we try to identify some causes of differential regional growth. The description of recent trends is based on real GDP per capita levels (both PPS adjusted and unadjusted) for each year over the period 1987-1997, covering 200 (principally NUTS 2) regions of the EU[3]. These data give us Figures 1.7 and 1.8 which are 'snapshots' of the spatial distribution[4] of relative GDP per capita levels in 1987 and 1997, and these clearly illustrate the marked and persistent differences between regions which have been the focus of EU and national regional policies for many decades. It appears that the forces that are maintaining these regional discrepancies are about as strong as those working to break them down.

---

[3] See Appendix, and also Chapter 2 for a critical appraisal of the construction of PPS series.

[4] The maps divide the distributions according to quantiles, with 8 groups each consisting of 25 regions.

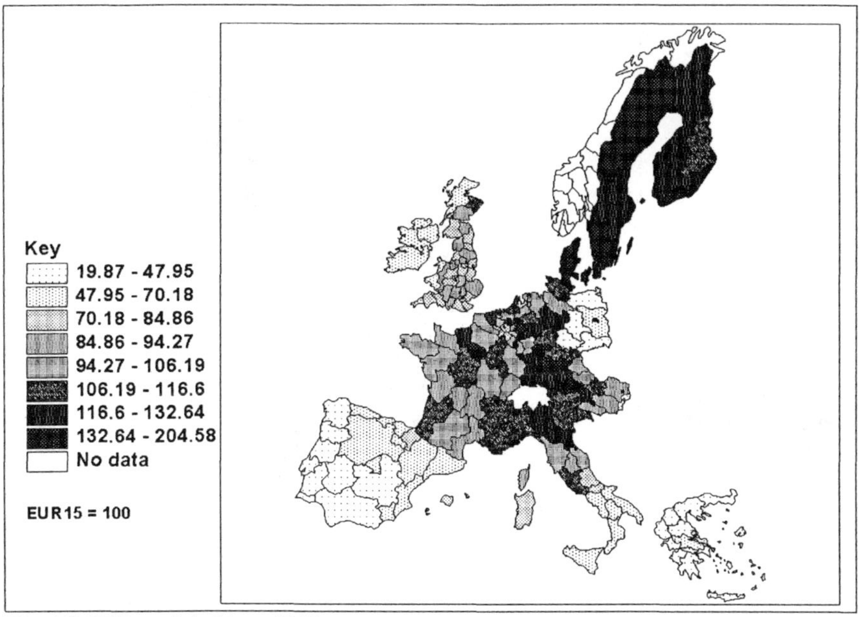

Fig. 1.7 GDP per inhabitant 1987

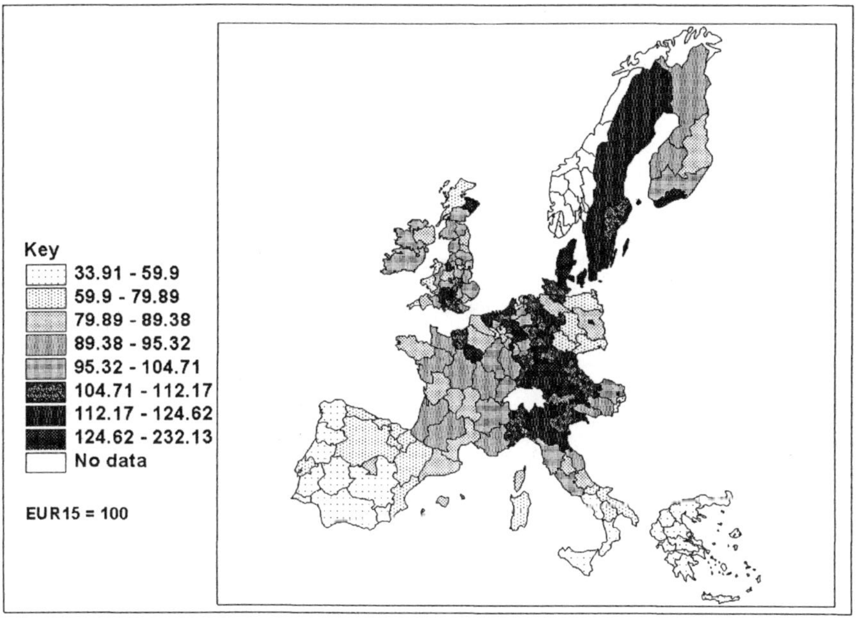

Fig. 1.8 GDP per inhabitant 1997

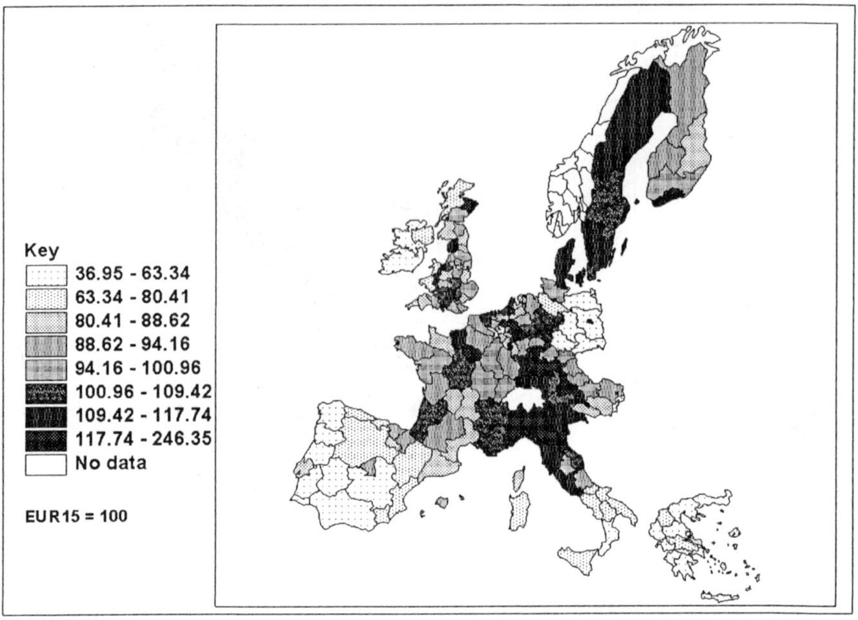

Fig. 1.9 PPS adjusted series 1987

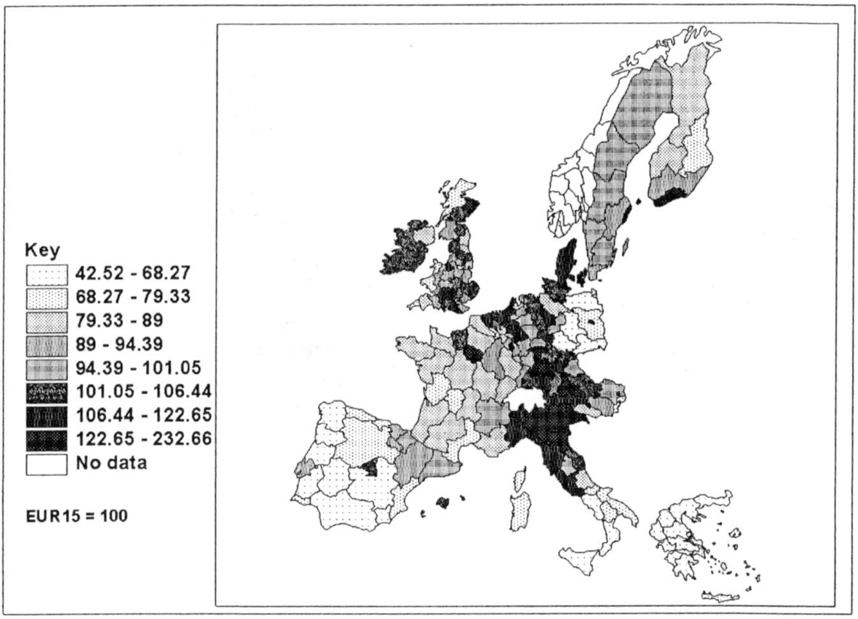

Fig. 1.10 PPS adjusted series 1997

However there are differences between the distributions which indicate that there has also been a degree of convergence in levels over the period. For instance, looking at the key to Figure 1.8, we can see that the lower bound for the richest group for 1997 is 124.62 and the upper bound for the poorest group is 59.9, whereas if we look at the 1987 data given in Figure 1.7 we see a wider gap between the bounds of the rich and poor regions. However this narrowing of interregional differences is not so apparent from Figures 1.9 and 1.10 which are based on the purchasing power standard (PPS) adjusted data.

1.3.1 The EU Regions – σ Convergence and Related Dispersion Measures

As a complement to the maps, and in order to provide more focus on the details of change over time, we use some simple numerical measures. An indication of the changing dispersion of income per capita is provided by the standard deviation (σ) of the distribution measured for each year (t), which we use in the form of the coefficient of variation σ/mean_t. In addition we use the Gini coefficient (G_t) to measure the level of inequality in the distribution at different points in time. We also obtain an indication of changing dispersion via the inter-quartile range (IQR), which is the difference between the upper and lower quartiles of the *ln* GDP per capita distribution. Clearly as the inter-quartile range narrows, the central 50% of regions become more similar in terms of GDP per capita. Compared to the coefficient of variation and the Gini coefficient, the inter-quartile range is robust because it is unaffected by (perhaps spurious) extreme values, but the corollary is that it is insensitive to changes in the upper and lower quartiles, whereas both $\sigma/mean_t$ and G_t are sensitive to transfers anywhere in the distribution. In contrast the range, since it is simply the difference between the maximum and minimum *ln* GDP per capita, is obviously sensitive to outliers.

Table 1.1 Measures of dispersion for GDP per capita (unadjusted)

year	cv	Gini coefficient	IQR	range
87	0.3628	0.0256	0.5040	2.3319
88	0.3466	0.0235	0.4388	2.0884
89	0.3435	0.0227	0.4169	2.0203
90	0.3399	0.0223	0.4103	1.9345
91	0.3367	0.0216	0.4226	1.9074
92	0.3299	0.0209	0.4372	1.9090
93	0.3390	0.0210	0.4163	1.9487
94	0.3406	0.0210	0.4063	1.9674
95	0.3530	0.0215	0.4489	1.9931
96	0.3314	0.0200	0.4275	1.9206
97	0.3120	0.0184	0.3393	1.9237

Table 1.2 Measures of dispersion for GDP per capita (PPS adjusted)

year	cv	Gini coefficient	IQR	range
87	0.2822	0.0177	0.3185	1.8972
88	0.2744	0.0168	0.2997	1.8872
89	0.2719	0.0165	0.2864	1.8761
90	0.3235	0.0189	0.3663	1.8246
91	0.2779	0.0166	0.3048	1.8992
92	0.2741	0.0160	0.3104	1.6833
93	0.2664	0.0151	0.2794	1.6627
94	0.2640	0.0148	0.2898	1.6499
95	0.2642	0.0147	0.2919	1.6635
96	0.2618	0.0144	0.2962	1.6549
97	0.2615	0.0143	0.2905	1.6997

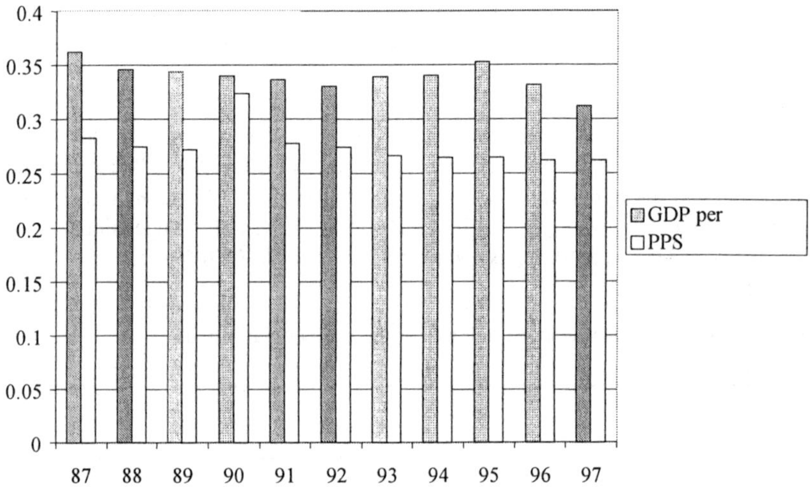

Fig. 1.11 Dynamics of the coefficient of variation, EU regions 1987-97

The time trends in these various measures of inequality and dispersion are given in Tables 1.1 and 1.2, and these support the general thesis of slightly reducing inter-regional disparities, but there are also some subtle variations in these data that are worth highlighting. Figure 1.11 shows that there is some visible evidence that there has been a small amount of convergence in GDP per inhabitant, although the changes over the decade are not dramatic. If we look at the PPS adjusted series, following the peak of 1990, there has been a reduction in dispersion but no evidence of convergence post-1992. In fact, although we have not displayed them, the Gini coefficient series are remarkably similar, with the exception that ongoing convergence is more evident through to 1997 for the PPS data. The inter-quartile range (IQR) for the unadjusted series provides only limited evidence of a systematic convergence process, and follows a fairly uneven path, although the IQR for 1987 is significantly larger than for 1997. The PPS adjustment again has the effect of reducing the dispersion in the data. Overall, there is very little visible evidence of a reduction in dispersion from 1994 onwards.

1.4 Neoclassical Growth Models

Neoclassical growth models are the natural starting point of our analysis of EU regional income disparities in this volume, and therefore in some Chapters the analysis is very much slanted towards this essentially mainstream approach to regional economics. Of course, this does not mean that I, or indeed my colleagues, uncritically accept the neoclassical model of how regional economies function[5], for example the assumption of diminishing returns that drives the neoclassical convergence process is one that is particularly questionable for regional economics. However, there are solid empirical reasons why it makes sense to fit models in which there is a significant convergence process, even if the reasons for this convergence might be debated. In other words, these mainstream models have a sound empirical track record and for that reason they are useful as tools for describing data trends and for simulating alternative scenarios. Later in the book, (see Chapter 13), I embody some of the more recent theory that incorporates increasing returns within a general equilibrium framework involving consistent assumptions at the micro-level, and build on recent work seeking to capture some of the additional technological externalities that are omitted from formal theory.

1.4.1 Theoretical and Empirical Basis[6]

Consider first a single economy with a production function involving two factors, labour and capital, and constant returns to scale (Solow 1956, Swan, 1956). Output from the single sector in this closed economy consists of a homogeneous good which is either consumed or invested, with the investment rate equal to an exogenously given savings rate. Population growth which we assume is equal to

---

[5] Chapter 13 summarizes the theory I find most attractive, at the present moment.

[6] See the Appendix to this Chapter for a fuller account of the theory.

workforce growth is also exogenously determined and for simplicity we assume that this is a constant. The economy inevitably converges on a stable equilibrium at which point the capital stock per worker produces output and hence savings and investment that are equal to the rate of depreciation, so that per worker capital stock is maintained at its equilibrium level. If the economy happens to be above the stable point, saving will be higher because output is higher, but because the marginal product of capital is diminishing, whereas depreciation is linear in capital, the savings schedule will have dropped below the depreciation schedule. The inevitable consequence is that the economy has to move to the stable point to restore the investment-depreciation balance, and any move in the opposite direction will have the effect of making the imbalance more pronounced and unsustainable. Likewise, below the equilibrium point, the savings schedule is above the depreciation schedule and capital stock increases depending on the distance between the two, falling to zero as the economy moves to the unique stable steady state. At this point capital stock, consumption and output each grow at the same rate as population, hence in the absence of technical progress per capita values are constant. This means that the economy cannot experience sustained growth of output per capita unless there are shifts in the production function resulting from exogenously determined technological progress. Moreover, the rate of technological progress determines the long-run rate of growth of the economy.

Equation (1.1) is an approximation to the rate of growth of output per capita which is derived from a linear approximation to the dynamics around the steady state using a Taylor series expansion.

$$p_t = -(1-\pi)(\theta+\phi+\delta)(\ln P_{t-1} - \ln P^*_{t-1}) + \phi \qquad (1.1)$$

In equation (1.1), p_t represents the growth of output per capita at time t, P_{t-1} denotes the level of output per capita at t-1 and P^*_{t-1} is the output per capita on the equilibrium path. Also π is the capital share parameter of the Cobb-Douglas production function, ϕ is the rate of technical progress, θ is the rate of growth of population (workforce) and δ is the depreciation rate. Integrating this equation leads to equation (1.2) which is the convenient reduced form which forms the basis of empirical modelling. In this the left hand side is the annual average exponential growth rate of output per capita over the period from year t-T to t, $\beta = (1-\pi)(\theta+\phi+\delta)$ is the annual rate of convergence towards the steady state, u represents the disturbance term and α is the constant to which the economy converges.

$$\ln(P_t / P_{t-T}) / T = \alpha - (1-\exp(-\beta T)\ln(P_{t-T}) / T + u_t \qquad (1.2)$$

As it stands this theory relates to a single economy or several identical economies, and it is necessary to extend it to a number of differentiated economies in order for it to be useful as a basis for modelling the heterogeneity typical of the

EU regions. Basic neoclassical growth theory assumes that regions tend to the same equilibrium growth path for capital and hence output per capita α, but from the perspective of European regional analysis, this is not very realistic, since to achieve this one has to assume that the capital share, rate of technical progress, the savings rate, depreciation rate, and the population and workforce growth rate are constants across regions. If instead we have, for instance, two regions with different population growth rates[7], the consequence is that the one with the faster population growth rate will tend to a lower level of capital per capita in equilibrium. It will more quickly reach an upper limit of capital per person at which all savings are used up offsetting the effects of depreciation and population growth.

It therefore seems realistic to assume that the European regions will be converging on different steady states. National or regional steady states are commonly modelled as a function of differences in variables such as investment-output ratios, research and development ratios, institutional differences, differences in levels of political stability, different tax policies, subsidies etc, which one might envisage are related to saving and investment and population growth rates, and which pick up discrepancies that might exist between workforce and population growth rates. Additionally, indicators of economic structure have been used (Armstrong 1995, Barro and Sala-i-Martin 1995) to control for the effects of asymmetrical shocks on regional growth rates.

Table 1.3 OLS Estimates of neoclassical reduced forms

Variable	Unconditional	Conditional
Constant	0.3161	0.4386
	(18.13)	(17.30)
LGDPPC87	-0.02832	-0.03954
Log level of GDP per inhabitant 1987	(-15.03)	(-15.31)
MAIRP		0.000497
air passengers per capita		(2.22)
PER_PPS		-0.0001965
peripherality index, circular regions		(-4.47)
EMP_AGRI		-0.000679
share of employment in agriculture		(-4.52)
EDUC_L		-0.0001187
share low educational attainment		(-2.04)
EDUC_H		0.000279
share high educational attainment		(2.76)
MUNLTR		-0.0997
long term unemployment		(-4.20)
PC_8993		0.01504
structural fund spending in Objective 1 regions 89-93		(2.96)

[7] We might also wish to assume that savings and technical progress rates vary across regions.

Table 1.3 (cont)

R^2(adjusted)	53.1%	71.8%
Residual standard error	0.0115	0.00894
β	0.03329	0.05031
(se β)	(0.002628)	(0.004272)
Half Life(LB)	17.98	11.78
Half Life	20.82	13.78
Half Life(UB)	24.72	16.59

For instance, price shocks may have differentiated effects because of regional differences in industrial structure. What this means in practice is that a range of variables have typically been introduced as covariates alongside the initial level of output per capita to explain variation in output per capita growth across regions or countries (see for example Mankiw et al 1992, Barro and Sala-i-Martin 1995, Armstrong and Vickerman, 1995, Levine and Renelt, 1992). While they do not influence the long run growth rate of output per capita, which depends only on the exogenously determined technical progress rate, they influence the growth rate in the short and medium term. Ultimately, since at steady state all economies grow at the same rate[8], the extra variables will cause the steady state level of output per capita to which regions are converging to differ between regions at any single time point. In this set up, the estimated conditional[9] β convergence rate is the rate of convergence of each region to its own steady state.

Introducing additional variables to control for different equilibrium output per capita levels across regions typically produces a negative partial correlation between growth rate and initial level of output per capita as is predicted by basic neo-classical theory. If on the other hand the necessary covariates are omitted, then the correlation is often weak, positive or non-existent.

With a mixed set of regions such as in the EU, a standard approach to the application of the neoclassical model is therefore to include in the reduced form, an appropriate set of covariates that account for the wide interregional diversity that exists. Taking the average annual GDP per capita growth for 200 NUTS 2 (or equivalent) regions of the EU as the dependent variable, we identify the significant variables[10] and include those identified in the initial preferred model which is summarised in Table 1.3. This conditional convergence model specification is further elaborated, but it is essentially the core model of this study. The assumption is that these variables are exogenous determinants of growth. This is justified in the case of educational attainment since although the data post-date (slightly) the period of analysis, it seems unlikely that it will be a response to GDP per capita growth over the period. In Chapter 13, in the context of manufacturing productivity growth, we give

[8] Technology is assumed to be a public good which is freely accessible and therefore invariant across regions.

[9] As opposed to the unconditional β convergence of basic neo-classical theory.

[10] The variables come from several sources, including the EC and Cambridge Econometrics.

some results from Hausman tests of exogeneity that suggest that educational attainment may be treated as exogenous and that any endogeneity or error-in-variables effects will have minimal effect on the estimates produced. Educational attainment is assumed to be primarily a response to external factors such as Government policy and broader social factors. Also, we assume that regional spending lags, rather than responds to, regional economic performance since it depends on indicators of past performance. The other variables (LGDPPC87, EMP_AGRI, PER_PPS) either pre-date the study period, are start-of-period measures and are assumed not to be responses, or are averages (MAIRP, MUNLTR) over a number of years and assumed to be stable indicators of earlier conditions rather than short-run responses to growth over 1987-97.

1.4.2 The Unconditional Convergence Model

The reduced form in equation (1.2) means that regressing annual average GDP per capita on the initial level gives

$$\hat{b}_0 = \hat{\alpha} \qquad \hat{b}_1 = (1 - \exp(-\hat{\beta}T))/T \tag{1.3}$$

so that

$$\hat{\beta} = -\ln(1 - T\hat{b}_1)/T \qquad s.e.(\hat{\beta}) = s.e.(\hat{b}_1)/\exp(-\hat{\beta}T) \tag{1.4}$$

The unconditional convergence model estimates $\hat{b}_0, \hat{b}_1$ are given in Table 1.3, together with the half life and the approximate 95% confidence interval, defined as

$$H = \ln(2)/\hat{\beta} \tag{1.5}$$

$$LB = \ln(2)/(\hat{\beta} + 2s.e.(\hat{\beta})) \qquad UB = \ln(2)/(\hat{\beta} - 2s.e.(\hat{\beta})) \tag{1.6}$$

The results indicate that regressing the average annual growth of GDP per capita on the initial level of GDP per capita (LGDPPC87) accounts for more than half of the variance of the dependent variable; the initial level is a highly significant variable. Regions with lower initial GDP per capita levels saw faster GDP per capita growth over the period 1987-97, as one would anticipate from basic neoclassical theory. We could interpret this catching-up as being due to diminishing returns, but it could also be a consequence of to the diffusion of innovations to poorer regions which are able to benefit more from access to new technology, or due to the impact of regional policies and structural funds.

1.4.3 Conditional Convergence Models

Two other models, fitted in order to attempt to allow for omitted variables, are also of interest. The first is the outcome of using a stepwise model selection procedure involving 172 different models and a large number of potential regressors, and is summarised in the second column of Table 1.3. The third model has the same structural specification, but also models residual autocorrelation and heteroscedasticity (Table 1.4, 1.5). It is evident from both the models that GDP per capita growth remains significantly negatively associated with the initial level of GDP per capita, but the additional covariates suggest that each region is converging to its own, rather than a common, steady state, with an interim effect on growth. The estimates indicate that the growth rate is positively related to the intensity of air passengers per capita (MAIRP) but there is a negative association with peripherality (PER_PPS), showing that increased peripherality tends to be associated with slower GDP per capita growth, controlling for other variables. Likewise, regions specialized in agriculture (EMP_AGRI) and with a comparatively large proportion with low educational attainment (EDUC_L) also grow more slowly, while those with a higher level of educational attainment (EDUC_H) grow more rapidly. The significance of long-term unemployment (MUNLTR) is taken alongside the significance of educational attainment as an additional indicator of how low levels of human capital reduce GDP per capita growth. Finally, structural fund spending over 1989-93 (PC_8993) also appears to contribute significantly to GDP per capita growth, although spending in 1994-99 is not significant. This might be due to a delayed impact, or the fact that the Pearson product moment correlation coefficient for the two spending variables is 0.919 so that its significance is obscured.

Table 1.4 Autoregressive errors model (ML estimates)

variable	**coefficient**	**s.e.**	**ratio**	**probability**
CONSTANT	0.407479	0.0267645	15.224645	0
LGDPPC87	-0.0369995	0.00267701	-13.821242	0
MAIRP	0.00057639	0.000197037	2.925286	0.003441
PER_PPS	-0.000220315	5.18E-05	-4.252084	0.000021
EMP_AGRI	-0.000319488	0.000140741	-2.270042	0.023205
EDUC_L	-4.12E-05	7.43E-05	-0.554143	0.579481
EDUC_H	0.00052411	0.000123369	4.24832	0.000022
MUNLTR	-0.117531	0.0231491	-5.077151	0
PC89_93_	0.010746	0.00520063	2.066289	0.038801
ρ	0.513115	0.0677566	7.572921	0
R^2	0.6825		Likelihood	682.223

Table 1.5 Groupwise heteroskedasticity (ML estimates)

variable	coefficient	s.e.	ratio	probability
CONSTANT	0.415602	0.0266564	15.591055	0
LGDPPC87	-0.0378113	0.00266596	-14.182996	0
MAIRP	0.000573459	0.000194059	2.955079	0.003126
PER_PPS	-0.00022702	5.26E-05	-4.313759	0.000016
EMP_AGRI	-0.000316091	0.000137213	-2.303646	0.021243
EDUC_L	-4.18E-05	7.40E-05	-0.564548	0.572381
EDUC_H	0.000524957	0.000123677	4.244569	0.000022
MUNLTR	-0.118496	0.0226656	-5.228011	0
PC89_93_	0.00991286	0.00509765	1.944594	0.051824
ρ	0.519317	0.0672681	7.720098	0
R^2	0.6966		Likelihood	690.459
σ^2 core	5.55E-05			
σ^2 periphery	6.47E-05			

1.5 Convergence

We measure the degree of convergence at an intermediate stage on the way to equilibrium, and at equilibrium, using three alternative indicators, namely the coefficient of variation, the Gini coefficient, and the share of the EU population at or below 75% of mean GDP per capita. The projected distributions are obtained from the regression models described above. To show how we obtain equilibrium[11] from a model, it is convenient to work with matrix notation, so that the expected value of the dependent variable is represented by

$$E(p) = E(\ln(P_t / P_{t-T}) / T) \tag{1.7}$$

The matrix of regressors is the n by k matrix X, and the regression coefficients are represented by the k by 1 vector b, so that

$$E(p) = Xb \tag{1.8}$$

[11] See equations 1.23-1.26 and Chapter 13 for related material on this topic.

The equation sequence (1.9) to (1.12) represents one round of iteration, and it is a simple matter to cycle through the sequence a number of times to see the evolution of the vector $E(p)$. In the sequence P denotes GDP per capita level and P^* represents the GDP per capita level of the leading region. In each cycle the matrix X changes as the variable $\ln(P)$, which forms one of the columns (v)of X, changes.

$$E(p_t) = X_t b \tag{1.9}$$

$$P_{t+1} = P_t \exp(E(p_t)) \tag{1.10}$$

$$P^*_{t+1} = P^*_t \exp(E(p^*_t)) \tag{1.11}$$

$$X_{t+1,v} = \ln(P_{t+1}) \tag{1.12}$$

The determining factor controlling the development of vector $E(p)$ is the coefficient denoted by b_v which corresponds to $\ln(P_{t+1})$. If $b_v < 0$ then a higher level produces slower growth and vice versa, and cycling through equations (1.9 to 1.12) for $t = 1$ to $t = 100$ (for example) has the effect that $E(p_t)$ tends to a constant across regions, and consequently $R_t = P_{t+1} / P^*_{t+1}$ tends to the constant vector R^e.

At equilibrium, there is no change in 'regional pecking order' so that the ratios, contained in vector R^e are constant with the highest level region[12] having a ratio of 1 and most other regions having ratios fixed at some value less than 1.

Assuming $b_v < 0$ we have an equivalent analytical solution, which is that

$$R^e = \exp\left[\left(X^* - X)b^\bullet\right)\left(b_v\right)^{-1}\right] \tag{1.13}$$

in which X^* has rows equal to the GDP per capita leader's row, and $b^\blacklozenge$ is b with the cell corresponding to b_v set to zero (see Fingleton, 2000, 2002, for related results). Thus the equilibrium vector depends on the difference between the level of the covariates in each region and the leading region, and on the estimated coefficients[13], so that for example faster convergence also causes regions to be closer to the leading region.

The three models discussed above provide input into the equation sequence, with the preferred estimates coming from the model in Table 1.5. Figure 1.15 shows how the coefficient of variation evolves, first through the historical period (the data here are identical to those in Table 1.1 for 1987 and 1997) and then via the iterations to equilibrium. Evidently, the equilibrium has a much less dispersed distribution than the period 1987-97. There is also a reduction in the share of the population living in regions with GDP per capita below 75% of the EU mean. In 1987 the share was 28.89% falling to 20.81% by 1997. Figure 1.15 shows that at

---

[12] In fact the initial leading region may be overtaken so that some values can exceed 1.

[13] The estimated ρ has no direct effect on the equilibrium vector.

equilibrium the share is approximately 15% of the EU population. We see also that the Gini coefficient converges to a stable value which also shows a less unequal set of GDP per inhabitant values in equilibrium. The dynamics of the three indicators in Figure 1.15 shows that equilibrium is effectively achieved by about 2040.

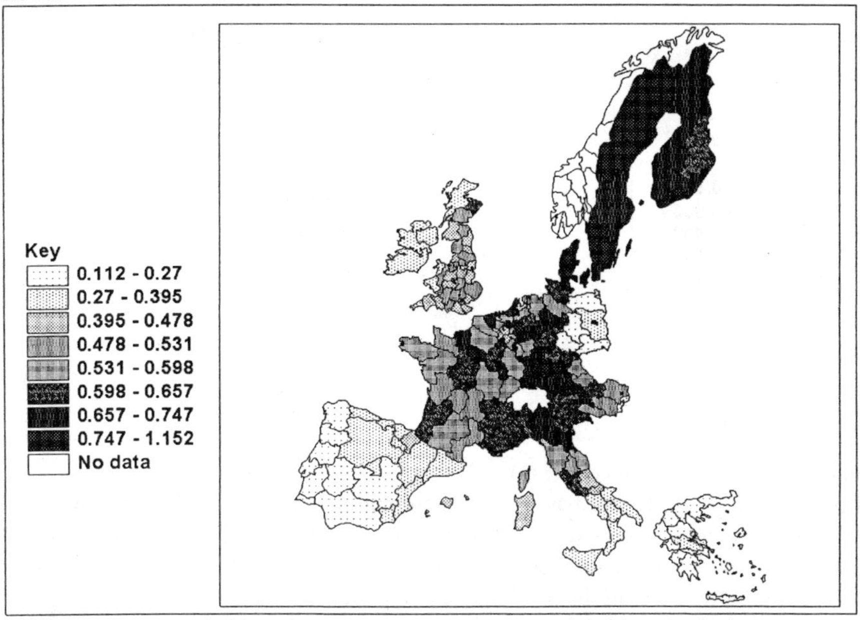

Fig. 1.12 GVA per inhabitant in 1987 as a ratio of GDP per inhabitant in Ile de France

The distributional changes implied by the model are also illustrated by the contrast between the distribution of the ratio *R* in 1987 (Figure 1.12) compared with R^e (Figure 1.13). A notable feature of these maps is that while the inequality as indicated by the coefficient of variation and share of the population living in the lowest income level regions reduces progressively through time to equilibrium, the spatial pattern at equilibrium is very polarized with an obvious core-periphery pattern. The per-region differences between the 1987 and equilibrium ratios are given by Figure 1.14, highlighting those regions projected to see comparatively large changes in relative GDP per capita. It is apparent from Figure 1.14 that the majority of regions see a positive difference between their 1987 ratio and the projected equilibrium ratio, with regions in Britain, the eastern Lander of Germany, the Low Countries, parts of Spain and Portugal and Greece showing the largest improvements in their GDP per capita relative to the Ile de France.

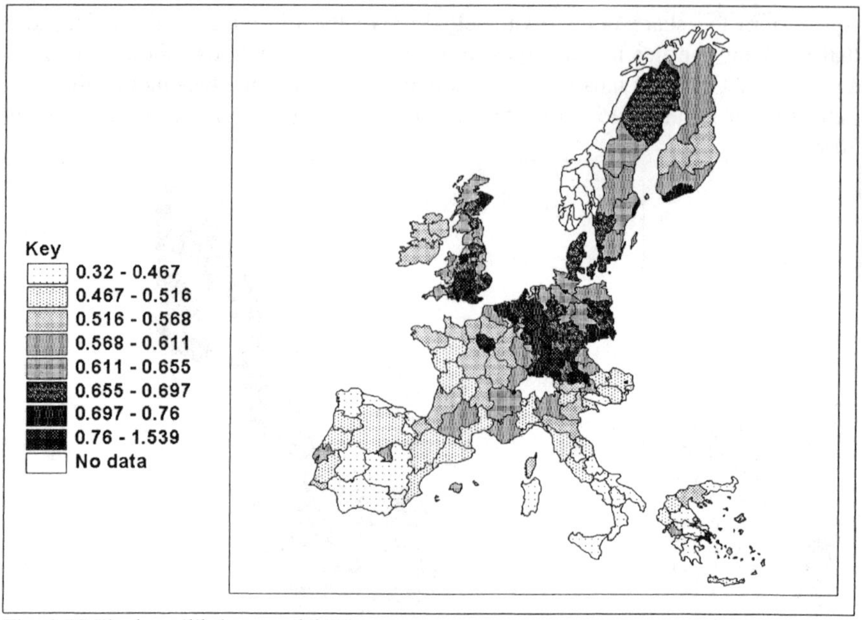

Fig. 1.13 Final equilibrium model (2)

Figures 1.13 and 1.15 are based on an assumption that the variables remain at their 1987-97 levels. The next section continues the study of the implications of this preferred model using simulations of equilibria, which are the outcome of alternative assumptions about the levels of key variables.

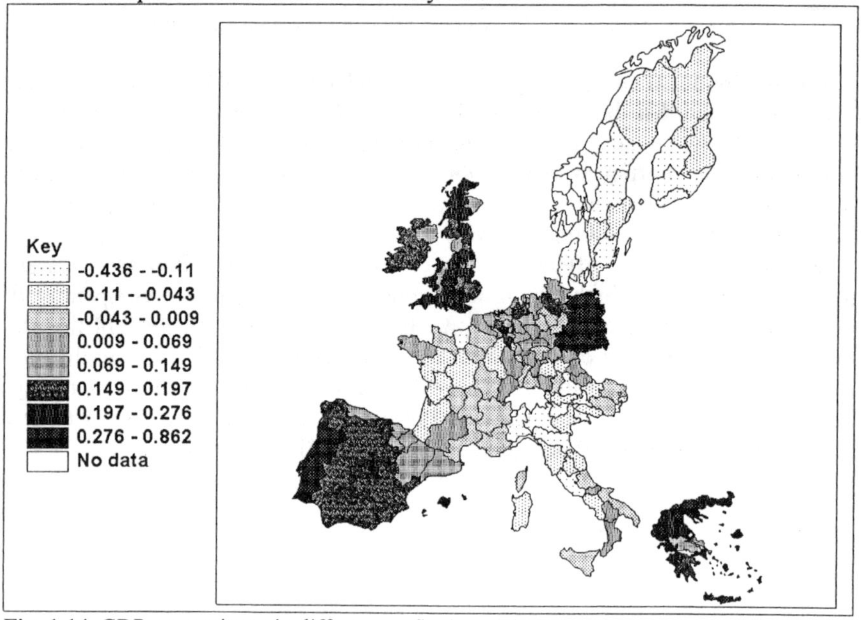

Fig. 1.14 GDP per capita ratio differences, final model minus 1987

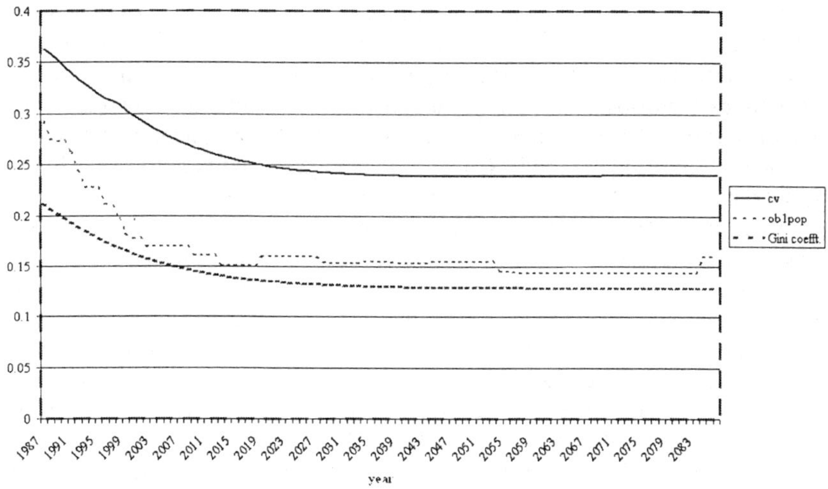

Fig. 1.15 Convergence indicators – final model (2)

1.6 Simulations

The evidence presented above suggests that regions will converge to some extent, although quite large differences are projected to remain at equilibrium. Of course the equilibrium that has been illustrated is based on an assumption that the current regional disparities in the driving variables will remain unchanged in the future. In reality this will not be the case, and in this section we therefore simulate some alternative equilibrium distributions by looking at the impact of alternative assumptions about variables that are, to some extent, under the control of policy makers. First we consider the impact of reduced spending in the regions that were in receipt of structural funds under Objective 1. Two alternative scenarios are played out, one involves spending levels reduced to 50% of the 1989-93 levels, the other looks at the consequences of cutting structural fund spending to zero[14]. Table 1.6 shows the results in terms of the three measures of inequality. It is not surprising that that the effect of the reduction in structural fund spending by 50% increases disparities between regions using all three measures, and that eliminating it en-

[14] This is not an unreal assumption. For 2000-2006, some regions in Ireland, Belgium, Netherlands, Scotland, France, Portugal, Spain, Sweden and Italy see a phasing-out of Objective 1 status.

tirely further widens disparities. For instance equilibrium with zero structural fund spending increases the share of the EU population living in poor regions to 17.54% and the standard deviations goes from 24.04 to 26.48. This is reinforced visually by the maps of the equilibrium distributions under these two assumptions. Figure 1.16 gives the equilibrium distribution of GDP per inhabitant ratios (vis-à-vis Ile de France) under model (2) with spending reduced to 50%, and Figure 1.17 is the outcome under the zero spending assumption.

In reality, structural fund spending is unlikely to be eliminated completely in this way, but the simulations remain useful as an indication that structural fund spending is of undoubted importance in limiting income disparities between regions. However, the impact of reducing or eliminating structural fund spending should be seen in the context of removing the effect of human capital differences between regions. To show this, for models (1) and (2) a scenario is created in which human capital levels are set equal across all regions, where human capital is represented by the levels of educational achievement and long-term unemployment. In order to eliminate differences between regions, both of these variables are set to their EU mean. Table 1.6 summarises the outcome, indicating that there would be a stronger effect on reducing the disparity in regional income levels and the proportion of the population in poor regions than the effect of eliminating structural fund spending. For example, under model (2), the proportion falls from 15.96% at equilibrium assuming current human capital disparities, to a negligible 0.53%. Likewise, there is a marked difference between the map, Figure 1.18, which is produced by this scenario, and Figures 1.16 and 1.17. While it is unlikely that we would ever be able to eliminate human capital differentials across the EU's regions, the indication is that stronger policies aimed at increasing levels of human capital in lagging regions could have a significant impact on reducing income per capita disparities.

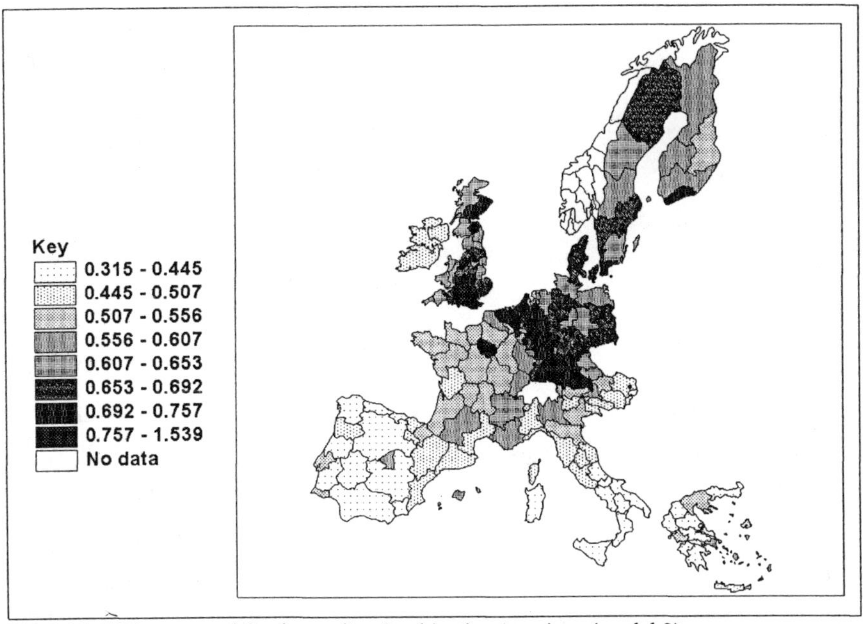

Fig 1.16 50% structural fund spending in objective 1 regions (model 2)

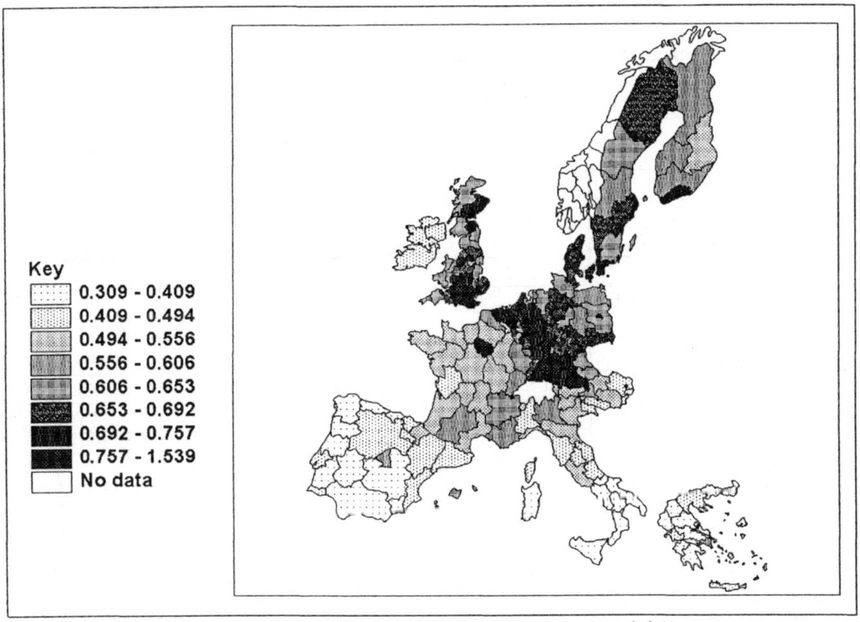

Fig 1.17 Zero structural fund spending in objective 1 regions (model 2)

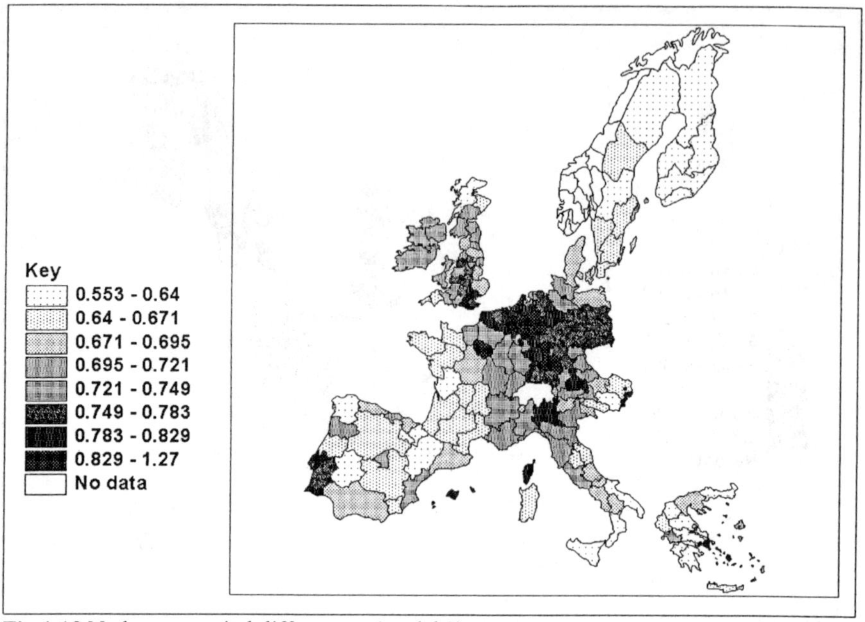

Fig 1.18 No human capital differences (model 2)

Table 1.6 Convergence and simulation analysis

model/period	half-life	σ	Gini	%.75% EU mean
1987 GDP (186 reg.)				25.58%
1997 GDP(186 reg.)				18.83%
1987 GDP (200 reg.)		36.28	0.2122	28.89%
1997 GDP (200 reg.)		31.20	0.1718	20.81%
CONVERGENCE BY 2007				
model (0)		23.91	0.1317	16.42%
model (1)		26.54	0.1462	16.04%
model (2)		27.24	0.1490	16.97%
EQUILIBRIUM				
model (0)	20.82	0	0	0
model (1)	13.78	21.79	0.1217	12.94%
model (2)	14.59	24.04	0.1288	15.96%

Table 1.6 (continued) Convergence and simulation analysis

model/period	half-life	σ	Gini	%.75% EU mean
SIMULATIONS				
(1) 50% structural fund spending		23.55	0.1326	15.88%
(1) zero structural fund spending		25.50	0.1446	17.03%
(1) no human capital differentials		14.33	0.0768	1.44%
(2) 50% structural fund spending		25.21	0.1369	15.67%
(2) zero structural fund spending		26.48	0.1456	17.54%
(2) no human capital differentials		13.67	0.0699	0.53%

model (0) is the unconditional convergence model (see Table 1.3).
model (1) is the conditional convergence model (Table 1.3).
model (2) is the SAR error model with groupwise heteroskedasticity (see Table 1.5).
1. The Gini coefficient here is a measure of the inequality in the ratios of GDP per capita to GDP per capita in the Ile de France. Its value approaches zero as the ratios approach unity across all regions.
2. The standard deviation is equal to 100 times the coefficient of variation in this case, since it is calculated from data with mean equal to 100.

1.7 Conclusion

This Chapter sets the scene for the rest of the book in the form of an empirical analysis inspired by mainstream, neoclassical growth theory and based on the methodology and tools of spatial econometrics. It therefore provides a background to subsequent Chapters which to varying degrees adopt, either explicitly or implicitly, a similar theoretical and empirical stance, or which take the critique and limitations of the neoclassical approach as their starting point. The emphasis on empirics extends to simulations intended to highlight possible geo-economic futures for the EU, although there are many caveats that have and could be introduced into the discussion that should be heeded, since explanations can be contested and forecasting is a risky business! With this proviso, the empirical analysis suggests a process of conditional convergence for the EU's regions, with convergence to region-specific trend growth paths involving still quite marked discrepancies between GDP per capita levels, although disparities will be less than is presently the case. The analysis also highlights the role of regional assistance in maintaining a degree of economic cohesion, and the importance of the role that could be played by advances in educational attainment and human capital in general in promoting the development of lagging regions.

1.8 APPENDIX

1.8.1 Neoclassical Growth Theory

In order to outline the basic theory, assume that the level of output at each point in time t in a given region, Y_t, is given by

$$Y_t = K_t^{\alpha}(A_t H_t)^{\beta} \tag{1.14}$$

in which K_t denotes the level of capital at time t, A_t is labour augmenting total factor productivity, and H_t is skilled labour. The coefficient α is capital's share in income; augmented labour's share is assumed to be β which is equal to 1-α so that we have diminishing returns to both inputs. In order to trace the dynamics, assume also that technology (A) evolves with an exogenously determined growth rate equal to g, and that skilled labour is a function of raw labour (L), number of years schooling (c) and the rate of return to a year of schooling (ϕ) (following Jones, 1997), with raw labour growing at a rate (n) equal to the population growth rate. The level of capital at time t is determined by the level of investment at t-1 and the depreciation rate d applied to K_{t-1}. The level of investment I_t equals the level of GDP multiplied by the investment or savings rate s.

$$A_t = A_{t-1}\exp(g) \tag{1.15}$$

$$L_t = L_{t-1}\exp(n) \tag{1.16}$$

$$H_t = L_t\exp(\phi c) \tag{1.17}$$

$$K_t = I_{t-1} + (1-d)K_{t-1} \tag{1.18}$$

$$I_t = sY_t \tag{1.19}$$

From this it follows that the region converges to the equilibrium or balanced growth path consistent with the equilibrium capital-labour ratio k^*, which is the solution to the expression

$$\dot{k}_t = sy_t - (n+d)k_t \tag{1.20}$$

in which sy_t is the (nonlinear) savings schedule and $(n + d)k_t$ is the (linear) depreciation schedule (the lower cases indicate variables normalised by the size of the labour force). The equilibrium level of output per worker (y^*) at each time point t is given by

$$\ln y^* = \ln A + (\alpha/(1-\alpha)) \ln s - (\alpha/(1-\alpha)) \ln(n+g+d) + \ln(H/L) \tag{1.21}$$

The effect of different assumptions about exogenous parameters (c, α, ϕ, d, s, g, and n) can be seen from the different paths for y in the region as a result of changing one parameter and holding everything else constant[15]. For example, one of the (perhaps counterintuitive) results from this model is that changing the investment rate s has exactly no effect on the equilibrium growth rate, although it does affect the level of output per worker. This is seen in Figure 1.19. In this, we describe the path followed to equilibrium assuming different investment rates, all other things being equal[16]. Similarly the effect of the region's population and labour force growing at 1% as opposed to 5%, all else equal, is seen in Figure 1.20. In this the other parameters are set to the values used to generate Figure 1.19, with $s = 0.05$. Note that faster population and labour force growth reduces the equilibrium level of output per worker, but again output per worker growth is unaffected. However, the level of output and investment at equilibrium grows at the rate $n + g$, so that a faster population growth rate means that the scale of the economy increases. It is easy to show that increasing the number of years schooling or the rate of return to a year's schooling has a similar effect of boosting the level of the path of output per worker but has no long-term impact on the rate of growth. Similarly, an increase in the rate of depreciation has a similar impact to an increase in the population growth rate, with the economy converging to a lower level trend path but with no sustained effect on the growth of output per worker. The path diagrams produced by these scenarios are very similar to Figures 1.19 and 1.20.

In contrast, if the region's technical progress rate is 1% rather than 5%, this slows not only the output growth rate but also the rate of growth of output per worker, as shown[17] by Figure 1.21. Also, we see from equation (1.17) that provided $\ln s > \ln (n + g + d)$, the higher is capital's share (α) then the higher is the trend path, but with our assumed parameter values the capital share makes no difference to the growth of output per worker in the long-run. Also, changing several parameters simultaneously makes no difference, the region converges to a new

[15] Below we give some results specific to particular parametric assumptions, without necessarily implying general rules.

[16] For this simulation, for visual clarity I have assumed that $g = 0.05$, $n = 0.05$, $A_0 = 110$, $K_0 = 10$, $L_0 = 20$, $Y_0 = 90$, $c = 9$, $\phi = 0.1$, $\alpha = 0.333$, $d = 0.025$, and $s = 0.01$ and 0.05 correspond to the alternative investment ratios.

[17] This uses the same set of parameter values as previously, except that $g = 0.01$ and 0.05 are the two alternative technical progress rates.

steady-state path and subsequently output per worker grows at the same rate as prior to the change.

Let us now consider what happens if we eliminate the assumption of constant returns to scale[18] by choosing $\beta > 1-\alpha$. Perhaps unsurprisingly, we see that with increasing returns the region deviates from the trend path it would have followed, and follows a steeper path commensurate with the faster steady state growth rate of output and output per capita. With constant returns we have seen convergence to a steady state at which investment, capital, output and effective labour (equal to AH) grow at the same rate $(n + g)$ and the capital/effective labour ratio is a constant. If we relax the constant returns assumption in favour of increasing returns to scale ($\beta > 1-\alpha$, $\alpha < 1$), with effective labour growing at one rate but investment, capital, output growing at a different (common) equilibrium rate, the capital/effective labour ratio is not constant but increases with time. We find that if the region's production function suddenly alters in this way, the region converges to a new equilibrium level and growth rate, with the growth of output equal to $(n + g)\beta/(1-\alpha)$ and output per capita equal to $(n + g)\beta/(1-\alpha) - g$. Figure 1.22 gives the simulated outcome[19] assuming $\beta = 0.75$

This all now seems a long way from neoclassical growth theory, and part of the reason we may want to preserve the constant returns assumption is that we can use results from the mainstream body of theory, for instance under a competitive market assumption the wage rate is equal to the marginal product of effective labour, and similarly for compensation to capital. In this we assume that all factors are privately owned, markets clear, all output is used as factor rewards with shares equal to α and $\beta = 1-\alpha$, and factor rewards are therefore consonant with their value as scarce resources. If however we have increasing returns to scale then we no longer have this interpretation. In the real world we know that wages are affected by numerous institutional factors and imperfectly competitive markets and although this could be accommodated within a constant returns framework, it is usual to move beyond the restrictive assumptions of pure neoclassical theory in general in the direction of endogenous growth theory, and beyond, when attempting to capture the general implications of institutional and political factors for growth.

These last two examples highlight the important feature of this model, which is that permanently changing the rate of growth of output per capita is not easy to accomplish; it can be achieved if there is some mechanism whereby the rate of growth of technical progress is altered. It can also be achieved if we abandon the assumption of constant returns to scale in favour of increasing returns.

---

[18] While constant returns to scale is a standard assumption, it is not essential to the theory. Solow (1994) points out that 'the model can perfectly well get along without constant returns to scale', although its presence facilitates simplification.

[19] The other parameters are set to the same values as for the other simulations.

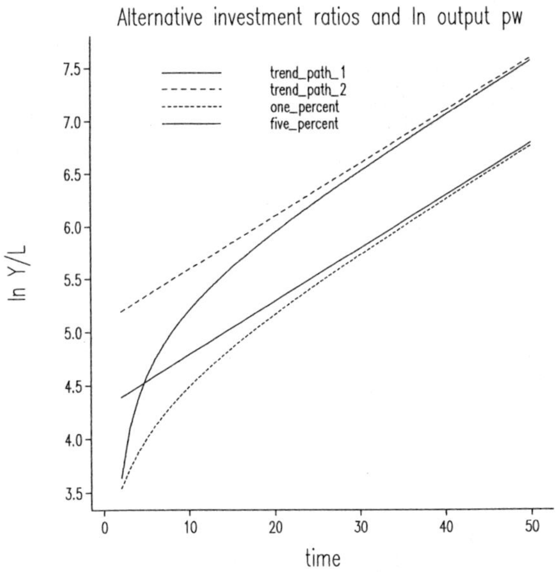

Fig. 1.19 The effect of different investment ratios

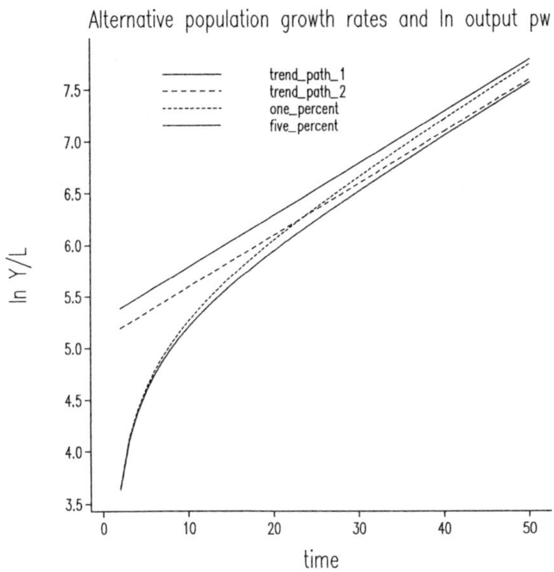

Fig. 1.20 The effect of different population/labour force growth rates

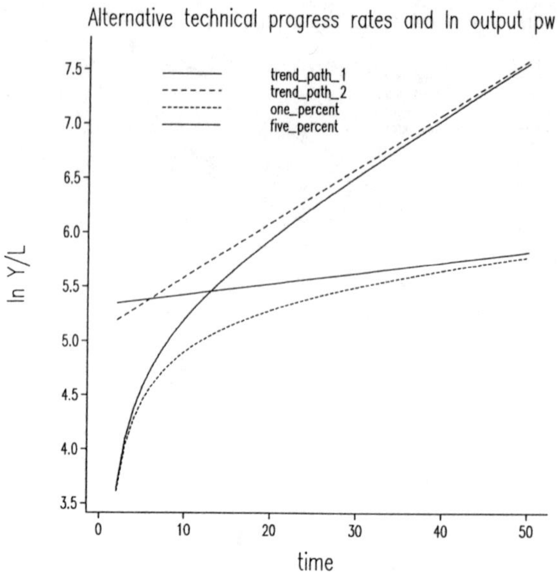

Fig. 1.21 The effect of different technical progress rates

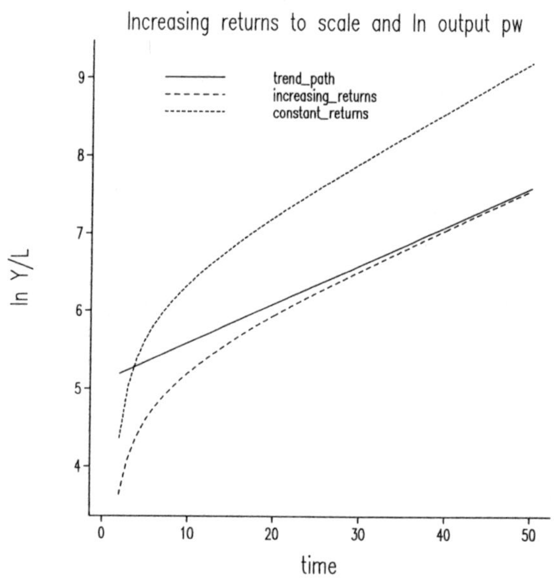

Fig. 1.22 Increasing versus constant returns to scale

So far we have considered the dynamics of a single region under alternative assumptions. We could of course just as easily be contemplating the dynamics of two regions that are identical in every respect apart from the parameter of interest. We can see from the above that the two regions will grow at different rates in the short term, as they move towards their different equilibrium paths, after which growth rates will be equal. Only when the rate of technical progress differs between the two regions, or where the returns to scale are not constant, is there a persistent output per worker growth rate discrepancy. Of course the same arguments also apply when there are more than two regions.

This theory therefore provides important insights into the causes of differences that are clearly apparent when one looks at the empirical data for the EU's regions. With regard to the EU regions' level of output per worker differences, these can be understood as a consequence of parameters differences. With regard to growth rate differences, our theory tells us that these are phenomena associated with a period of transition to trend paths, although the transition period could be lengthy. On the other hand they may also be a more permanent feature, perhaps resulting from differences in technical progress rates between regions, and as we have seen in Figure 1.21 the outcome is then an ever-widening gap between productivity levels. While the rate of technical progress is an important determinant of regional differences, since it has such wide ranging and permanent effects on both the level and growth of output per worker, in fact the theory we have outlined says nothing more about it. An obviously important issue in regional research is therefore the matter of whether technical progress rates do differ according to region, and if they do, what are the causes.

One cause of technical progress rate differentials might be that technical progress depends on the scale of the economy, so that for instance larger regions see faster technical progress. This leads us into the area of endogenous growth theory and the early work by Arrow (1962), Romer (1986) and Lucas (1988). We follow the approach typical of Romer (1990), Grossman and Helpman (1991, 1994) and Aghion and Howitt (1992), and assume that the level of resources devoted to research and development increases the technical progress rate. It takes only a minor modification of our neoclassical set-up to capture the impact and implications of making technical progress endogenous. Previously, technical progress grew at the exogenously determined rate g, but now we assume that it depends on the number of workers (L_A) involved in the search for new knowledge, where the total labour force is equal to $L = L_A + L_Y$ and there are L_Y workers involved directly in production. We retain the assumption that the labour force grows at an exogenously determined rate n and introduce an assumption that a constant share of the labour force at time t is devoted to research and development. The rate of technical progress is now simply proportional to the size of this section of the workforce, and therefore increases as L_A increases[20]. Hence

---

[20] For this simulation, for I assume that $n = 0.05$, $A_0 = 110$, $K_0 = 10$, $L_0 = 20$, $Y_0 = 90$, $c = 9$, $\phi = 0.1$, $\alpha = 0.333$, $\beta = 1 - \alpha$, $d = 0.025$, $s = 0.05$ and either $g = 0.05$ or technical progress is endogenously determined. Also, $L_A = L/2$ and $\kappa = 0.1$.

$$\dot{A}/A = \kappa L_A \qquad A_t = A_{t-1} + A_{t-1}\kappa L_A \tag{1.22}$$

Putting equation (1.22) in place of equation (1.15) produces Figure 1.23. Here we see the contrast between the dynamics associated with exogenously determined technical progress, and a technical progress rate that increases as the level of research and development employment increases. Endogenising technical progress produces accelerating output per worker growth, and there is no prospect of a steady-state with productivity converging to a linear trend-path and constant steady-state growth rate. The outcome is one of increasing returns to scale.

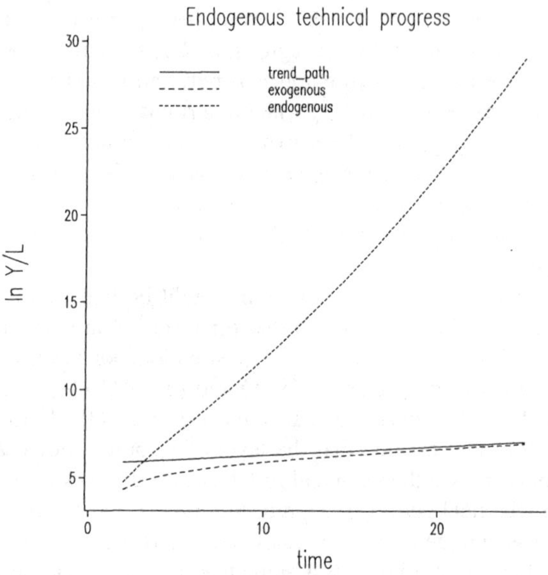

Fig. 1.23 Endogenous versus exogenous technical progress

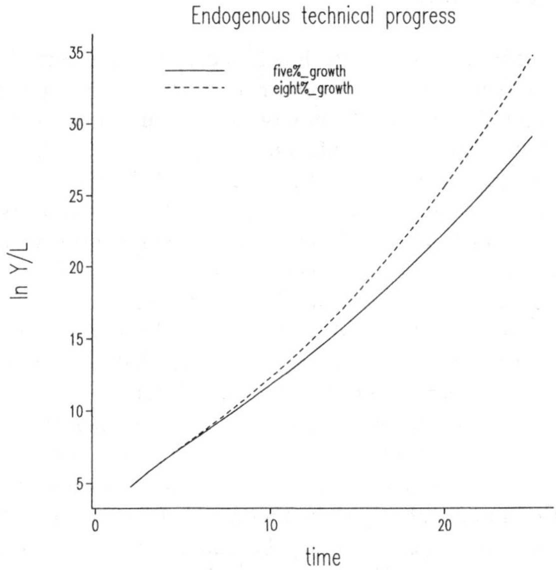

Fig. 1.24 Two regions with different workforce growth rates

Figure 1.24 shows the paths of two regions which are identical, with the same initial values and exogenous coefficients, except that one region has workforce growth equal to 5% and the other grows at 8%. The fact that the growth of workers is faster in one region means that the number of workers involved in knowledge creation becomes larger and therefore the rate of technical progress is faster. The outcome is not convergence to equilibrium paths whereupon the ratio of the levels of output per worker is constant over time, but instead the ratio increases with time. The increasing scale of the region with the larger workforce and hence larger level of resources devoted to knowledge creation boosts the rate of growth of technology and therefore the growth of output per worker. This kind of scale effect has brought criticism from some (for example Jones, 1995) who argues that this type of growth path is unrealistic and that the increase in the number of knowledge workers has not been matched by a commensurate increase in total factor productivity in the USA. However, from the European perspective, with some exceptions, it does seem to be the case that the regions devoting most to research and development do seem to be the wealthiest, and it appears that success breeds success, with rich regions becoming increasingly rich and economic activity concentrating where it already exists. Within the context of the European regions, the work of Cheshire and Carbonaro (1995) has a basis in the Romer (1990) model (see also Fingleton, 2001).

1.8.2 Spatial Effects in Regression

Spatial effects can occur for various reasons. For instance, we have seen the effect of different population growth rates on the trend-path of regions, and it is reasonable to expect that regions that are neighbours or are in some way connected to each other will tend to have correlated population growth rates[21]. Equally, some of the other determining parameters highlighted in Section 1.8.1 may be correlated across regions, for example there is reason to suggest that the spillovers which characterise endogenous growth theory will carry across regional boundaries and impact on the growth rates in neighbouring regions. In other words, the premise is that space, which is absent from the theories outlined above, matters.

Regression analysis will produce spatially autocorrelated residuals when there are omitted variables that are spatial autocorrelated, given that the omitted variables are relevant ones and the dependent variable is itself spatially autocorrelated. One group that comes into this category consists of those variables that are constant across regions within each country, but differ from country to country. The most obvious of these is variables representing national policy instruments which have a common effect on groups of regions within national boundaries. Another candidate is the effect of national economic cycles causing a nation's regions to grow either faster or slower than the EU average and thus show up as a cluster of similar valued residuals. For some of these effects, it should be possible to model the spatial dependence in a regression by incorporating suitable explanatory variables. For example, we may wish to include country-specific dummy variables, equal to 1 in all regions that are within a country and zero in other regions. However, in general only a limited set of data is available and we may be compelled by lack of data to omit one or more significant variables, and thus leaves open the possibility that the model will suffer from specification error. For example, if a significant missing variable is correlated with the initial level of GDP per capita, the initial level will to an extent capture the effect of the missing variable and the consequence will be a biased estimate of the rate of convergence.

In order model residual spatial autocorrelation and we make use of a so-called standardised weights matrix. We commence with the unstandardised matrix W^* which is a 200 by 200 matrix of zeros and ones, defined so that the row i and column j cell initially takes a value 1 if regions i and j are contiguous (share a common boundary), otherwise $W^*_{ij} = 0$. This is then standardised to give the matrix W by dividing each cell entry in W^* by the sum of the respective row, in other words $W_{ij} = W^*_{ij}/\Sigma_j W^*_{ij}$. Clearly there are many ways in which a standardised W matrix can be constructed, with a wide variety of optional 'distance' measures such as

[21] Commuting across area boundaries is likely to negatively correlate growth rates in neighbouring areas, since if a central city region attracts workers from its neighbours, then the population growth in surrounding areas will add to GDP per inhabitant growth in the central city, and may detract from it in surrounding areas

great circle distances or some measure of economic distance based on transport costs. However it is often the case that detection of residual autocorrelation is quite robust to difference weighting systems, for example when weights are based quantitative distances embodying a steep decay effect rather than contiguity, although of course the details of the analyse will be different and so we would not wish to be too strict regarding significance thresholds. However, if residuals are highly significantly autocorrelated using contiguity based weights, there is a strong possibility that they would also be significant under a different, but not too dissimilar, weighting scheme.

Given the W matrix, it is possible to use both in order to carry out a test for residual spatial autocorrelation based on Moran's I statistic. The version[22] of Moran's I (I_R) appropriate to OLS regression residuals is defined as

$$I_R = \sum_i \sum_j W_{ij} (Y_i - \mu)(Y_j - \mu) / \sum_i (Y_i - \mu)^2 \tag{1.22}$$

in which Y is the dependent variable and μ here denoted the fitted values from the regression model. Interpretation should be cautious though, since I_R has power against a range of alternatives, including non-normality, heteroscdasticity and spatial lag dependence (see Anselin 1988, Fingleton 1999). Other tests are also available for residual autocorrelation, notably Lagrange multiplier (LM) test, which we also report below. Testing the residuals of the conditional convergence model using I_R suggests that they are significantly positively autocorrelated, with a standardised value equal to 6.14 which well outside the range of plus and minus 1.96 which one would anticipate in the absence of residual autocorrelation. This is supported by the results of the LM test (Burridge, 1980), where the test statistic (equal to 30.04) is very atypical of the appropriate null distribution ($\chi2$ with 1 degree of freedom). The Kelejian-Robinson test, which is robust to non-normality, also indicates significant error autocorrelation. It may be the case that there is also error heteroscedasticity, although in practice it is difficult to disentangle both sources of non-sphericity. The White test statistic, which has power against unspecified sources, takes a value of 67.49 which has a p-value of only 0.0129 in the asymptotic null distribution[23].There is a strong suggestion that the conditional model summarised in Table 1.3 is incorrectly specified, with both residual dependence and heterogeneity a possibility.

1.8.3 Spatial Regression Models

Let us denote our dependent variable by the vector Y, the set of independent variables (such as LGDPPC87, PER_PPS etc) by the matrix X, the regression coefficients by the vector b and the residuals by $e = Y - Xb$. Residual autocorrelation is introduced in the form of an autoregressive process in which the vector e is a func-

---

[22] This is simplified because of the use of a standardized W matrix.

[23] χ^2 with 44 degrees of freedom.

tion of the weighted average of the residuals in contiguous regions plus a (heterogeneous) random shock, in other words

$$Y = Xb + e \tag{1.23}$$

$$e = \rho We + \xi \tag{1.24}$$

$$\xi \sim N(0, \sigma_i^2 I) \tag{1.25}$$

$$E(Y) = Xb \tag{1.26}$$

This specification allows for spillover effects and seeks to represent the effects of missing variables which are manifest as autocorrelated residuals. Haining (1990) also supports modelling the effect of omitted variables by an autoregressive error process, arguing that when the set of explanatory variables is not clear-cut, and extra independent variables are not identified or not capable of removing the residual correlation, it is helpful to adopt the autocorrelated errors specification to soak up spatial effects and help to minimise biased estimation.

1.8.4 Empirical Results from Spatial Regressions

Table 1.4 summarizes the results of estimating via maximum likelihood the autoregressive errors specification given by equation 1.23-1.26 using the reduced model in Table 1.3 as a guide as to the set of variables in X. In this initial estimation, we assume error homogeneity. It is apparent that ρ is significantly greater than zero, reinforcing the suggestion of significant residual autocorrelation mentioned above. This is reaffirmed by the Likelihood ratio test statistic, which is equal to 36.06 and is highly significant when referred to χ^2 with 1 degree of freedom. However, the LM test for a spatial lag in the model with error autocorrelation (see Anselin and Florax 1995) gives a value of 1.27 which is not significant when referred to χ^2 with 1 degree of freedom. While some of the coefficients estimates have changed in value as a result of modelling error dependence, they are not fundamentally different from those in Table 1.3, although we now see that low educational attainment is not a significant variable. However, the diagnostics from the spatial regression show that we have not picked up the residual heterogeneity that we also touched upon. Inspection of the data in fact points to higher residual variance in the 'peripheral' regions of the EU than in the core. Evidently some peripheral regions such as Ireland growing much faster than expected under the model while other peripheral regions grow much more slowly, while the more spatially compact core regions tend to deviate less from expectation. In order to accommodate error heterogeneity on a core-periphery[24] basis we fit a groupwise

[24] Using a dummy variable with 500km distance from Luxembourg defining core and periphery regions.

error heterogeneity specification[25], thus simultaneously allowing for residual autocorrelation and the apparently significantly higher error variance in the periphery. On fitting the groupwise error heterogeneity specification by maximum likelihood, the likelihood ratio test on error homogeneity gives a test statistic[26] of 16.47 which is highly significant in the chi-squared distribution with 1 degree of freedom, indicating that we should indeed reject the null of error homogeneity. Table 1.5 shows that the model generally supports the earlier results and affirms the significance of the variables apart from low educational attainment.

While we have argued the case for using our autoregressive error specification in an attempt to capture spatial effects that are actually due to omitted variables, on the grounds that omitted variables are unavailable, we have also already mentioned one set of easy-to-construct variables that could be invoked in place of our somewhat complex error specification. This is the set of national dummy variables (14 of them, excluding one so as to avoid the dummy variable trap) to represent nation-specific policies and possibly nationally differentiated cyclical patterns. Previously this has been found to be effective as a way of eliminating residual autocorrelation in the European context (Armstrong 1995).

However national dummies may not suffice, and to test this we include them jointly with the autoregressive error specification, the outcome being the estimates of Table 1.7. Evidently this signals that the complex error process we have used is unnecessary, since the ρ estimate is insignificantly different from zero, and the test of groupwise heteroscedasticity is null (the likelihood ratio test statistic is 2.71, which is less than the upper 5% point of the χ^2 distribution with 1 degree of freedom). However, the disadvantage of this specification is that it also eliminates the significance of some of the key policy-related variables, which become subsumed by national effects. Thus, peripherality become of marginal significance, while the share of employment in agriculture and structural spending in Objective 1 regions are insignificant. Fitting the same specification via OLS assuming spatially independent and homoscedastic errors produces the estimates in Table 1.8, which are similar although the significance levels are lower for the marginal variables. This last model highlights some difficulties associated with using the national dummy approach. First, it may not entirely eliminate residual autocorrelation. In the present case the standardised value of Moran's I statistic for the residuals is equal to 1.85 and a one tailed p-value (which is appropriate since the natural alternative hypothesis is positive residual autocorrelation) in the normal distribution of 0.032.

Also, the multicollinearity condition number for this model is equal to 231.8, which is about 10 times larger than the maximum acceptable value. This reminds us that the presence of the dummy variables may nullify the significance of some variables because of their correlation. In the extreme case, consider a (hypothetical) variable which takes the same value in all regions within a given country, and which takes another value, which is constant across all regions in a second coun-

[25] An option available in SPACESTAT.

[26] Equal to twice the difference in the likelihoods of the autoregressive errors and groupwise heterogeneity models.

try, and so on across countries. It is then impossible to include this variable in the presence of country dummies because of aliasing due to perfect multicollinearity. Hence while country dummies may control for spatial autocorrelation, it may be preferable to represent the effects of what are presumed to be country-specific policies and diverse other spatial effects via the error specification, as has been done in the model summarized by the Table 1.5. This may avoid obscuring the significance of variables of substantive interest and more properly capture subtle and localized spatial interaction effects.

Table 1.7 The groupwise heteroscedasticity spatially autocorrelated error model with national dummy effects

variable	**coefficient**	**s.e.**	**ratio**	**probability**
CONSTANT	0.39973	0.0270577	14.773212	0
LGDPPC87	-0.0326698	0.00272185	-12.002771	0
MAIRP	0.00062496	0.000164951	3.788762	0.000151
PER_PPS	-9.21E-05	4.66E-05	-1.976481	0.0481
EMP_AGRI	-6.09E-05	0.000126964	-0.479882	0.631312
EDUC_L	-0.000276164	0.000142205	-1.942011	0.052136
EDUC_H	0.000651677	0.000160362	4.063796	0.000048
MUNLTR	-0.11033	0.0214535	-5.142739	0
PC89_93	0.00616316	0.00481763	1.279291	0.200794
ρ	0.0373576	0.0931496	0.40105	0.688383
Austria	-0.0292391	0.00737095	-3.966808	0.000073
Belgium	-0.0369923	0.0070094	-5.277533	0
Germany	-0.0397563	0.00672885	-5.908334	0
Denmark	-0.0362192	0.00744269	-4.866415	0.000001
Spain	-0.0296414	0.00794269	-3.731912	0.00019
Finland	-0.0622924	0.00716211	-8.697488	0
France	-0.0427115	0.00651506	-6.555808	0
Greece	-0.0392821	0.00768079	-5.114336	0
Ireland	-0.0045642	0.00813357	-0.561156	0.574691
Italy	-0.0292524	0.00675784	-4.328663	0.000015
Netherlands	-0.0376566	0.00664322	-5.668423	0
Portugal	-0.0117011	0.00930352	-1.257704	0.208499
Sweden	-0.0509744	0.00693258	-7.352873	0
Britain	-0.0383257	0.00661266	-5.795802	0
R^2	0.8796		Likelihood	742.213
σ^2 core	3.29E-05			
σ^2 periphery	3.98E-05			

Table 1.8 The table 1.7 specification with spherical errors

variable	coefficient	s.e.	ratio	probability
CONSTANT	0.394225	0.0282578	13.951025	0
LGDPPC87	-0.0321961	0.00287117	-11.213594	0
MAIRP	0.000607155	0.000178696	3.397705	0.000839
PER_PPS	-8.56E-05	4.81E-05	-1.780517	0.076707
EMP_AGRI	-6.45E-05	0.000139093	-0.463543	0.643545
EDUC_L	-0.000267325	0.000153026	-1.746919	0.082386
EDUC_H	0.000665311	0.000170473	3.902737	0.000135
MUNLTR	-0.107401	0.0231905	-4.631244	0.000007
PC89_93_	0.00611813	0.00528376	1.15791	0.248461
Austria	-0.0289929	0.00751264	-3.85922	0.000159
Belgium	-0.0371599	0.00709332	-5.23872	0
Germany	-0.0397336	0.00680308	-5.840526	0
Denmark	-0.0364438	0.00762836	-4.777408	0.000004
Spain	-0.0299174	0.00818672	-3.654377	0.00034
Finland	-0.0627437	0.00728772	-8.609518	0
France	-0.0426049	0.0065605	-6.494161	0
Greece	-0.0391023	0.00790608	-4.945854	0.000002
Ireland	-0.00454946	0.00842006	-0.540312	0.589661
Italy	-0.0294023	0.00684074	-4.298119	0.000028
Netherlands	-0.0376717	0.00666195	-5.654759	0
Portugal	-0.0113994	0.00971866	-1.172939	0.242396
Sweden	-0.0511978	0.00702836	-7.284467	0
Britain	-0.0382734	0.00667395	-5.734746	0
R2-adj	0.8584		Likelihood	740.737
σ^2	4.01E-05			
σ^2 (ML)	3.55E-05			

1.8.5 GDP per Inhabitant

The Chapter is based mainly on the growth of real GDP[27] per inhabitant over the period 1987-97, using raw data initially provided by the European Commission. The analysis uses a set of 200 regions, and to achieve this detailed level of coverage, some data reconstruction has been necessary. One important factor has been the accession to the EU of the eastern Lander following the unification of Germany, and this created a problem because in 1987, the start of the period of analysis, this region was outside the EU and comparable official statistics were unavailable. In addition, boundary changes, for example involving regions of the United Kingdom, again interfere with across-time consistency. In order to achieve this consistency, simple interpolation methods have been used, simply involving the application of the following formula,

$$Y_{it} = Y_{iT} Y_{It} / Y_{IT} \tag{1.27}$$

in which Y is the level of GDP per inhabitant, i denotes region, t denotes year (with missing data), T denotes the first year with data, and I is a higher level region such as a national territory. Thus, in the case of the new Lander, I is Germany and T is 1991. In the case of the UK, boundary changes with the introduction of new regions also meant that series had to be extended back to 1987. Hence the inner and outer London series were reconstructed based on the complete series for the South east region and national series were the basis of the regional series for Wales and Scotland. In the case of Ireland and Denmark, we do not in effect distinguish between the regions. The same basic method was also used to reconstruct the PPS adjusted series.

Although the results are largely unreported, apart from briefly in Table 1.6, in fact parallel analyses were carried out on a similar set of data which made fewer assumptions, as a check on the robustness of the results obtained. The analyses performed on this second data series are very similar to those reported here, indicating that the conclusions we have drawn are not influenced by the method of data construction described above. This second series consists of 186 rather than 200 regions. The differences between the two series are listed below.

regions of Ireland (2) merged to form 1 region
inner and outer London (2) merged to form 1 region
East and West Wales (2) merged to form 1 region
Scottish regions (4) amalgamated to form 1 region
New Lander and Berlin (6) eliminated
Danish regions (3) merged to form 1 region

The outcome is that 19 regions have been replaced by 5 regions.

[27] Some analysis of PPS adjusted data has also been carried out.

1.8.6 Other Variables

i) Data for the variable PAVE are available for only 169 of the 200 regions, but data on IPC patents per capita (MPATPC, 1985-95 average) are available for all regions. In this case the variable PAVE was regressed on the MPATPC data and the resulting model used to predict, given the MPATPC value, the PAVE data for the missing regions. The regression accounted for 85.8% of the variance of PAVE, and the *t* ratio for the regression was equal to 31.88.

ii) For Berlin, the population growth rate for 1987-97 was assumed to be same as in the surrounding Lander (Brandenburg), thus allowing the 1987 population level to be calculated.

iii) Spending on structural funds in the Objective 1 regions is the total structural fund spending by region 1989-93 (measured in current ECU's) and spending for 1994-99 (1994 ECU's), on a per capita basis.

iv) The peripherality index, which was constructed by Copus[28], combines spatial accessibility and the mass of each region. Because it takes into account how a region is linked to the interregional transport network, this index distinguishes between peripheral regions in terms of different levels of accessibility, as in the case of Athens versus the rest of Greece.

A fuller description of these variables and various caveats is provided in Pinelli et. al. (1998).

References

Aghion P. and Howitt P. (1992) A model of growth through creative destruction, *Econometrica*, 60, 323-351

Anselin L. (1988) *Spatial Econometrics: Methods and Models* Dordrecht: Kluwer

Anselin L. and Florax R. (1995) Small sample properties of tests for spatial dependence in regression models: some further results, in L. Anselin and R. Florax (eds) *New Directions in Spatial Econometrics* Berlin: Springer-Verlag, 21-75

Armstrong H.W. (1995) Convergence among regions of the European Union 1950-1990, *Papers in Regional Science* 74, 143-152

Armstrong H.W. and Vickerman R.W.(eds.) (1995) *Convergence and Divergence among European Regions*, London: Pion

[28] Copus, 1997. The present Chapter uses the index calculated by Copus using total 1984 GDP in PPS as the mass variable, hence emphasising demand variation. Copus provides two alternative measures for this index: the first assuming circular regions, the second assuming mininum bounded rectangular self potential estimation for the area of the regions. They are in fact very similar and we use the circular definition in this chapter.

Arrow, K.J. (1962) The economic implications of learning by doing, *Review of Economic Studies* 29, 155-173

Barro R. and Sala-i-Martin X. (1995) *Economic Growth* New York: McGraw Hill Inc

Burridge P. (1980) On the Cliff-Ord test for spatial autocorrelation, *Journal of the Royal Statistical Society B*, 42, 107-8

Cheshire P. and Carbonaro G. (1995) Convergence-divergence in regional growth rates: an empty black box?, in H.W. Armstrong and R.W. Vickerman (eds) *Convergence and Divergence among European Regions* European Research in Regional Science, Pion, London, 89-111

Copus A. K. (1997) *A new peripherality index for European regions* a report for the Highlands and Islands European Partnership

Fingleton B. (1999) Estimates of time to economic convergence: an analysis of regions of the European Union, *International Regional Science Review*, 22, 5-35

Fingleton B. (2000) Convergence: international comparisons based on a simultaneous equation model with regional effects, *International Review of Applied Economics*, 143, 285-305

Fingleton B. (2001) Equilibrium and economic growth: spatial econometric models and simulations, *Journal of Regional Science*, 41, 117-148

Fingleton B. (2002) Regional economic growth and convergence : insights from a spatial econometric perspective, forthcoming in L. Anselin and R. Florax (eds) *New Advances in Spatial Econometrics*, Heidelberg: Springer-Verlag

Fingleton B., Lewney R. and Pinelli D. (1997) *Regional Growth and Convergence*, published as Vol 1 of Subseries VI of the EC's Single Market Review, London: Kogan Page

Grossman G.M. and Helpman E. (1991) *Innovation and Growth in the Global Economy*, Cambridge: MIT Press

Grossman G.M. and Helpman E. (1994) Endogenous innovation in the theory of growth, *Journal of Economic Perspectives*, 8, 23-44

Haining R. P. (1990) *Spatial Data Analysis in the Social and Environmental Sciences*, Cambridge: Cambridge University Press

Jones C. I. (1995) R&D-based models of Economic growth, *Journal of Political economy*, 103 759-784

Jones C. I. (1997) Convergence revisited, *Journal of Economic Growth*, 2, 131-153

Levine R. and Renelt D. (1992) A sensitivity analysis of cross-country growth regressions, *American Economic Review*, 82, 942-963

Lucas R.E. Jr (1988) On the mechanics of development planning, *Journal of Monetary Economics*, 22, 3-42

Mankiw N. G., Romer P., and Weil D. N. (1992) A contribution to the empirics of growth, *Quarterly Journal of Economics*, 107, 407-438

Pinelli D., Giacometti R., Lewney R., and Fingleton B. (1998) *European Regional Competitiveness Indicators*, Discussion Paper 103, Department of Land Economy, Cambridge University

Romer P.M. (1986) Increasing returns and long-run growth, *Journal of Political Economy*, 94, 1002-1037

Romer P.M. (1990) Endogenous Technical Change, *Journal of Political Economy*, 98, 71-102

Romer P.M. (1994) The origins of endogenous growth, *Journal of Economic Perspectives*, 8, 3-22

Solow R.M. (1956) A contribution to the theory of economic growth, *Quarterly Journal of Economics*, 70, 65-94

Solow R.M. (1994) Perspectives on growth theory, *Journal of Economic Perspectives,* 8, 45-54

Swan T. W. (1956) Economic Growth and capital accumulation, *Economic Record*, 32, 334-361

2 An Exploratory Spatial Data Analysis of European Regional Disparities, 1980-1995

Cem Ertur and Julie Le Gallo

University of Burgundy, France

2.1 Introduction

European integration has stimulated numerous studies of regional economic convergence within the European Union in recent macroeconomic and regional science literature (e.g. Abraham and Von Rompuy 1995; Armstrong 1995a; Neven and Gouyette 1995; Martin 2001). Most of the time, the empirical methods that have been used are identical to the methods employed in international studies. However, spatial effects, particularly spatial autocorrelation and spatial heterogeneity, must be taken into account when analyzing the convergence process at regional scale. There are number of factors - trade between regions, technology, knowledge diffusion and more generally regional spillovers - that lead to geographically dependent regions. Because of spatial interactions between regions, geographical location is important in accounting for the economic performances of regions. Yet for all its importance, the role of spatial effects in convergence processes has only been recently examined using the appropriate spatial statistic and econometric methods (Armstrong 1995b; Fingleton 1999; López-Bazo et al. 1999, for European regions; Rey and Montouri 1999; Rey 2001, for US states).

This Chapter aims therefore at investigating more closely the space-time dynamics of the distributions of regional per capita GDP and average growth rate in Europe over the 1980-1995 period in order to show that spatial autocorrelation and spatial heterogeneity are indeed unavoidable features. More specifically, *spatial autocorrelation* can be defined as the coincidence of value similarity with locational similarity (Anselin 2001). *Spatial heterogeneity* means that economic behavior is not stable over space and may generate characteristic spatial patterns of economic development under the form of spatial regimes: a cluster of rich regions (the core) being distinguished from a cluster of poor regions (the periphery).

To this end, we use the recently developed methods of Exploratory Spatial Data Analysis. By identifying global and local spatial autocorrelation, we can characterize the way economic activities are located in the European Union and the way this pattern of location has changed over time. Moreover, local spatial statistics can provide an insight into spatial heterogeneity within the sample and its persistence over time.

In Section 2.2, we briefly set out the methods of Exploratory Spatial Data Analysis (ESDA). Section 2.3 presents the sample of 138 European regions over the 1980-1995 period as well as the spatial weight matrices used in this Chapter. Section 2.4 gives detailed empirical results for log per capita GDP: we show that its distribution is indeed characterized by significant positive global spatial autocorrelation. Next, we compute local spatial statistics - Getis-Ord statistics (Ord and Getis 1995) - in order to detect clusters of high and low per capita GDP. The occurrence of such clusters during the period is an indication of the persistence of spatial disparities between European regions. Local Moran autocorrelation statistics (Moran scatterplot and LISA: Anselin 1995, 1996) are also computed and reveal in addition the existence of "atypical regions". The robustness of the results is investigated with an original application of transition probability matrices used for example by Quah (1996a) in a different context. In Section 2.5, the spatial pattern of regional growth is examined in order to study the dynamism of European regions in relation to their initial per capita GDP level. Major implications of the results for estimating and testing standard β-convergence models are finally emphasized in the conclusion.

2.2 Exploratory Spatial Data Analysis (ESDA)

Exploratory Spatial Data Analysis (ESDA) is a set of techniques aimed at describing and visualizing spatial distributions, at identifying atypical localizations or spatial outliers, at detecting patterns of spatial association, clusters or hot spots, and at suggesting spatial regimes or other forms of spatial heterogeneity (Haining 1990; Bailey and Gatrell 1995; Anselin 1998a,b). These methods provide measures of global and local spatial autocorrelation.

2.2.1 Global Spatial Autocorrelation

Spatial autocorrelation can be defined as the coincidence of value similarity with locational similarity (Anselin 2001). Therefore, there is positive spatial autocorrelation when high or low values of a random variable tend to cluster in space and there is negative spatial autocorrelation when geographical areas tend to be surrounded by neighbors with very dissimilar values.

The measurement of global spatial autocorrelation is based on Moran's I statistic, which is the most widely known measure of spatial clustering (Cliff and Ord 1981; Upton and Fingleton 1985). For each year of the 1980-1995 period, this statistic is written as following:

$$I_t = \frac{n}{S_0} \cdot \frac{\sum_i \sum_j w_{ij}(x_{i,t} - \mu_t)(x_{j,t} - \mu_t)}{\sum_t (x_{i,t} - \mu_t)^2} \qquad t = 1,...,16 \tag{2.1}$$

where $x_{i,t}$ is the observation in region i and year t; μ_t is the mean of the observations across regions in year t; n is the number of regions and w_{ij} is one element of the spatial weight matrix W. This matrix contains the information about the relative spatial interdependence between the n regions i. The elements w_{ii} on the diagonal are set to zero whereas one element w_{ij} indicates the way region i is spatially connected to the region j. S_0 is a scaling factor equal to the sum of all the elements of W.

Noting z_t the vector of the n observations for year t in deviation from the mean μ_t, (2.1) can be written in the following matrix form:

$$I_t = \frac{n}{S_0} \cdot \frac{z_t' W z_t}{z_t' z_t} \qquad t = 1,...,16 \tag{2.2}$$

In order to normalize the outside influence upon each region, the spatial weight matrix is row-standardized such that the elements w_{ij} in each row sum to 1. In this case, (2.2) simplifies since for row-standardized weights $S_0 = n$.

Various spatial weight matrices can be considered: a simple binary contiguity matrix, a binary spatial weight matrix with a distance-based critical cut-off, above which spatial interactions are assumed negligible, and more sophisticated generalized distance-based spatial weight matrices with or without a critical cut-off. The critical cut-off may be the same for all regions or may be defined to be specific to each region leading in the latter case, for example, to k-nearest neighbors weight matrices. In these nearest neighbors matrices, the critical cut-off for each region is determined so that each region has the same number of neighbors. The critical cut-off may thus be expressed as a fixed distance d or as a fixed number k of nearest neighbors. The appropriate choice of the spatial weight matrix is one of the most difficult and controversial methodological issues in exploratory spatial data analysis and spatial econometrics (Anselin 1988; Florax and Rey 1995; Anselin and Bera 1998).

Moran's I statistic gives a formal indication on the degree of linear association between the vector z_t of observed values and the vector Wz_t of spatially weighted averages of neighboring values, called the spatially lagged vector. Values of I larger than the expected value $E(I) = -1/(n-1)$ indicate positive spatial autocorrelation, whereas values smaller than the expected indicate negative spatial autocorrelation.

Inference for Moran's I statistics may be based on the normal approximation or on the randomization approach. In this latter case, it is assumed that, under the null hypothesis, each observed value could have occurred at all locations with equal likelihood (Cliff and Ord 1981). Alternatively, inference may be based on the permutation approach: instead of using theoretical mean and standard deviation of the randomization assumption, a reference distribution is empirically generated for I, from which the mean and standard deviation are computed. In practice, this is carried out by permuting the observed values over all locations and by re-computing I for each new sample. The mean and standard deviation for I are then computed moments for the reference distribution for all permutations (Anselin 1995).

2.2.2 Local Spatial Autocorrelation

Moran's I statistic is a global statistic and does not allow to assess the regional structure of spatial autocorrelation. However, it may be asked whether there are *local* spatial clusters of high or low values, which regions contribute more to the global spatial autocorrelation, and finally to what extent the global evaluation of spatial autocorrelation masks atypical localizations or "pockets of local nonstationarity", i.e. respectively regions or groups of neighboring regions, which deviate from the global pattern of spatial autocorrelation.

In this respect, the analysis of local spatial autocorrelation is analyzed with three tools. First, $G_i(d)$ statistics (Getis and Ord 1992; Ord and Getis 1995), based on spatial accumulations, are used to assess the significance of local spatial clustering around individual locations. Second, Moran scatterplots (Anselin 1996) allow visualizing local spatial instability and outliers. Third, Local Indicators of Spatial Association (Anselin 1995), based on spatial covariances, are also used to test the assumption of random distribution by comparing the values of each specific localization with the values in the neighboring locations.

2.2.2.1 Getis-Ord (1992, 1995) Statistics

Getis and Ord (1992) suggest the use of the $G_i(d)$ statistic to detect local "pockets" of dependence that may not show up when using global statistics. This statistic for each region i and year t can then be written as following (Getis and Ord 1992):

$$G_{i,t}(d) = \sum_{j \neq i} w_{ij}(d) x_{j,t} / \sum_{j \neq i} x_{j,t} \tag{2.3}$$

where $w_{ij}(d)$ are the elements of a symmetric binary spatial weight matrix equal to one for all links within distance d of a given region i and equal to zero for all other links including the link of region i to itself. The variable x has a natural origin and is positive.

Once standardized, a positive value of $G_{i,t}(d)$ indicates a spatial cluster of *high* values, whereas a negative value indicates clustering of *low* values around region i and year t. This statistic has been extended to variables that do not have a natural origin and to non binary standardized weight matrices (Ord and Getis 1995). In this case, it has the following expression:

$$G_{i,t}(d) = \frac{\sum_{j \neq i} w_{ij}(d) x_{j,t} - W_i \mu_t}{\sigma_t \left\{ \left[(n-1) \; S_{1i} - W_i^2 \right] \; /(n-2) \right\}^{1/2}} \tag{2.4}$$

where $W_i = \sum_j w_{ij}$; $S_{1i} = \sum_j w_{ij}^2$ for $j \neq i$; μ_t and σ_t are the usual sample mean and standard deviation for the sample of size n - 1 excluding region i for year t. The sign of this statistic is interpreted in the same way as the preceding one.

These statistics are based on spatial accumulations and can thus help to deepen the analysis for the detection of spatial clusters around each region i without being affected by the value taken by the variable in that region i.[1] Moreover, they may help reveal problems with the spatial scale of the observational units by incrementing d.

2.2.2.2 Moran Scatterplot

Local spatial instability is studied by means of the Moran scatterplot (Anselin 1996), which plots the spatial lag Wz_t against the original values z_t. The four different quadrants of the scatterplot correspond to the four types of local spatial association between a region and its neighbors: HH a region with a high[2] value surrounded by regions with high values, LH a region with a low value surrounded by regions with high values, LL a region with a low value surrounded by regions with low values, HL a regions with a high value surrounded by regions with low values. Quadrants HH and LL refer to positive spatial autocorrelation indicating spatial clustering of *similar* values, whereas quadrants LH and HL represent negative spatial autocorrelation indicating spatial clustering of *dissimilar* values. The Moran scatterplot may thus be used to visualize atypical localizations, i.e. regions in quadrants LH or HL. Moreover, the use of standardized variables makes the Moran scatterplots comparable across time.

Global spatial autocorrelation may also be visualized on this graph since, from (2.2) Moran's I is formally equivalent to the slope coefficient of the linear regression of Wz_t on z_t using a row-standardized weight matrix. The detection of outliers and sites, which exert strong influence on Moran's I, is then based on standard regression diagnostics (Anselin 1995): studentized residuals and leverage measures are used to detect outliers, and Cook's distance is an influence measure (Belsley et al. 1980; Donald and Maddala 1993). More precisely, the studentized residual is a measure of the extreme character of an observation along the dependent variable domain and is calculated as the studentized difference between the actual value and the predicted value. The leverage quantifies the extreme nature of an observation in the range of the independent variable and is assessed using the diagonal elements of the hat matrix (Haoglin and Welsch 1978).[3] Finally, the Cook's distance combines the two previous diagnostics and measures the extent to which regression coefficients are changed by the deletion of a particular observation (Cook 1977; Weisberg 1985).

Moran scatterplots do not give any indications of *significant* spatial clustering and therefore, they cannot be considered as a LISA in the sense defined by Anselin (1995).

---

[1] Note that statistics, which include the value taken by the variable in region i have also been suggested by Getis and Ord (1992, 1995).

[2] High (resp. low) means above (resp. below) the mean.

[3] The hat matrix is defined as $H = X(X'X)^{-1}X'$ where X is the matrix of observations on the explanatory variables in a regression.

2.2.2.3 Anselin (1995) Local Indicators of Spatial Association (LISA)

Anselin (1995) defines a Local Indicator of Spatial Association (LISA) as any statistics satisfying two criteria: first, the LISA for each observation gives an indication of significant spatial clustering of similar values around that observation; second, the sum of the LISA for all observations is proportional to a global indicator of spatial association.

The local version of Moran's I statistic for each region i and year t is written as:

$$I_{i,t} = \frac{(x_{i,t} - \mu_t)}{m_0} \sum_j w_{ij}(x_{j,t} - \mu_t) \text{ with } m_0 = \sum_i (x_{i,t} - \mu_t)^2 / n \tag{2.5}$$

where x_{it} is the observation in region i and year t; μ_t is the mean of the observations across regions in year t and where the summation over j is such that only neighboring values of j are included. It is straightforward to see that the sum of local statistics can be written:

$$\begin{aligned} \sum_i I_{i,t} &= \frac{1}{m_0} \sum_i (x_{i,t} - \mu_t) \sum_j w_{ij}(x_{j,t} - \mu_t) \\ &= \frac{1}{m_0} \sum_i \sum_j w_{ij}(x_{i,t} - \mu_t)(x_{j,t} - \mu_t) \end{aligned} \tag{2.6}$$

From (2.1), it follows that the global Moran's I statistic is proportional to the sum of local Moran's statistics:

$$I_t = \sum_t I_{i,t} / S_0 \tag{2.7}$$

For a row-standardized weight matrix, $S_0 = n$ so that $I_t = (1/n)\sum_t I_{i,t}$: the global Moran's I equals the mean of the local Moran's statistics.

A positive value for $I_{i,t}$ indicates spatial clustering of *similar* values (high or low) whereas a negative value indicates spatial clustering of *dissimilar* values between region i and its neighbors in year t. Note that this statistic is based on spatial covariances rather than spatial accumulation and measures a different concept of local spatial association compared to Getis and Ord (1992, 1995) statistics. It is therefore interesting to consider them in conjunction with $G_i(d)$ statistics.[4]

Anselin (1995) gives two interpretations of LISA. They can be used first as indicators of significant local spatial clusters ("hot spots") as the $G_i(d)$ statistics and second as diagnostics for local instability (atypical localizations or "pockets of nonstationarity"), significant outliers and spatial regimes. This second interpretation is similar to the use of a Moran scatterplot to identify outliers and leverage points for Moran's I: since there is a link between the local indicators and the global statistic, LISA outliers will be associated with the regions, which exert the

---

[4] Following Anselin (1995, p.101), the $G_i(d)$ statistic cannot be considered as LISA "since its individual components are not related to a global statistic of spatial association".

most influence on Moran's *I*. Finally, combining the information in a Moran scatterplot and the significance of LISA yields to the Moran significance map, showing the regions with significant LISA and indicating by a color code the quadrants in the Moran scatterplot to which these regions belong (Anselin and Bao 1997).

2.3 Data and Spatial Weight Matrix

We apply ESDA to European regional per capita GDP in logarithms expressed in ECUs, the former European Currency Unit, replaced by the Euro since 1999. The data are extracted from the EUROSTAT-REGIO databank. Our sample includes 138 regions in 11 European countries over the 1980-1995 period: United Kingdom (11) at NUTS1 level and Belgium (11), Denmark (1), France (21), Germany (30), Greece (13), Luxembourg (1), Italy (20), Netherlands (9), Portugal (5) and Spain (16) at NUTS2 level (see Appendix 1 for more details).

We are aware of all the shortcomings of the database we use, especially with regard to the adequacy of the regional breakdown adopted, which can raise a form of ecological fallacy problem (King 1997; Anselin and Cho 2000) or "modifiable areal unit problem" well known to geographers (Openshaw and Taylor 1979; Arbia 1989). The choice of the NUTS2 level as our spatial scale of analysis may appear quite arbitrary and may have some impact on our inference results. Regions in NUTS2 level may be too large in respect to the variable of interest and the unobserved heterogeneity may create an ecological fallacy, so that it might have been more relevant to use NUTS3 level. Conversely, they may be too small so that the spatial autocorrelation detected could be an artifact that comes out from slicing homogenous zones in respect to the variable considered, so that it might have been more relevant to use NUTS1 level. Even if, ideally, the choice of the spatial scale should be based on theoretical considerations, we are constrained in empirical studies by data availability. Moreover, our choice to prefer NUTS2 level to NUTS1 level, when data is available, is based on European regional development policy considerations: indeed it is the level at which eligibility under Objective 1[5] of Structural Funds is determined since their reform in 1989. Our empirical results are conditioned by this choice and could be affected by different levels of aggregation and even by missing regions.

Figs. 2.1 and 2.2 display per capita GDP relative to the sample average in 1980 and in 1995. Poorer regions appear here with per capita GDP less than 75% of the sample average: they are mainly southern regions. Richer regions appear with per capita GDP more than 125% of the sample average and are mainly northern regions. We must note that the specific geographical configuration of European regions will indeed have some consequences in the choice of the weight matrix to which we turn now.

---

[5] For regions where development is lagging behind (in which per capita GDP is generally below 75% of the EU average). More than 60% of total EU resources used to implement structural policies are assigned to Objective 1.

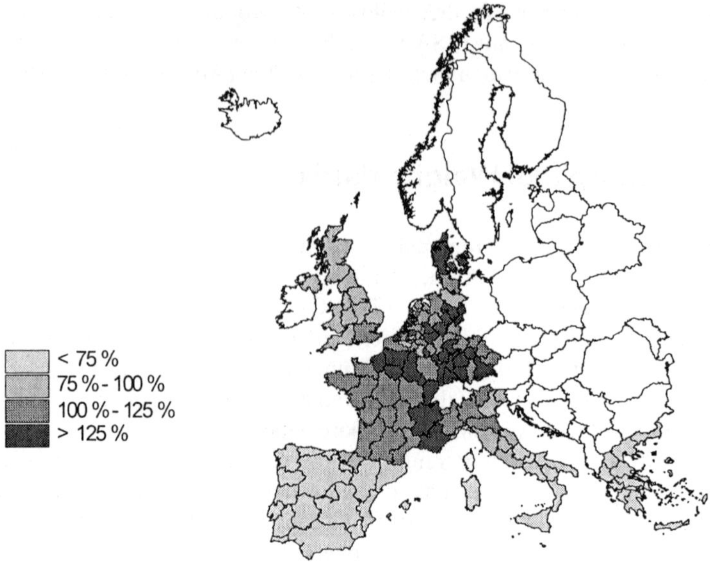

Fig. 2.1 Per capita GDP relative to the sample average in 1980

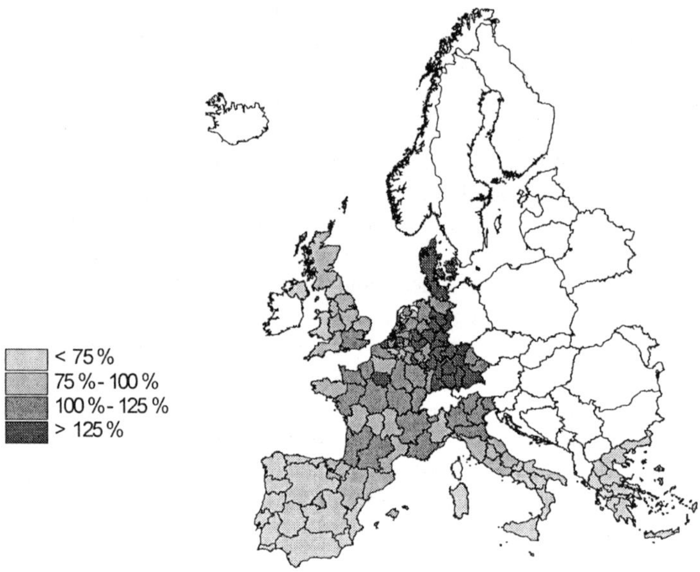

Fig. 2.2 Per capita GDP relative to the sample average in 1995

Given the characteristics of our sample, the spatial weight matrix W we use in this study is based on the k-nearest neighbors computed from the great circle distance between region centroids. Nearest neighbors matrices are also used by Pace and Barry (1997), Pinkse and Slade (1998) and Baller et al. (2001) in different contexts. It is important to keep in mind that all subsequent analyses are conditional upon the choice of the spatial weight matrix. Indeed, the results of diagnostic tests are dependent on spatial weights. We use therefore k = 10,15,20,25 to check for robustness of our results (see Sect. 2.4.5 below). The general form of the k-nearest neighbors weight matrix $W(k)$ is defined as following:

$$\begin{cases} w^*_{ij}(k) = 0 \text{ if } i = j \\ w^*_{ij}(k) = 1 \text{ if } d_{ij} \le d_i(k) \\ w^*_{ij}(k) = 0 \text{ if } d_{ij} > d_i(k) \end{cases} \quad \text{and} \quad w_{ij}(k) = w^*_{ij}(k) \Big/ \sum_j w^*_{ij}(k) \tag{2.8}$$

where $d_i(k)$ is a critical cut-off distance defined for each region i. More precisely, $d_i(k)$ is the kth order smallest distance between regions i and j such that each region i has exactly k neighbors. These matrices are preferred to the simple contiguity matrix, as used for example by López-Bazo et al. (1999), for various reasons. First, they connect a number of islands such as Sicilia, Sardegna, and Baleares to continental Europe thus avoiding rows and columns in W with only zero values. With a simple contiguity matrix, unconnected observations are indeed implicitly eliminated from the computed spatial autocorrelation statistics but this leads to a change in the sample size and must be explicitly accounted for in statistical inference. Second, with $k = 10$, the United Kingdom is connected to continental Europe and Greece to Italy, therefore avoiding the block-diagonal structure of the simple contiguity matrix when ordered by country. The spatial connections between regions belonging to different countries are thus guaranteed. Actually, 24.28% of the 10-nearest neighbors belong to a different country. This ratio increases when we increment the number of nearest neighbors in our robustness analysis. Third, by choosing a fixed number of nearest neighbors, we avoid specific methodological problems that may occur when the number of neighbors is allowed to vary. This problem arises in the case of a simple contiguity matrix or a distance based weight matrix with the same fixed distance critical cut-off for all regions. These features are of particular interest when working on a sample of European regions, which are less closely connected and compact than US states.[6]

---

[6] There must be at least 10 neighbors for United Kingdom and Greece to be connected with continental Europe so that the weight matrix is not block-diagonal. However, using k = 10,15,20 and 25 implies many neighbors in the weighting scheme as pointed out by an anonymous referee. For example 10 means that almost all UK regions are connected and 25 that almost all German regions are connected. However, we only use k = 10,15,20 and 25 for studying the robustness of our results (see Sect. 2.4.5). The choice of nearest neighbors matrices was also based on the fact that it was not possible to obtain inference using the permutation approach for LISA statistics using a distance-based weight matrix.

2.4 ESDA on Regional per Capita GDP, 1980-1995

The first step of ESDA is to evaluate global spatial autocorrelation in the sample to determine if spatial concentration of similar values in terms of per capita GDP in logarithms exists. Then, we explicitly describe these spatial concentrations of poor or rich regions with three tools : Getis-Ord statistics, Moran scatterplots and LISA statistics. Finally, we evaluate the robustness of all our results with an original application of transition probability matrices.

2.4.1 Global Spatial Autocorrelation

Table 2.1 displays the evolution of Moran's I statistic of log per capita GDP over the 1980-1995 period for the 138 European regions of the sample. Inference is based on the permutation approach with 10 000 permutations (Anselin 1995). It appears that per capita regional GDPs are positively spatially autocorrelated since the statistics are significant with $p = 0.0001$ for every year.[7] This result suggests that the distribution of per capita regional GDP is by nature clustered over the whole period. In other words, the regions with relatively high per capita GDP (resp. low) are localized close to other regions with relatively high per capita GDP (resp. low) more often than if their localizations were purely random. We note that the standardized values of Moran's I statistic appear to be very high possibly indicating a spatial scale problem. This could imply that part of the autocorrelation detected derives from the NUTS2 regional breakdown adopted by Eurostat.

Considering the evolution of Moran's I statistic over time shows that the standardized values of the statistic remain approximately the same over the whole period. Therefore, it indicates a global significant tendency toward geographical clustering of similar regions in terms of log per capita GDP.

Moran's I statistics computed using 15, 20 and 25 nearest neighbors lead to the same results for the sign and significance of global spatial autocorrelation, highlighting the robustness of our results with regard to the choice of the spatial weight matrix.

2.4.2 $G_i(k)$ Statistics (Ord and Getis 1995)

Since Moran's I yields a single result for the entire data set, it cannot discriminate between a spatial clustering of high values and a spatial clustering of low values in the case of global positive spatial autocorrelation. Such clusters are detected by computing the extended Ord and Getis (1995) statistic for k nearest neighbors.

---

[7] All computations are carried out using SpaceStat 1.90 (Anselin 1999) and Arcview 3.2 (Esri).

Table 2.1 Moran's *I* statistics for log per capita GDP over 1980-1995

	W(10)			*W*(15)		
Year	**Moran's I**	**St. dev.**	**St. Value**	**Moran's I**	**St. dev.**	**St. value**
1980	0.774	0.033940	23.024	0.742	0.027106	27.643
1981	0.760	0.033971	22.574	0.727	0.027127	27.045
1982	0.746	0.033956	22.161	0.710	0.027084	26.478
1983	0.779	0.034083	23.060	0.737	0.027186	27.385
1984	0.757	0.034019	22.446	0.710	0.027128	26.447
1985	0.766	0.034077	22.692	0.722	0.027184	26.808
1986	0.785	0.034126	23.213	0.733	0.027220	27.178
1987	0.789	0.034164	23.289	0.733	0.027250	27.171
1988	0.773	0.034196	22.802	0.719	0.027259	26.632
1989	0.750	0.034221	22.113	0.689	0.027206	25.594
1990	0.762	0.034242	22.461	0.698	0.027248	25.860
1991	0.754	0.034311	22.174	0.688	0.027292	25.479
1992	0.770	0.034323	22.651	0.708	0.027316	26.192
1993	0.790	0.034272	23.259	0.746	0.027306	27.567
1994	0.799	0.034267	23.514	0.759	0.027304	28.043
1995	0.802	0.034222	23.653	0.766	0.027268	28.357

	W(20)			*W*(25)		
Year	**Moran's *I***	**St. dev.**	**St. value**	**Moran's *I***	**St. dev.**	**St. value**
1980	0.688	0.022623	30.713	0.602	0.019475	27.643
1981	0.672	0.022635	30.011	0.583	0.019488	27.045
1982	0.653	0.022624	29.185	0.561	0.019483	26.478
1983	0.674	0.022728	29.985	0.578	0.019558	27.385
1984	0.644	0.022697	28.698	0.549	0.019524	26.447
1985	0.653	0.022752	29.032	0.558	0.019574	26.808
1986	0.654	0.022783	29.037	0.556	0.019617	27.178
1987	0.651	0.022837	28.827	0.553	0.019659	27.171
1988	0.637	0.022855	28.194	0.542	0.019667	26.632
1989	0.602	0.022836	26.678	0.506	0.019661	25.594
1990	0.604	0.022865	26.724	0.506	0.019690	25.860
1991	0.594	0.022938	26.205	0.497	0.019738	25.479
1992	0.618	0.022946	27.229	0.521	0.019734	26.192
1993	0.675	0.022895	29.805	0.588	0.019689	27.567
1994	0.694	0.022872	30.648	0.608	0.019662	28.043
1995	0.708	0.022829	31.308	0.626	0.019626	28.357

Note: The expected value for Moran's I statistic is constant for each year: $E(I) = -0.007$. All statistics are significant at $p = 0.0001$.

In this case, inference is based on the normal asymptotic approximation as suggested by Ord and Getis (1995) even though they concede there might be a problem in presence of global spatial autocorrelation (Ord and Getis 1995, pp. 298-299).[8] Inference is further complicated by the fact that these local statistics will be correlated when the neighborhood sets of two regions contain common elements (Anselin 1995; Ord and Getis 1995). This is actually a problem of multiple statistical comparison.[9] As pointed out by Anselin (1995, p. 96): "This means that when the *overall* significance associated with the multiple comparisons (correlated tests) is set to α, and there are m comparisons, then the individual significance α_i should be set to α/m (Bonferroni) or $1-(1-\alpha)^{1/m}$ (Sidák)". The second procedure requires the variables to be multivariate normal, which is asymptotically the case for $G_i(k)$. With $m = n$ (the number of regions of the sample), as in most of the empirical papers (Pereira et al. 1998), these procedures can be overly conservative in evaluating the significance of the $G_i(k)$ statistics (Anselin 1995; Ord and Getis 1995). However, using k-nearest neighbors spatial weight matrices, we note that the number of comparisons cannot exceed k because two given regions cannot have more than k common neighbors. In this respect, we suggest using $m = k$ and we present the results obtained with both the usual 5% significance level, which may be too liberal, and the 5% Sidák pseudo-significance level with $m = k$, which is indeed less conservative than using $m = n$. These two significance levels can therefore be considered as the two extreme bounds for inference.

The results of this procedure for $k = 10$ are summarized in the Table of Appendix 2. Columns 2 and 3 report the number of years during which $G_i(10)$ is significantly positive or negative and the corresponding years are displayed in the columns 4 and 5.[10] Using the 5% significance level, we mainly detect significant local clusters of high values around northern French regions, Belgian, Dutch and German regions (57.41% of significant statistics). Northern Italian regions are mostly significant in the middle of the period. However, only northern German regions remain significant at the 5% Sidák pseudo-significance level. These clusters are highly persistent over the whole period. Local clusters of low values are also highly persistent and concern the poorest regions of the European Union: most of the Spanish regions, some southern Italian regions, all the Greek and Portuguese regions.[11] We note that these poor regions entered the EU in 1986 and belong to the "cohesion countries". They have benefited since 1989 from regional aid to the Structural Funds Objective 1 regions but over the period as a whole, the per capita GDP of these regions has remained lower than the average.

---

[8] Recently, Ord and Getis (2001) have suggested a new local spatial autocorrelation statistic in presence of global autocorrelation.

[9] More about this problem can be found in Savin (1984).

[10] 54.12% of these indicators are significant at the 5% significance level (1 195 versus a total of 2 208) and only 29.30% at the 5% Sidák pseudo-significance level (647 versus a total of 2 208).

[11] All the Greek and Portuguese regions are even significant using the 5% Sidák pseudo-significance level.

Globally, these results reveal the high persistence of spatial inequality among the European regions over time following the well-known North-South polarization, to the extent that only the neighborhood of these regions is considered.[12] This specific polarization type can be interpreted as a core-periphery scheme developed in economic geographic models (Krugman 1991). We therefore detect a characteristic form of spatial heterogeneity between European regions.

The 10 highest and 10 smallest extreme values for $G_i(10)$ statistics are shown in Table 2.3 below. In 1980 and 1995, clusterings of very high values are only to be found in Germany. In 1980, clusterings of very low values mainly concern Portuguese and Greek regions while in 1995 all the Portuguese regions are replaced by some more Greek regions.

Usually, the Getis-Ord statistics are used to determine the appropriate scale of analysis by incrementing the critical cut-off distance d (in our case, it would mean incrementing the number of neighbors k). One possibility is to choose the d (or the k) value that maximizes the absolute sum of the $G_i(d)$ (or the $G_i(k)$) statistics for all the observations (Getis 1995). However, in our case, the absolute sum never decreases indicating the non-cessation of spatial effects when the number of nearest neighbors taken into account is incremented. We therefore think that it is more informative to check the robustness of the empirical results using different k values (see below Sect. 2.4.5).

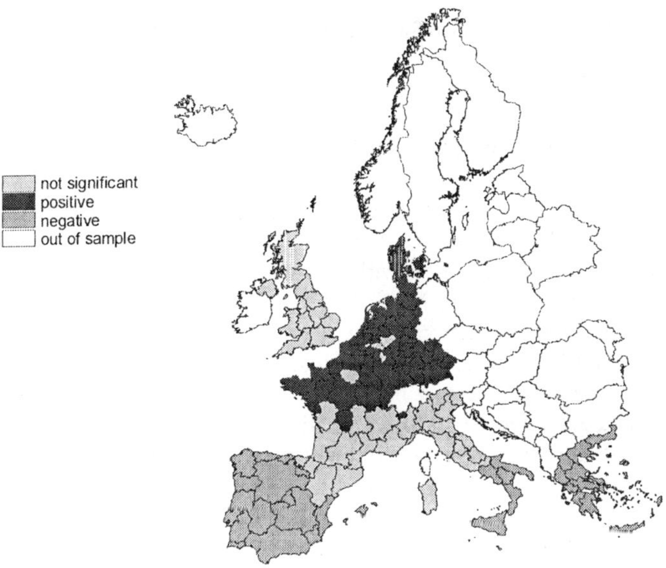

Fig. 2.3 Getis-Ord significance map for log per capita GDP in 1980 (5% significance level)

[12] Detailed results for the initial and final years of our sample are displayed in figures 2.3, 2.4 and in Table 2.8.

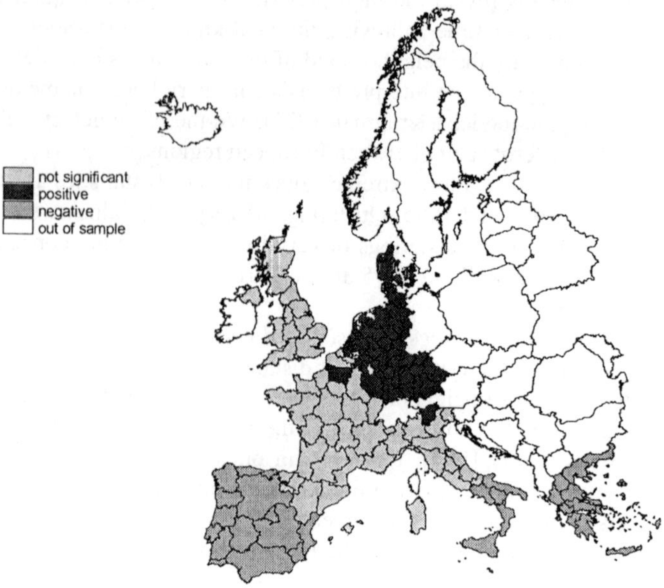

Fig. 2.4 Getis-Ord significance map for log per capita GDP in 1995 (5% significance level)

2.4.3 Moran Scatterplots (Anselin 1996)

Another way to detect local spatial clusters but also to analyze local instability in the form of atypical localizations, spatial outliers and spatial regimes is to use Moran scatterplots in conjunction with LISA (Anselin 1995). In the presence of global positive autocorrelation, Moran's I statistic may indeed mask regions that deviate from this global pattern.

Figures 2.5 and 2.6 display the Moran scatterplots for the initial and final years of our sample and $k = 10$ nearest neighbors.[13] It can be seen that most European regions are characterized by positive spatial association. More specifically, in 1980, 97.8% of European regions exhibited association of similar values (65.2% in quadrant HH and 32.6% in quadrant LL). In 1995, 94.9% of the European regions exhibited the same positive association (56.5% in quadrant HH and 32.4% in quadrant LL). It is interesting to note that some English regions are the only "northern" regions belonging to quadrant LL in 1995. These results therefore highlight the difficulties of old industrial regions in United-Kingdom like North, North-West or Yorkshire and Humberside.

[13] Detailed results for the initial and final years of our sample are displayed in Table 2.8.

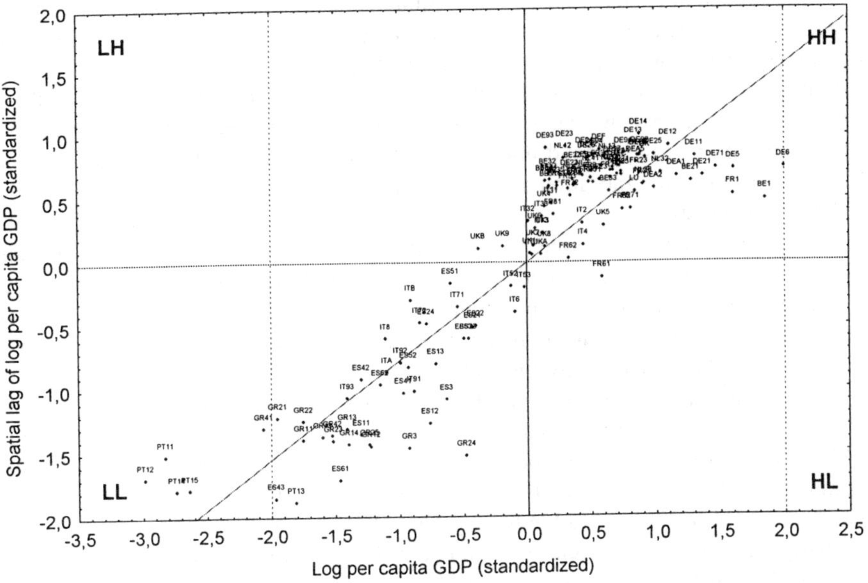

Fig. 2.5 Moran scatterplot for log per capita GDP in 1980

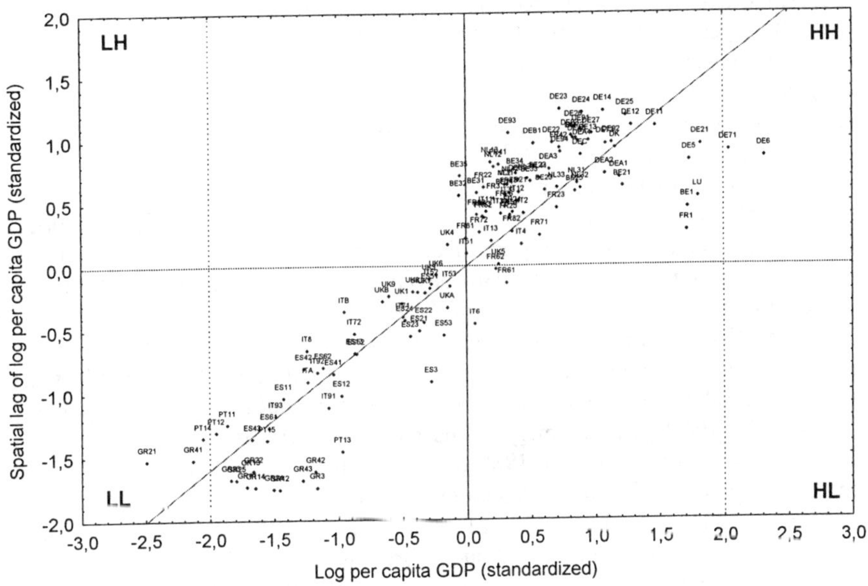

Fig. 2.6 Moran scatterplot for log per capita GDP in 1995

Furthermore, the Moran scatterplots can help to identify atypical regions, i.e. regions deviating from the global pattern of positive autocorrelation. In 1980, only 3 regions displayed association of dissimilar values: Wales (United-Kingdom) and

Northern Ireland in quadrant LH and Aquitaine (France) in quadrant HL. In 1995, there are 7 atypical regions: Hainaut and Namur (Belgium), Languedoc-Roussillon (France), East Anglia (United Kingdom) in quadrant LH and Aquitaine, Midi-Pyrénées (France) and Lazio (Italy) in quadrant HL.[14] The results therefore favor the existence of only two regimes, the first one corresponding to the regions in the HH quadrant and the second one to the regions in the LL quadrant. These results are consistent with the simple North-South polarization pattern found with the Getis-Ord statistics.

All these results are based on the analysis of the initial and the final periods. However, they can be extended to the whole period. More insight into the evolution of Moran's scatterplots over time is then provided by a newly introduced measure of space-time transitions based on the classification over time of a region and its neighbors in four groups (Rey 2001). The first group includes the transitions with a relative move of only the region, for example a HH region in the first period becoming a LH region in the following period. The other cases are HL-LL, LH-HH and LL-HL. The second group contains the transitions of the neighbors only: HH-HL, HL-HH, LH-LL and LL-LH while the transitions of both a region and its neighbors belong to the third group: HH-LL, HL-LH, LH-HL and LL-HH. Finally, the 4 cases in which the region and its neighbors remain at the same level are in the fourth group. High stability in the types of transitions is reflected by a large number of type 4 transitions and low values of the flux (or instability) measure, which is defined as the frequency of the first and second types of transitions over all 15 years of transitions. For time intervals of 1, 5 and 10 years, type 4 transitions are always the most common ones (95.6%, 89.9% and 85.3%) and the flux measure is respectively equal to 4%, 7.9% and 8.8%. These results denote a high stability between European regions and a low rate of mobility, increasing slowly with the transition interval.

Finally, the Moran scatterplot can also be used to assess the presence of outliers, which are defined as the points further than two units away from the origin. For example, in 1995, Hamburg and Darmstadt (Germany) were outliers with per capita GDPs more than two standard deviations above the mean (Fig. 2.6). Note however that the production activity of these city-regions is overstated due to large commuter flows, while in surrounding regions, where commuters live, it is understated. Some poor objective 1 regions are outliers with per capita GDP less than two standard deviations below the mean. There are no outliers on the vertical axis.

Table 2.2 displays a summary of the most extreme observations according to the regression diagnostics for 1980 and 1995: first, studentized residuals larger than 2 in absolute value and second, observations associated with leverages higher than $2p/n$ (where p is the number of explanatory variables in the regression, i.e. $p = 2$ and $n = 138$). There are 12 such observations in 1980 and 1995, most of them being located in Portugal, Greece and Germany. Finally, no region can be considered to be influential according to the Cook's distance criterion. This latter result suggest that, although some regions have high leverages and studentized re-

---

[14] The Moran scatterplots computed with the other nearest neighbors matrices lead to sensibly the same picture.

siduals, no region appears to be particularly influential in the sample. Therefore, Moran's I cannot be considered as being strongly influenced by any given region and appears to be a general feature of the sample.

Table 2.2 Moran scatterplot outliers: initial and terminal years for log per capita GDP

	1980		1995	
	Region	**Studentized Residual**	**Region**	**Studentized Residual**
Studentized Residuals Exceeding 2 in absolute Value	Sterea Ellada (1)	-3.440	Ile-de-France	-3.139
	Bruxelles	-2.893	Hamburg	-2.886
	Hamburg	-2.500	Bruxelles	-2.654
	Attiki (1)	-2.298	Luxembourg (Lux)	-2.612
	Ile-de-France	-2.226	Attiki (1)	-2.337
	Asturias (1)	-2.099	Darmstadt	-2.130
	Lüneburg	2.073	Madrid	-2.005

	1980		1995	
	Region	**Leverage**	**Region**	**Leverage**
Leverage Exceeding 4/n	Centro (1)	0.072	Ipeiros (1)	0.052
	Norte (1)	0.066	Hamburg	0.046
	Alentejo (1)	0.062	Voreio Aigaio (1)	0.040
	Algarve (1)	0.058	Alentejo (1)	0.038
	Voreio Aigaio (1)	0.038	Darmstadt	0.038
	Hamburg	0.036	Centro (1)	0.035
	Extremadura (1)	0.035	Norte (1)	0.033
	Ipeiros (1)	0.035	Dyptiki Ellada (1)	0.032
	Bruxelles	0.032	Oberbayern	0.031
	Lisboa (1)	0.031	Luxembourg (Lux)	0.031
	Ionia Nisia (1)	0.029	Peloponnissos (1)	0.031
	Anatoliki Makedonia (1)	0.029	Bremen	0.029

Note: (1) denotes Objective 1 regions since 1989

Moran scatterplots allow capturing the persistence of economic disparities. However, since they do not provide any significance result, these different patterns should be assessed by the use of local indicators of spatial association. All these findings will therefore be refined by the study of LISA statistics.

2.4.4 Lisa (Anselin 1995)

LISA were computed to further examine these results (consistent with the periodic Economic and Social Cohesion reports of the European Commission, e.g. 1996). With the aim of identifying the spatial movements that occurred during the whole 1980-1995 period, we retain only the phenomena of local spatial clusters for which the local Moran's statistics are significant.

Due to the presence of global spatial autocorrelation, inference must be based on the conditional permutation approach with 10 000 permutations (Anselin 1995). It should be stressed that p-values obtained for the local Moran's statistics are actually pseudo-significance levels. In addition, as normality is unlikely to be the case with LISA we use the Bonferroni 5% pseudo-significance level with $m = k$ instead of the Sidák 5% pseudo-significance level.

The number of years over the whole period with significant LISA is displayed in column 6 of the Table in the Appendix 2.[15] The number of years during which the region falls into a certain quadrant of the Moran scatterplot with a significant LISA are displayed in the following columns and the corresponding years are finally displayed in columns 11 and 12 in the Table. The associated Moran significance maps at the 5% pseudo-significance level for the initial and final years are presented in Figs. 2.7 and 2.8.[16] Several points can be highlighted.

First, the local pattern of spatial association reflects the global trend toward positive spatial autocorrelation since 98.84% of the significant LISA, using the 5% pseudo-significance level, fall either into quadrant HH or in quadrant LL of the scatterplot. We note, however, that the distribution between associations of the HH and LL types is uneven since 62.27% of the regions fall into quadrant HH: we thus detect mainly regions or sets of regions with high per capita GDP surrounded by other regions with high per capita GDP.[17]

Second, deviations of the global trend are extremely limited and are dominated by the LH type of negative spatial association (1.17% of the significant LISA). Only Madrid (Spain) is significantly HL ("diamond in the rough") for 1991 and 1992. The "doughnuts" or LH regions are Brabant Wallon and Namur (Belgium) for 3 years, Hainault (Belgium) for 2 years, Friesland (The Netherlands) for 6 years and Drenthe (The Netherlands) for just one year.[18]

Third, four regional clusters persist in time. The first is a significant LL type of clustering between all the Portuguese regions and almost all the Spanish regions. Two other LL form of clusterings are located between some Italian Objective 1 regions (Campania, Puglia, Basilicata, Calabria, Sicilia) and between all the Greek regions (the Greek and the Portuguese regions are even significant using the 5% Bonferroni pseudo-significance level). The last clustering, of the HH type, mainly concerns German, French, Belgian and Dutch regions. All these results show a high persistence of spatial disparities between European regions across time: rich regions surrounded by rich neighbors still benefit from their environment whereas poor regions with poor neighbors remain negatively affected.

---

[15] 66.26% of these indicators are significant at the 5% pseudo-significance level (1 463 versus a total of 2 208) and only 41.35% at the 5% Bonferroni pseudo-significance level (913 versus a total of 2 208).

[16] Detailed results for the initial and final years of our sample are displayed in Table 2.8.

[17] Using the Bonferroni 5% pseudo-significance level, 53.9% of significant LISA fall in quadrant HH and 46.1% of significant LISA fall in quadrant LL.

[18] No atypical localization is found when the Bonferroni 5% pseudo-significance level is used.

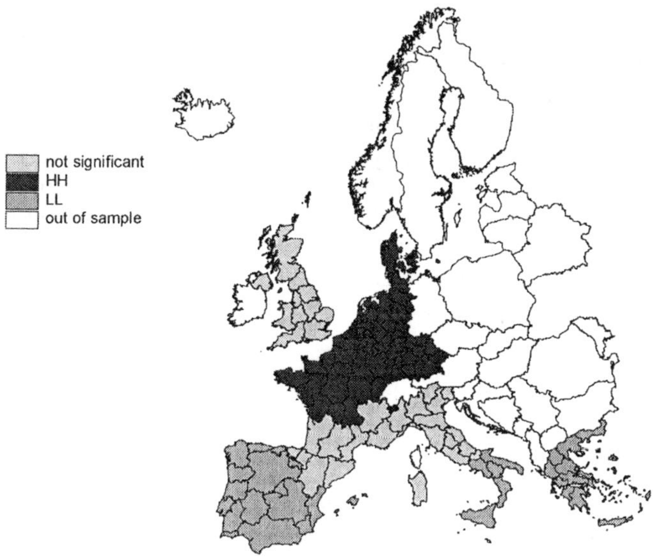

Fig. 2.7 Moran significance map for log per capita GDP in 1980 (5% pseudo-significance level)

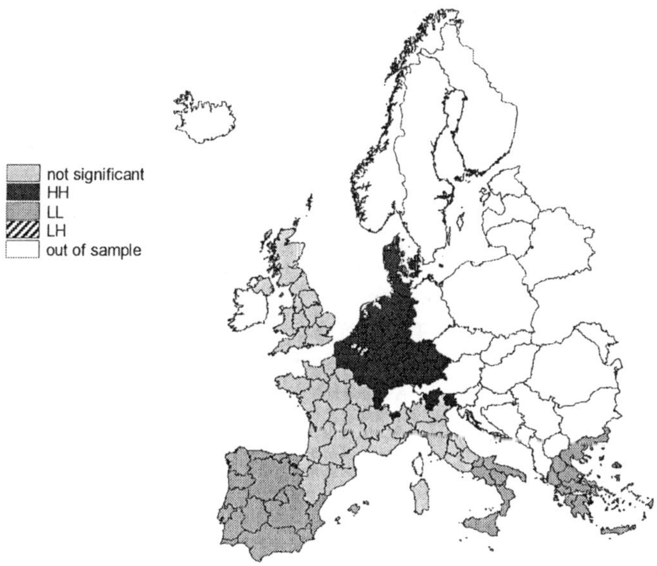

Fig. 2.8 Moran significance map for log per capita GDP in 1995 (5% pseudo-significance level)

These results strongly indicate the existence of only two regimes, the first one corresponding to the HH scheme and the second one to the LL scheme, both of them representing positive spatial association. The North-South polarization pattern appears therefore to be the most relevant feature in our sample with regard to spatial heterogeneity. Furthermore, these results are consistent with those obtained with the Getis-Ord statistics and Moran scatterplots.

The spatial outliers identified by the 2-sigma rule are shown in Table 2.3. In 1980, all Portuguese regions as well as the Spanish region Extremadura formed clusters with very similar values. The situation in 1995 was different since all the Portuguese regions, except Alentejo, were replaced by Greek regions, indicating the good performance in terms of economic development of the former regions after joining the European Union.

Table 2.3 Getis-Ord extreme values and LISA outliers : initial and terminal years for log per capita GDP

	1980	
Getis-Ord 10 largest values	**Getis-Ord 10 smallest values**	**LISA outliers 2-sigma rule**
Tübingen	Lisboa (1)	Extremadura (1)
Freiburg	Extremadura (1)	Norte (1)
Karlsruhe	Alentejo (1)	Centro (1)
Oberpfalz	Algarve (1)	Lisboa (1)
Lüneburg	Centro (1)	Alentejo (1)
Schleswig-Holstein	Andalucia (1)	Algarve (1)
Hannover	Norte (1)	
Oberfranken	Sterea Ellada (1)	
Braunschweig	Attiki (1)	
Weser-Ems	Kentriki Makedonia (1)	

	1995	
Getis-Ord 10 largest values	**Getis-Ord 10 smallest values**	**LISA outliers 2-sigma rule**
Oberpfalz	Kentriki Makedonia (1)	Anatoliki Makedonia (1)
Tübingen	Sterea Ellada (1)	Kentriki Makedonia (1)
Oberfranken	Thessalia (1)	Dytiki Makedonia (1)
Mittelfranken	Anatoliki Makedonia (1)	Thessalia (1)
Karlsruhe	Attiki (1)	Ipeiros (1)
Stuttgart	Peloponnisos (1)	Ionia Nisia (1)
Unterfranken	Dytiki Ellada (1)	Dytiki Ellada (1)
Braunschweig	Kriti (1)	Sterea Ellada (1)
Schwaben	Dytiki Makedonia (1)	Peloponnisos (1)
Lüneburg	Ionia Nisia (1)	Voreio Aigaio (1)
		Alentejo (1)

Note: (1) denotes Objective 1 regions since 1989

2.4.5 Robustness Analysis Using Transition Probability Matrices

In order to study the robustness of our results with respect to the choice of the weight matrix, we suggest an original application of transition probability matrices. The results are robust if a region in a particular state with the 10-nearest neighbor matrix (i.e. not significant, positive or negative for $G_i(k)$ statistics; not significant, HH, LL, HL or LH for LISA statistics), remains in this state for the other weight matrices (15, 20 or 25 neighbors). Transition probability matrices are therefore an appropriate tool for studying this issue since they can be used to describe the transitions between two different distributions *for a given point in time.*[19]

Table 2.4 reports three different transition probability matrices for studying the robustness of the $G_i(k)$ statistics with respect to the choice of the nearest neighbor matrix. The first matrix expresses the transitions of the regions between the three different possible $G_i(k)$ states with 10 neighbors and the three possible $G_i(k)$ states with 15 neighbors. For example, with the 5% Sidák pseudo-significance level [20], there were 1 561 instances of a region with a non-significant $G_i(k)$ a given year with 10 neighbors, 77.5% of these same regions had a $G_i(k)$ statistic still not significant the same year with 15 neighbors, 16.3% had a significant $G_i(k)$ that became positive and 6.1% had a significant $G_i(k)$ that became negative. The other two matrices display similar transitions with 20 and 25 neighbors. Robustness of the results to the choice of neighbors would imply that these three matrices are fundamentally diagonal: indeed, a region is in the same $G_i(k)$ state whatever the number of neighbors chosen.

In this respect, Table 2.4 provides two interesting results. First, the more neighbors there are, the more regions there are with significant $G_i(k)$ statistics. This is reflected by the first line of each matrix. The percentage of regions with non-significant $G_i(k)$ for 10 neighbors that are still associated with a non-significant $G_i(k)$ for 15 neighbors is 77.5%. This probability drops to 58.5% for 20 neighbors and 48.8% for 25 neighbors. When a region is associated with a $G_i(k)$ statistic that becomes significant with a larger number of neighbors, then the associated $G_i(k)$ statistic mainly becomes positive (16.3% with 10 neighbors, 32.4% with 15 neighbors and 41.5% with 25 neighbors). This is the case for most of the German regions, three French regions close to Germany, two regions in the North of Italy. The $G_i(k)$ statistic becomes significantly negative for all the remaining Spanish regions when adding neighbors.

Second, the sub-matrices corresponding to the two $G_i(k)$ states (positive and negative) are diagonal, which means that a region with a significant $G_i(k)$ keeps the same sign when different numbers of neighbors are chosen.

---

[19] These particular kinds of transition probability matrices have been used to study the impact of spatial conditioning of regional income distributions in Europe (Quah 1996b; Le Gallo 2001) and in the United States (Rey 2001).

[20] All the robustness results remain similar using the 5% asymptotically normal significance level.

Table 2.4 Robustness analysis for $G_i(k)$ statistics

		10 to 15 neighbors		
	Total number	Not significant	Positive	Negative
Not significant	1561	0.775	0.163	0.061
Positive	228	0.000	1.000	0.000
Negative	419	0.000	0.000	1.000

		10 to 20 neighbors		
	Total number	Not significant	Positive	Negative
Not significant	1561	0.585	0.324	0.091
Positive	228	0.000	1.000	0.000
Negative	419	0.000	0.000	1.000

		10 to 25 neighbors		
	Total number	Not significant	Positive	Negative
Not significant	1561	0.488	0.415	0.097
Positive	228	0.000	1.000	0.000
Negative	419	0.000	0.000	1.000

Table 2.5 reports similar transition probability matrices for the LISA statistic using 5% Bonferroni significance level.[21] Again, a higher number of neighbors implies that there are more regions with a LISA statistic falling in the first quadrant (17% for 10 neighbors, 28.3% for 15 neighbors and 40.3% for 25 neighbors). The regions concerned are the same as those associated with a positive G_i statistic. The submatrices corresponding to the four states of the LISA statistic (HH, LH, LL, HL) are diagonal as well. Therefore, a region associated with a significant LISA statistic remains in the same state whatever the number of neighbors chosen.

All these findings therefore highlight the robustness of our results with respect to the choice of the weight matrix and the number of neighbors.

2.5 Spatial Patterns of Average Growth Rates

The preceding results on per capita GDP levels indicated high persistence of spatial inequality between northern and southern regions during the 1980-1995 period. However, if poor regions grow faster than rich regions, this spatial inequality may decrease in the future.

---

[21] All the robustness results remain similar using the 5% pseudo-significance level.

Table 2.5 Robustness analysis for LISA statistics

10 to 15 neighbors						
	Total number	Not significant	HH	LH	LL	HL
Not significant	1295	0.754	0.170	0.072	0.002	0.002
HH	492	0.002	0.998	0.000	0.000	0.000
LH	421	0.000	0.000	1.000	0.000	0.000
LL	0	0.000	0.000	0.000	0.000	0.000
HL	0	0.000	0.000	0.000	0.000	0.000

10 to 20 neighbors						
	Total number	Not significant	HH	LH	LL	HL
Not significant	1295	0.586	0.283	0.110	0.002	0.019
HH	492	0.000	1.000	0.000	0.000	0.000
LH	421	0.000	0.000	1.000	0.000	0.000
LL	0	0.000	0.000	0.000	0.000	0.000
HL	0	0.000	0.000	0.000	0.000	0.000

10 to 25 neighbors						
	Total number	Not significant	HH	LH	LL	HL
Not significant	1295	0.432	0.403	0.115	0.002	0.049
HH	492	0.000	1.000	0.000	0.000	0.000
LH	421	0.000	0.000	1.000	0.000	0.000
LL	0	0.000	0.000	0.000	0.000	0.000
HL	0	0.000	0.000	0.000	0.000	0.000

Therefore, in a refinement of the analysis, we apply the ESDA techniques to the average growth rates of per capita GDP in order to study the geographical patterns of the growth process.

Computation of Moran's I statistics on the annual average growth rates of per capita GDP between 1980 and 1995 of the various regions reveals a positive spatial autocorrelation (12.709 with a p-value of 0.0001). We note that its standardized value is well below the values computed for per capita GDP levels. This means though that regions with relatively high per capita GDP average growth rate (resp. low) are localized close to other regions with relatively high per capita GDP average growth rate (resp. low) more often than if these localizations were purely random.

The procedure of evaluating local spatial autocorrelation applied to the average growth rates (Fig. 2.9 and Table 2.8) shows that only 34.8% of the Getis-Ord G_i

statistics are significant at the 5% significance level.[22] Significant local spatial clusters of high values are found mainly in Portugal, Spain and Southern German regions, which are the only northern regions belonging to a local clustering of high growth rate values. Inversely, British, French and some Greek regions belong to local concentrations of low growth rate values. Extreme values are displayed in Table 2.6.

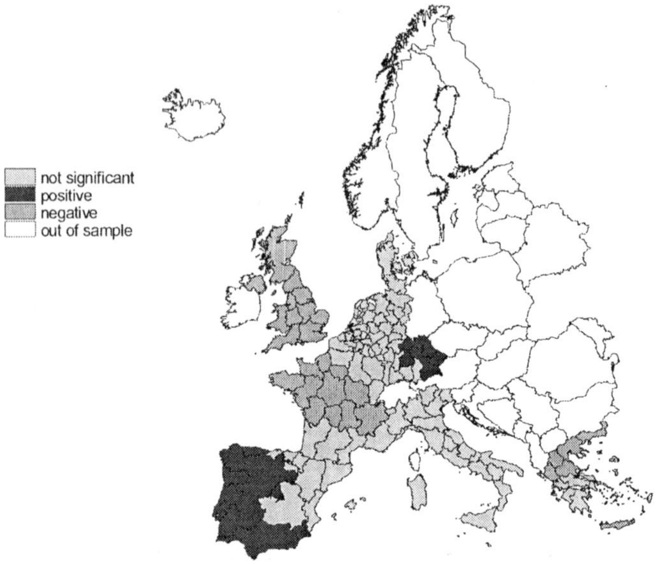

Fig. 2.9 Getis-Ord significance map for average growth rate of per capita GDP over 1980-1995 (5% significance level)

Table 2.6 Getis-Ord extreme values: average growth rates for log per capita GDP

Getis-Ord extreme values	
10 largest values	**10 smallest values**
Extremadura (1)	Northern Ireland (1)
Alentejo (1)	Scotland
Lisboa (1)	Wales
Algarve (1)	North
Andalucia (1)	South West
Centro (1)	West Midlands
Galicia (1)	Yorkshire and Humberside
Norte (1)	East Midlands
Asturias (1)	North West
Oberfranken	Bourgogne

Note: (1) denotes Objective 1 regions since 1989

[22] 13.8% of these statistics are positively significant and 21% are negatively significant.

The Moran scatterplot for average growth rates is displayed in Fig. 2.10. Compared with the scatterplots for per capita GDP in 1980 and 1995, there is much more spatial instability. Indeed, much more regions belong to the HL or the LH quadrants: only 73.19% of the European regions exhibit association of similar values (33.33% in quadrant HH and 39.86% in quadrant LL) while 26.81% of the regions are negatively associated (11.59% in quadrant LH and 15.22% in quadrant HL). All Portuguese regions and Luxembourg have growth rates more than two standard deviations above the mean while only Sterea Ellada (Greece) has a growth rate less than two standard deviations below the mean. We saw that some of these regions were outliers in the opposite quadrant in 1980 for log per capita GDP levels (see Table 2.2). We will come back to this inverse relationship between the per capita GDP in 1980 and average growth rates below.

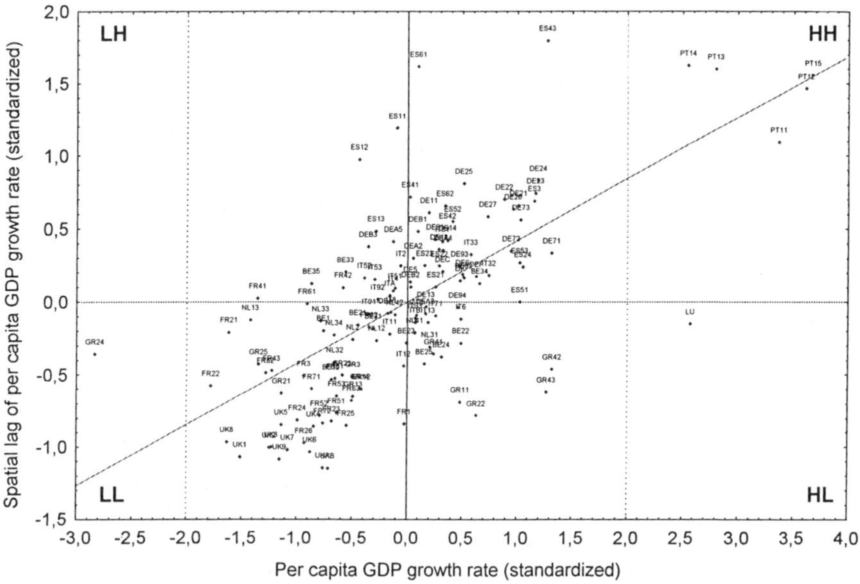

Fig. 2.10 Moran scatterplot for average growth rate of per capita GDP over 1980-1995

Finally, the most extreme observations according to the Moran regression diagnostics are shown in Table 2.7. Observations with high studentized residuals and leverage are mostly situated in Objective 1 regions in Portugal, Spain and Greece. Luxembourg is also an outlier is this respect. However, as for per capita GDP in 1980 and 1995, there was no influential region according to the Cook's distance criterion, so that the global spatial autocorrelation detected in the average growth rate appears to be a general feature of the sample and is not due to the influence of some isolated regions.

Table 2.7 Moran scatterplot outliers: average growth rates for log per capita GDP

Moran scatterplot outliers			
Studentized Residuals		**Leverage**	
Andalucia (1)	3.497511	Algarve (1)	0.105763
Extremadura (1)	2.822284	Centro (1)	0.102492
Galicia (1)	2.745314	Norte (1)	0.089878
Luxembourg (Lux)	-2.666020	Sterea Ellada (1)	0.065942
Asturias (1)	2.591420	Lisboa (1)	0.064531
Kriti (1)	-2.436728	Luxembourg (Lux)	0.055558
Ionia Nisia (1)	-2.195220	Alentejo (1)	0.054656
Notio Agaio (1)	-2.142425	Picardie	0.030390

Note: (1) denotes Objective 1 regions since 1989

Inspection of LISA statistics for average growth rates (Table 2.8) shows that the patterns of spatial association remain dominated by clustering of LL or HH types.[23] These results are therefore consistent with those previously found with Getis-Ord statistics: the dynamism of Spanish, Portuguese and Southern German regions contrasts with the behaviour of French, British and Northern Greek regions. Galicia and Asturias (Spain) are significantly of the HL type, while Anatoliki Makedonia, Ionia Nisia and Kriti (Greece) are significantly of the LH type. The LISA outliers according to the 2-sigma rule are the 5 Portuguese regions. The significant LISA at the 5% level are shown in a Moran significance map (Fig. 2.11).

All these results are robust with respect to the choice of the weight matrix. However, they must be interpreted with caution since they are dependent of the start and end dates of the time period chosen. The start and end dates can indeed be at different points in the business cycle as also pointed out by Dowrick and Nguyen (1989) and Fingleton et al. (1997) introducing therefore a possible bias in the computation of the average growth rate. However, correcting for business cycles distortion by estimating the average growth rate using a time trend regression for the log of per capita GDP for each of the regions in the sample can also raise some problems, relating to the potential presence of unit roots and the reliability of statistical inference for local and global statistics based on these *estimated* growth rates, but these are beyond the scope of this Chapter.

Let us now look more closely at the inverse relationship between log per capita GDP in 1980 and average growth rates to study the dynamism of the European regions depending on their starting point (Table 2.8). In only 42.75% of cases, the regions that were in a certain quadrant of the Moran scatterplot for log per capita GDP level in 1980 are in the opposite quadrant for their average growth rate.[24]

[23] 42.75% (resp. 21.01%) of the LISA computed are significant at the 5% pseudo-level (resp. 5% Bonferroni pseudo-level).

[24] 57.25% of the $G_i(k)$ statistics that were positive (resp. negative) in per capita GDP in 1980 are negative (resp. positive) for average growth rates.

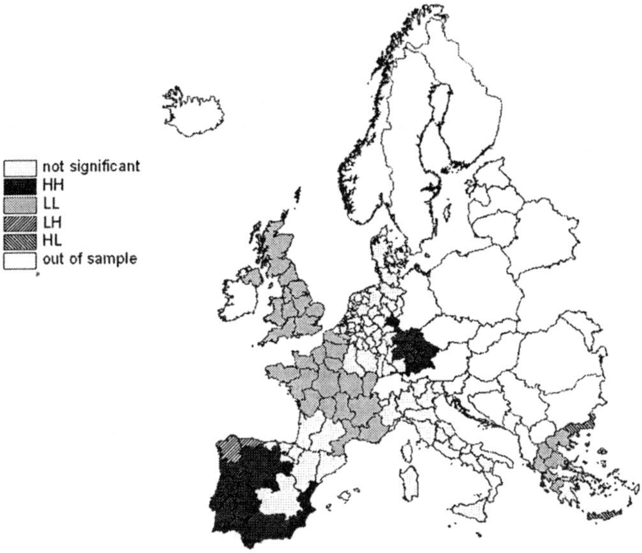

Fig. 2.11 Moran significance map for average growth rate of per capita GDP over 1980-1995 (5% pseudo significance level)

It appears that, in only 42.75% of cases, the regions that were in a certain quadrant of the Moran scatterplot for log per capita GDP level in 1980 are in the opposite quadrant for their average growth rate.[25] However, this global feature masks different behaviors. Indeed, the regions of Portugal and some Spanish regions exhibited a low per capita GDP in 1980 and were surrounded by regions with low per capita GDP (clustering of the LL type) although their average growth rate was, as for their neighbors, higher than the average (clustering of the HH type). Local spatial autocorrelation indicators therefore highlight the dynamism of these regions, whose economic performances within the group of the Southern regions of Europe were often underlined. On the contrary, almost all of the British regions, some regions in Belgium and in France, are characterized by a configuration of the initial per capita GDP of HH type and a configuration of the average growth rates of the LL type.

Other characteristics between the patterns of spatial association can be highlighted. On the one hand, within the group of the Southern regions, some poor regions of Spain, Italy and Greece fail to take off, just like their neighbors (configurations of the LL type for the initial per capita GDP and growth rates) or in spite of the dynamism of their neighbors (configuration of the LL type for the initial per capita GDP and of LH type for the growth rates). For example, this is the case for Galicia, Asturias, Cantabria in Spain; Umbria and Marche in Italy.

[25] 57.25% of the $G_i(k)$ statistics that were positive (resp. negative) in per capita GDP in 1980 are negative (resp. positive) for average growth rates.

Table 2.8 Spatial association patterns: $G_i(k)$ and LISA statistics for 10-nearest neighbors. Initial year, final year and average growth rates for log per capita GDP (1980-1995)

Code	Region	G_i(10) 1980	LISA 1980	G_i(10) 1995	LISA 1995	G_i(10) growth	LISA growth
	BELGIUM						
Be1	Bruxelles	+	HH*	+	HH	-	LL
Be21	Anvers	+*	HH*	+*	HH*	-	LL
Be22	Limburg (B)	+*	HH**	+*	HH*	-	HL
Be23	Oost Vlaanderen	+*	HH*	+*	HH*	-	LL
Be24	Vlaams Brabant	+*	HH*	+	HH*	-	HL
Be25	West Vlaanderen	+*	HH*	+	HH*	-	HL
Be31	Brabant wallon	+	HH*	+	HH*	-	LL
Be32	Hainaut	+*	HH**	+	LH*	-	LL*
Be33	Liège	+	HH*	+*	HH*	+	LH
Be34	Luxembourg (B)	+*	HH*	+*	HH*	+	HH
Be35	Namur	+*	HH*	+*	LH*	+	LH
	GERMANY						
De11	Stuttgart	+*	HH**	+**	HH**	+*	HH*
De12	Karlsruhe	+**	HH**	+**	HH**	+	HH
De13	Freiburg	+**	HH**	+**	HH**	-	HL
De14	Tübingen	+**	HH**	+**	HH**	+	HH
De21	Oberbayern	+*	HH**	+**	HH**	+*	HH*
De22	Niederbayern	+*	HH*	+**	HH**	+*	HH*
De23	Oberpfalz	+**	HH**	+**	HH**	+*	HH*
De24	Oberfranken	+**	HH**	+**	HH**	+*	HH*
De25	Mittelfranken	+*	HH**	+**	HH**	+*	HH*
De26	Unterfranken	+*	HH**	+**	HH**	+*	HH*
De27	Schwaben	+*	HH**	+**	HH**	+	HH*
De5	Bremen	+*	HH**	+*	HH**	+	HH
De6	Hamburg	+*	HH**	+**	HH**	+	HH
De71	Darmstadt	+*	HH**	+**	HH**	+	HH
De72	Giessen	+*	HH**	+**	HH**	+	HH
De73	Kassel	+*	HH**	+**	HH**	+	HH*
De91	Braunschweig	+**	HH**	+**	HH**	+	HH
De92	Hannover	+**	HH**	+**	HH**	+	HH
De93	Lüneburg	+**	HH**	+**	HH**	+	HH
De94	Weser-Ems	+*	HH**	+**	HH**	-	HL
Dea1	Düsseldorf	+*	HH*	+*	HH*	-	LL
Dea2	Köln	+	HH*	+*	HH*	+	HH
Dea3	Münster	+*	HH**	+*	HH**	-	HL
Dea4	Detmold	+*	HH**	+**	HH**	+	HH
Dea5	Arnsberg	+*	HH**	+**	HH**	+	LH
Deb1	Koblenz	+*	HH**	+**	HH**	+	HH
Deb2	Trier	+*	HH*	+*	HH*	+	HH
Deb3	Rheinhessen-Pfalz	+*	HH**	+**	HH**	+	LH
Dec	Saarland	+*	HH**	+**	HH**	+	HH
Def	Schleswig-Holstein	+**	HH**	+**	HH**	+	HH

Table 2.8 (cont.)

Code	**Region**	G_i **(10) 1980**	**LISA 1980**	G_i **(10) 1995**	**LISA 1995**	G_i **(10) growth**	**LISA growth**
Dk	***DENMARK***	+*	HH**	+**	HH**	+	HH
	SPAIN						
Es11	Galicia (1)	-**	LL**	-**	LL**	+**	LH**
Es12	Asturias (1)	-**	LL**	-**	LL**	+**	LH**
Es13	Cantabria	-*	LL*	-*	LL*	+	LH
Es21	Pais Vasco	-	LL	-	LL*	+	HH
Es22	Navarra	-	LL	-	LL	+	HH
Es23	La Rioja	-*	LL*	-	LL*	+	HH
Es24	Aragon	-	LL	-	LL	+	HH
Es3	Madrid	-**	LL**	-**	LL**	+*	HH*
Es41	Castilla-Leon (1)	-**	LL**	-**	LL**	+*	HH*
Es42	Castilla-la Mancha (1)	-**	LL**	-*	LL*	+	HH
Es43	Extremadura (1)	-**	LL**	-**	LL**	+**	HH**
Es51	Cataluna	-	LL	-	LL	+	HL
Es52	Valenciana (1)	-*	LL*	-*	LL*	+	HH*
Es53	Islas Baleares	-*	LL*	-	LL*	+	HH
Es61	Andalucia (1)	-**	LL**	-**	LL**	+**	HH**
Es62	Murcia (1)	-**	LL**	-*	LL**	+*	HH*
	FRANCE						
Fr1	Ile-de-France	+	HH*	+	HH	-*	LL**
Fr21	Champagne-Ardenne	+*	HH*	+	HH*	-	LL
Fr22	Picardie	+*	HH**	+*	HH*	-	LL*
Fr23	Haute-Normandie	+*	HH**	+	HH	-*	LL**
Fr24	Centre	+*	HH*	+	HH	-*	LL**
Fr25	Basse-Normandie	+*	HH*	+	HH	-*	LL**
Fr26	Bourgogne	+*	HH**	+	HH	-**	LL**
Fr3	Nord-Pas-De-Calais	+*	HH*	+	HH*	-	LL*
Fr41	Lorraine	+*	HH**	+*	HH**	+	LH
Fr42	Alsace	+*	HH**	+**	HH**	+	LH
Fr43	Franche-Comté	+*	HH**	+	HH*	-	LL*
Fr51	Pays de la Loire	+*	HH*	+	HH	-*	LL**
Fr52	Bretagne	+*	HH*	+	HH	-*	LL**
Fr53	Poitou-Charentes	+	HH*	+	HH	-*	LL*
Fr61	Aquitaine	-	HL	-	HL	-	LL
Fr62	Midi-Pyrénées	+	HH	-	HL	-	LL
Fr63	Limousin	+*	HH*	+	HH	-*	LL*
Fr71	Rhône-Alpes	+	HH	+	HH	-*	LL*
Fr72	Auvergne	+	HH*	+	HH	-*	LL**
Fr81	Languedoc-Roussillon	+	HH	+	LH	-	LL*
Fr82	PACA	+	HH	+	HH	-	LL*
	GREECE						
Gr11	Anatoliki Makedonia (1)	-**	LL**	-**	LL**	-*	HL*
Gr12	Kentriki Makedonia (1)	-**	LL**	-**	LL**	-*	LL*
Gr13	Dytiki Makedonia (1)	-**	LL**	-**	LL**	-*	LL*
Gr14	Thessalia (1)	-**	LL**	-**	LL**	-*	LL*

Table 2.8 (cont.)

Code	Region	G_i (10) 1980	LISA 1980	G_i (10) 1995	LISA 1995	G_i (10) growth	LISA growth
Gr21	Ipeiros (1)	-**	LL**	-**	LL**	-*	LL*
Gr22	Ionia Nisia (1)	-**	LL**	-**	LL**	-*	HL**
Gr23	Dytiki Ellada (1)	-**	LL**	-**	LL**	-	LL*
Gr24	Sterea Ellada (1)	-**	LL**	-**	LL**	-	LL
Gr25	Peloponnisos (1)	-**	LL**	-**	LL**	-	LL
Gr3	Attiki (1)	-**	LL**	-**	LL**	-	LL*
Gr41	Voreio Aigaio (1)	-**	LL**	-**	LL**	-	HL
Gr42	Notio Aigaio (1)	-**	LL**	-**	LL**	-	HL
Gr43	Kriti (1)	-**	LL**	-**	LL**	-*	HL*
	ITALY						
It11	Piemonte	+	HH	+	HH	-	LL
It12	Valle d'Aosta	+*	HH*	+	HH*	-	LL
It13	Liguria	+	HH	+	HH	-	HL
It2	Lombardia	+	HH	+	HH	+	LH
It31	Trentino – Alto Adige	+	HH	+*	HH*	+	HH
It32	Veneto	+	HH	+	HH	+	HH
It33	Friuli – Venezia Giulia	+	HH	+	HH*	+	HH
It4	Emilia – Romagna	+	HH	+	HH	+	LH
It51	Toscana	+	HH	+	HH	+	LH
It52	Umbria	-	LL	-	LL	+	LH
It53	Marche	-	LL	-	LL	+	LH
It6	Lazio	-	LL	-	HL	-	HL
It71	Abruzzo (1)	-	LL	-	LL	-	HL
It72	Molise (1)	-	LL	-	LL*	-	HL
It8	Campania (1)	-*	LL*	-*	LL*	-	LL
It91	Puglia (1)	-**	LL**	-**	LL**	-	LL
It92	Basilacata (1)	-*	LL*	-*	LL**	+	LH
It93	Calabria (1)	-**	LL**	-**	LL**	-	HL
Ita	Sicilia (1)	-**	LL**	-**	LL**	+	LH
Itb	Sardegna (1)	-	LL	-	LL	-	HL
Lu	***LUXEMBOURG***	+	HH*	+	HH*	-	HL
	NEDERLAND						
Nl12	Friesland	+*	HH**	+*	HH**	-	LL
Nl13	Drenthe	+*	HH**	+*	HH**	-	LL
Nl2	Oost Nederland	+*	HH*	+*	HH*	-	LL
Nl31	Utrecht	+*	HH**	+*	HH*	-	HL
Nl32	Noord-Holland	+*	HH**	+*	HH*	-	LL
Nl33	Zuid-Holland	+*	HH*	+*	HH*	-	LL
Nl34	Zeeland	+*	HH**	+*	HH*	-	LL
Nl41	Noord-Brabant	+*	HH*	+*	HH*	-	HL
Nl42	Limburg (NL)	+*	HH*	+*	HH*	-	LL
	PORTUGAL						
Pt11	Norte (1)	-**	LL**	-**	LL**	+**	HH**
Pt12	Centro (1)	-**	LL**	-**	LL**	+**	HH**

Table 2.8 (cont.)

Code	Region	G_i (10) 1980	LISA 1980	G_i (10) 1995	LISA 1995	G_i (10) growth	LISA growth
Pt13	Lisboa e vale do Tejo (1)	-**	LL**	-**	LL**	+**	HH**
Pt14	Alentejo (1)	-**	LL**	-**	LL**	+**	HH**
Pt15	Algarve (1)	-**	LL**	-**	LL**	+**	HH**
	UNITED-KINGDOM						
Uk1	North	+	HH	-	LL	-**	LL**
Uk2	Yorkshire and Humberside	+	HH	-	LL	-**	LL**
Uk3	East Midlands	+	HH	-	LL	-**	LL**
Uk4	East Anglia	+	HH	+	LH	-**	LL**
Uk5	South East	+	HH	+	HH	-**	LL**
Uk6	South West	+	HH	-	LL	-**	LL**
Uk7	West Midlands	+	HH	-	LL	-**	LL**
Uk8	North West	+	HH	-	LL	-**	LL**
Uk9	Wales	+	LH	-	LL	-**	LL**
Uka	Scotland	+	HH	-	LL	-**	LL**
Ukb	Northern Ireland (1)	+	LH	-	LL	-**	LL**

Notes: (1) denotes Objective 1 regions since 1989.
* for $G_i(10)$ statistics: significant at 5% (** significant at 5% Sidák) significance level based on Normal asymptotic approximation.
* for LISA statistics: significant at 5% (** significant at 5% Bonferroni) pseudo-significance level based on 10 000 permutations.

Moreover, the situation of some Greek regions continued to deteriorate for the whole period compared to the rest of the EU. Indeed, lots of them were initially poor and surrounded by other poor regions and they experienced growth rates significantly lower than the average, like their neighbors (Kentriki Makedonia, Dytiki Makedonia, Attiki,...) Thus, these regions show clear signs of delayed development. On the other hand, while some German regions are very dynamic since they, like their neighbors, started with high levels and still had a HH type form of clustering for their average growth rates, some other German regions were HH regions in 1980 but are LH regions for their growth rates, showing less dynamism than their neighbors.

2.6 Conclusion

The study of the spatial distribution of regional per capita GDP in Europe over 1980-1995 using Exploratory Spatial Data Analysis (ESDA) highlights the importance of spatial interactions and geographical location in regional growth issues. Despite progress in communication technologies and economic restructuration, traditional spatial disparities persist: economic activities are not randomly distributed in space and growth remains geographically concentrated.

ESDA appears as a powerful tool to finely reveal the characteristics of economic development of each region in relation to those of its geographical environment. The overall picture is one of spatial autocorrelation and spatial heterogeneity in the distribution of regional per capita GDP. Note that this result may be consistent with economic geography models that usually lead to core-periphery patterns in presence of regional integration and decrease of transport costs (Krugman 1991; Fujita et al. 1999). We have not attempted to explain in more depth specific results for each region from an economic perspective: here, ESDA should be considered as a first descriptive step before suggesting factors to explain the spatial patterns highlighted and before estimating and testing more sophisticated econometric models.

From the applied econometric perspective, ESDA results may have important implications for the proper estimation of standard β-convergence models (Barro and Sala-I-Martin 1991, 1995; Armstrong 1995a). Usually, in these cross-sectional models, average growth rates of per capita GDP are regressed on initial levels by using Ordinary Least Squares (OLS). Sometimes other variables that maintain constant the steady state of each region are added (conditional β-convergence). A significant negative regression coefficient indicates that poor regions tend to grow faster than rich regions so that the poor regions catch up in the long run the level of per capita GDP of the rich regions. Nevertheless, it should be stressed that OLS estimators are inefficient and statistical inference is unreliable if the basic assumptions about the error term are not met, i.e. absence of correlation and homoskedasticity.

First, ESDA reveals significant positive global spatial autocorrelation, which persists throughout the period: regions with relatively high (resp. low) per capita GDP are and remain localized close to other regions with relatively high (low) per capita GDP. This is also the case for average growth rates of per capita GDP on the 1980-1995 period. Spatial autocorrelation of the error term should then systematically be tested for in cross section specifications of β-convergence models (as in Armstrong 1995b) and if detected, an appropriate spatial specification (spatial lag model or spatial error model as presented for example in Anselin 1988; LeSage 1999; Anselin 2001) should be estimated using the proper econometric tools to achieve reliable statistical inference (e.g. Moreno and Trehan 1997; Fingleton 1999; Rey and Montouri 1999). Moreover, using such spatial specifications, it is possible to properly estimate geographical spillover effects affecting the growth process. The spatial diffusion process of random shocks can also be highlighted using simulation experiments (Le Gallo et al. 2002, see Chap.3).

Second and more interesting, Getis-Ord statistics show clustering of high and low values, persistent over the whole period, confirming the North-South polarization of European regions. Moran scatterplots and LISA confirm this initial finding and moreover reveal the existence of atypical regions, mainly poor regions surrounded by rich neighbors (LH type). These results highlight some kind of spatial heterogeneity hidden in the global positive spatial autocorrelation pattern and may indicate the co-existence of at least two distinct spatial regimes, which could be interpreted as spatial convergence clubs. The club convergence concepts as used by Durlauf and Johnson (1995) or Quah (1996a) in a non spatial framework is indeed appealing in respect to the heterogeneity problem faced by the standard β-convergence model. The convergence process, if it exists, may be different across these regimes. Moreover, spatial effects could also perform differently in each of these regimes. Once again from the applied econometrics perspective, this result suggests that the potential for distinct spatial regimes should also be carefully considered in the estimation of β-convergence models, which should be tested for spatial heterogeneity (Anselin 1990a, 1990b): structural instability (varying coefficients across regimes) and/or heteroskedasticity (varying error variances across regimes). This therefore suggests that the empirical methodology elaborated by Durlauf and Johnson (1995) should be extended to take into account spatial effects in the framework of a formal test of club convergence. These latter aspects have not yet been fully examined in the empirical literature and are studied in Baumont et al. (2002, see Chap. 4).

References

Abraham F, Von Rompuy P (1995) Regional convergence in the European Monetary Union. *Papers in Regional Science* 74:125-142

Anselin L (1988) *Spatial econometrics: methods and models.* Kluwer Academic Publishers, Dordrecht

Anselin L (1990a) Spatial dependence and spatial structural instability in applied regression analysis. *Journal of Regional Science* 30:185-207

Anselin L (1990b) Some robust approach to testing and estimating in spatial econometrics. *Regional Science and Urban Economics* 20:141-163

Anselin L (1995) Local indicators of spatial association-LISA. *Geographical Analysis* 27:93-115

Anselin L (1996) The Moran scatterplot as an ESDA tool to assess local instability in spatial association. In: Fisher M, Scholten HJ, Unwin D (eds) *Spatial analytical perspectives on GIS.* Taylor & Francis, London

Anselin L (1998a) Interactive techniques and exploratory spatial data analysis. In: Longley PA, Goodchild MF, Maguire DJ, Wind DW (eds) *Geographical information systems: principles, techniques, management and applications.* Wiley, New York

Anselin L (1998b) Exploratory spatial data analysis in a geocomputational environment. In: Longley PA, Brooks SM, McDonnell R, Macmillan B (eds) *Geocomputation, a primer.* Wiley, New York

Anselin L (1999) *SpaceStat, a software package for the analysis of spatial data, Version 1.90.* BioMedware, Ann Arbor

Anselin L (2001) *Spatial econometrics.* In: Baltagi B (ed) Companion to econometrics. Basil Blackwell, Oxford

Anselin L, Bao S (1997) Exploratory spatial data analysis linking SpaceStat and ArcView. In: Fischer M, Getis A (eds) *Recent developments in spatial analysis.* Springer, Berlin

Anselin L, Bera A (1998) Spatial dependence in linear regression models with an application to spatial econometrics. In: Ullah A, Giles DEA (eds) *Handbook of Applied Economics Statistics.* Springer, Berlin

Anselin L, Cho WKT (2000) Spatial effects and ecological inference. Working Paper, University of Illinois, Urbana-Champaign.

Arbia G (1989) *Spatial data configuration in statistical analysis of regional economic and related problems*. Kluwer Academic Publishers, Boston

Armstrong H (1995a) Convergence among the regions of the European Union. *Papers in Regional Science* 74:143-152

Armstrong H (1995b) An appraisal of the evidence from cross-sectional analysis of the regional growth process within the European Union. In: Armstrong H, Vickerman R (eds) *Convergence and divergence among European regions*. Pion, Londres

Bailey T, Gatrell AC (1995) *Interactive spatial data analysis*. Longman, Harlow

Baller RD, Anselin L, Messner SF, Deane G, Hawkins DF (2001) Structural covariates of U.S. County homicide rates: incorporating spatial effects. *Criminology* 39:561-590

Baumont C, Ertur C, Le Gallo J (2002) Spatial convergence clubs and the European growth process, 1980-1995. In: Fingleton B (ed) *European regional growth*. Springer, Berlin

Barro RJ, Sala-i-Martin X, (1991) Convergence across states and regions. *Brooking Papers on Economic Activity* 2:107-182

Barro RJ, Sala-i-Martin X (1995) *Economic growth*. McGraw-Hill, New York

Belsley D, Kuh E, Welsch R (1980) *Regression diagnostics: identifying influential data and sources of collinearity*. Wiley, New York

Cheshire P, Carbonaro G (1995) Convergence-divergence in regional growth rates: an empty black box? In: Armstrong H, Vickerman R (eds) *Convergence and divergence among European regions*. Pion, London

Cliff AD, Ord JK (1981) *Spatial processes: models and applications*. Pion, London

Cook R (1977) Detection of influential observations in linear regression. *Technometrics* 19:15-18

Donald SG, Maddala GS (1993) Identifying outliers and influential observations in econometric models. In: Maddala GS, Rao CR, Vinod HD (eds) *Handbook of statistics 11*. Elsevier Science Publishers B.V. , North-Holland

Dowrick S, Nguyen DT (1989) OECD comparative economic growth 1950-85: catch-up and convergence. *American Economic Review* 79:1010-1030

Durlauf SN Johnson PA (1995), Multiple regimes and cross-country growth behaviour. *Journal of Applied Econometrics* 10:365-384

European Commission (1996) *First report on economic and social cohesion*. Official Publication Office, Luxembourg

European Commission (1999) *The European regions: sixth periodic report on the socio-economic situation in the regions of the European Union*. Official Publication Office, Luxembourg

Eurostat (1999) *Regio database, user's guide, methods and nomenclatures*. Official Publication Office, Luxembourg

Fingleton B (1999) Estimates of time to economic convergence: an analysis of regions of the European Union. *International Regional Science Review* 22:5-34

Fingleton B, Lewney R, Pinelli D (1997) Regional growth and convergence. *The Single Market Review*, subseries 6, Aggregate and Regional Impact, Vol. 1. Kogan Page, London

Florax R, Rey S (1995) The impact of misspecified spatial interaction in linear regression models. In: Anselin L, Florax R (eds) *New directions in spatial econometrics*. Springer, Berlin

Fujita M, Krugman PR, Venables A (1999) *The spatial economy: cities, regions, and international trade*. MIT Press, Cambridge

Getis A (1995) Spatial filtering in a regression framework: examples using data on urban crime, regional inequality, and government expenditures. In: Anselin L, Florax R (eds) *New directions in spatial econometrics*. Springer, Berlin

Getis A, Ord JK (1992) The analysis of spatial association by use of distance statistics. *Geographical Analysis* 24:189-206

Haining R (1990) *Spatial data analysis in the social and environmental sciences*. Cambridge University Press, Cambridge

Hoaglin D, Welsch R (1978) The hat matrix in regression and ANOVA. *The American Statistician* 32:17-22

King G (1997) *A solution to the ecological inference problem: reconstructing individual behavior from aggregate data*. Princeton University Press, Princeton

Krugman PR (1991) *Geography and trade*. Cambridge, MIT Press

Le Gallo J (2001) Space-time analysis of GDP disparities among European regions: a Markov chains analysis. Working paper n°2001-06, LATEC, University of Burgundy, Dijon, France

Le Gallo J (2002) *Disparités géographiques et convergence des régions européennes: une approche par l'économétrie spatiale*. Ph.D. Thesis. Université de Bourgogne, Dijon.

Le Gallo J, Ertur C (2002) Exploratory Spatial Data Analysis of the distribution of regional per capita GDP in Europe, 1980-1995. *Papers in Regional Science*, forthcoming

Le Gallo J, Ertur C, Baumont C (2002) Spatial econometric analysis of convergence across European regions, 1980-1995. In: Fingleton B (ed) *European regional growth*. Springer, Berlin

LeSage JP (1999) The theory and practice of spatial econometrics. Unpublished manuscript available at: http://www.spatial-econometrics.com

López-Bazo E, Vayá E, Mora A, Suriñach J (1999) Regional economic dynamics and convergence in the European union. *Annals of Regional Science* 33:343-370

Martin R (2001) EMU versus the regions? Regional convergence and divergence in Euroland. *Journal of Economic Geography* 1:51-80

Moreno R, Trehan B (1997) Location and growth of nations. *Journal of Economic Growth* 2:399-418

Neven D, Gouyette C (1995) Regional convergence in the European community. *Journal of Common Market Studies* 33:47-65

Openshaw S, Taylor P (1979) A million or so correlation coefficient: three experiments on the modifiable areal unit problem. In: Wrigley N (ed) *Statistical applications in the spatial science*, London, Pion

Ord JK, Getis A (1995) Local spatial autocorrelation statistics: distributional issues and an application. *Geographical Analysis* 27:286-305

Ord JK, Getis A (2001) Testing for local spatial autocorrelation in the presence of global autocorrelation. *Journal of Regional Science* 41:411-432

Pace RK, Barry R (1997) Quick computation of spatial autoregressive estimators. *Geographical Analysis* 29:232-246

Pereira J, Carreiras J, Perestrello de Vasconcelos M (1998) Exploratory data analysis of the spatial distribution of wildfires in Portugal, 1980-1989. *Geographical Systems* 5:355-390

Pinkse J, Slade E (1998) Contracting in space: an application of spatial statistics to discrete-choice models. *Journal of Econometrics* 85:125-154

Quah D (1996a) Empirics for economic growth and convergence. *European Economic Review* 40:1353-1375

Quah D (1996b) Regional convergence clusters across Europe. *European Economic Review* 40:951-958

Rey S (2001) Spatial empirics for economic growth and convergence. *Geographical Analysis* 33:195-214

Rey S, Montouri B (1999) US regional income convergence: a spatial econometric perspective. *Regional Studies* 33:143-1564

Savin NE (1984) Multiple hypotheses testing. In: Griliches Z, Intriligator MD (eds) *Handbook of Econometrics, volume II*. Elsevier Science Publishers, North-Holland

Upton GJG, Fingleton B (1985) *Spatial data analysis by example*. Wiley, New York

Vanhoudt P, Matha T, Smith B (2000) How productive are capital investments in Europe. *EIB Papers* 5:81-106

Weisberg S (1985) *Applied linear regression*. Wiley, New York

Appendix 1. Data appendix

The data are extracted from the EUROSTAT-REGIO database. This database is widely used in empirical papers dealing with European regions (e.g. Neven and Gouyette 1995; Quah 1996b; López-Bazo *et al.* 1999).

Eurostat is the Statistical Office of the European Communities. Its task is to provide the European Union with statistics at European level that enable comparisons between countries and regions. These statistics are used by the European Commission and other European Institutions so that they can define, implement and analyze Community policies. The REGIO database is the official source of harmonized annual data at the regional level throughout the 1980-1995 period for the European Union.

In this Chapter, we use Eurostat 1995 nomenclature of statistical territorial units, which is referred to as NUTS (Nomenclature of Territorial Units for Statistics). The aim is to provide a single uniform breakdown of territorial units for the production of regional statistics for the European Union. In this nomenclature, NUTS1 means European Community Regions while NUTS2 means Basic Administrative Units. For practical reasons to do with data availability and the implementation of regional policies, this nomenclature is based primarily on the institutional divisions currently in force in the Member States following "normative criteria". Eurostat defines these criteria as following: "normative regions are the expression of political will; their limits are fixed according to the tasks allocated to the territorial communities, according to the size of population necessary to carry out these tasks efficiently and economically, and according to historical and cultural factors" (Eurostat 1999, p. 7). It excludes territorial units specific to certain fields of activity or functional units (Cheshire and Carbonaro 1995) in favor of regional units of a general nature. The regional breakdown adopted by Eurostat appears therefore as one of the major shortcomings of the Regio database.

We use the series E2GDP based on ESA79 and expressed in ECUs per inhabitant over the 1980-1995 period for 138 regions in 11 European countries mentioned in the text. For the United Kingdom, NUTS1 is used because there is no official counterpart to NUTS2 units, which are drawn up only for European

Commission use as groups of counties. This explains data non-availability at NUTS2 level throughout the period for this country. Luxembourg and Denmark may be considered as NUTS2 regions according to Eurostat. Our preference to prefer NUTS2 level to NUTS1 level, when data is available, is based on European regional development policy considerations: this is the level at which eligibility under Objective 1 of Structural Funds is determined (European Commission 1999). Our empirical results are certainly conditioned by this choice and could be affected by missing regions and different levels of aggregation. They must therefore be interpreted with caution.

We exclude Groningen in the Netherlands from the sample because of anomalies related to North Sea Oil revenues, which substantially increase its per capita GDP (as in Neven and Gouyette 1995). We also exclude the Canary Islands and Ceuta y Mellila, which are geographically isolated. Corse, Austria, Finland, Ireland and Sweden are excluded because data is not available for the whole 1980-1995 period in the EUROSTAT-REGIO database. Berlin and East Germany are also excluded for well-known historical and political reasons.

Some authors use per capita GDP expressed in purchasing power standards (PPS), i.e. adjusted for purchasing power parity (e.g. Armstrong 1995; López-Bazo et al. 1999), whereas we restrict our analysis to per capita GDP expressed in ECUs. The choice between per capita GDP expressed in ECUs or in PPS is rather complicated. Indeed, for international and interregional comparisons, per capita GDP values expressed in national currencies should be converted to a common currency (ECU for European countries). This conversion is carried out by means of official exchange rates but, for different reasons, these exchange rates don't mirror the real purchasing power of a given currency in the economic area of a country and their use don't give a reliable indication on the volume of goods and services produced and consumed in different countries. To overcome this drawback, an exchange rate based on purchasing power parity is often used. Values obtained this way are called purchasing power standards (PPS). However, it is worth stressing that the construction of regional accounts in purchasing power parity that are comparable across space and time is very complicated and can raise serious problems. First, this conversion should be based on regional purchasing power parity but, due to data non-availability, this adjustment is calculated on the basis of national price levels and so does not take into account regional differences in prices, which can be significant particularly when there are wide variations in income between regions. Second, per capita GDP expressed in PPS can change in one economy relative to another not only because of a difference in the rate of GDP growth in real terms but also because of a change in relative price levels. This complicates the analysis of changes over time insofar as a relative increase in per capita GDP, which arises from a reduction in the relative price level or from a re-estimation of the PPS adjustment might have slightly different implications than one which results from a relative growth in real GDP. Third, as pointed out by Vanhoudt et al. (2000) data obtained using this method are primarily designed to compare countries within the same year. Using these data in time-series comparisons could be problematic yielding "implausible results for annual average growth rates. This is often overlooked when analyzing determinants of economic growth in empirical exercises" (p.83). This problem does not seem to be taken into account, nor that the definition of the PPP has changed with every enlargement of the EU. Results obtained for growth rates using PPS should therefore be interpreted with caution.

Another problem concerning the use of per capita GDP for assessing regional disparities is that commuter flows affect comparisons between regions. In the case of city regions, surplus commuters ensure that production activity in these regions is higher than it would be with resident workers only. As a result, per capita GDP in these regions is generally overestimated, and that of the regions in which the commuters live is generally underestimated. However, note that this effect is significant only in a few cases (Brussels, Luxembourg, Hamburg, Darmstadt, Bremen, Ile-de-France), that poor regions are generally not affected by this problem and that these commuter flows are globally negligible at the NUTS2 level of regional breakdown.

Note that we also used per capita GDP measured in PPS (Le Gallo 2002; Le Gallo and Ertur 2002). This measure decreases regional disparities. Indeed with the PPS adjustment, low per capita GDP in less rich regions tend to be partly offset by the lower cost of living. However, the overall spatial pattern is similar. The main differences are the detection of a larger number of atypical regions in Moran scatterplots with the PPS adjustment and less significant local spatial autocorrelation statistics for average growth rates.

Appendix 2 Spatial association patterns: Gi(k) and LISA statistics for 10-nearest neighbors for log per capita GDP (1980-1995).

	$G_i(10)$ statistics				LISA statistics						
Code	**Signif+**	**Signif-**	**Years 5% s.**	**Years 5% Sidák. s.**	**Signif**	**HH**	**LH**	**LL**	**HL**	**Years 5% s.**	**Years 5% Bonf. s.**
Be1	0 (0)	0 (0)			1 (0)	1				80	80
Be21	2 (0)	0 (0)	80;95		6 (0)	6				80-81;87; 93-95	
Be22	5 (0)	0 (0)	80-81;93-95		12 (1)	12 (1)				80-83;85-88;92-95	
Be23	2 (0)	0 (0)	80;95		4 (0)	4				80-81;94-95	
Be24	1 (0)	0 (0)	80		4 (0)	4				80-81;94-95	
Be25	1 (0)	0 (0)	80		5 (0)	5				80-81;93-95	
Be31	0 (0)	0 (0)			5 (0)	2	3			80;95 / 81;93-94	
Be32	2 (0)	0 (0)	80-81		4 (1)	2 (1)	2			80-81 / 94-95	80
Be33	2 (0)	0 (0)	94-95		4 (0)	4				80;93-95	
Be34	4 (0)	0 (0)	80;93-95		9 (0)	9				80-81;86-88;92-95	
Be35	3 (0)	0 (0)	80;94-95		5 (0)	2	3			80-81 / 93-95	
De11	16 (14)	0 (0)	80-95	82-95	16 (16)	16 (16)				80-95	80-95
De12	16 (16)	0 (0)	80-95	80-95	16 (16)	16 (16)				80-95	80-95
De13	16 (16)	0 (0)	80-95	80-95	16 (16)	16 (16)				80-95	80-95
De14	16 (16)	0 (0)	80-95	80-95	16 (16)	16 (16)				80-95	80-95
De21	16 (5)	0 (0)	80-95	91-95	16 (15)	16 (15)				80-95	80; 82-95
De22	16 (9)	0 (0)	80-95	87-95	16 (14)	16 (14)				80-95	82-95
De23	16 (16)	0 (0)	80-95	80-95	16 (16)	16 (16)				80-95	80-95
De24	16 (15)	0 (0)	80-95	80;82-95	16 (16)	16 (16)				80-95	80-95
De25	16 (14)	0 (0)	80-95	82-95	16 (16)	16 (16)				80-95	80-95
De26	16 (12)	0 (0)	80-95	83-84;86-95	16 (16)	16 (16)				80-95	80-95
De27	16 (12)	0 (0)	80-95	83-84;86-95	16 (16)	16 (16)				80-95	80-95
De5	15 (0)	0 (0)	80-88;90-95		16 (8)	16 (8)				80-95	80-84; 93-95
De6	16 (1)	0 (0)	80-95	95	16 (13)	16 (13)				80-95	80-88; 91; 93-95
De71	16 (3)	0 (0)	80-95	93-95	16 (15)	16 (15)				80-95	80-91; 93-95
De72	16 (9)	0 (0)	80-95	83;86-88;91-95	16 (16)	16 (16)				80-95	80-95
De73	16 (3)	0 (0)	80-95	93-95	16 (14)	16 (14)				80-95	80; 82-91; 93-95
De91	16 (7)	0 (0)	80-95	80;83;91-95	16 (16)	16 (16)				80-95	80-95
De92	16 (7)	0 (0)	80-95	80;82-83;92-95	16 (15)	16 (15)				80-95	80-91;93-95
De93	16 (11)	0 (0)	80-95	80-85;91-95	16 (16)	16 (16)				80-95	80-95
De94	16 (3)	0 (0)	80-95	82;94-95	16 (14)	16 (14)				80-95	80-88;90-95

Appendix 2 (cont.)

	$G_i(10)$ statistics				LISA statistics						
Code	**Signif+**	**Signif-**	**Years 5% s.**	**Years 5% Sidák. s.**	**Signif**	**HH**	**LH**	**LL**	**HL**	**Years 5% s.**	**Years 5% Bonf. s.**
Dea1	4 (0)	0 (0)	80;93-95		14 (0)	14				80-90;93-95	
Dea2	3 (0)	0 (0)	93-95		14 (0)	14				80-81;83;85-95	
Dea3	11 (0)	0 (0)	80-87;93-95		16 (4)	16 (4)				80-95	80; 83; 94-95
Dea4	16 (3)	0 (0)	80-95	93-95	16 (12)	16 (12)				80-95	80-85; 87-88; 91; 93-95
Dea5	16 (3)	0 (0)	80-95	93-95	16 (15)	16 (15)				80-95	80-91; 93-95
Deb1	16 (3)	0 (0)	80-95	93-95	16 (8)	16 (8)				80-95	80; 86-88; 91; 93-95
Deb2	4 (0)	0 (0)	80;93-95		12 (0)	12				80-81;85-90;92-95	
Deb3	16 (7)	0 (0)	80-95	86-87;91-95	16 (16)	16 (16)				80-95	80-95
Dec	16 (1)	0 (0)	80-95	95	16 (11)	16 (11)				80-95	80-81;85-88;90-91;93-95
Def	16 (10)	0 (0)	80-95	80-85;92-95	16 (16)	16 (16)				80-95	80-95
Dk	16 (3)	0 (0)	80-95	93-95	16 (14)	16 (14)				80-95	80-88;90-91; 93-95
Es11	0 (0)	16 (16)	80-95	80-95	16 (15)			16 (15)		80-95	80-91; 93-95
Es12	0 (0)	16 (16)	80-95	80-95	16 (15)			16 (15)		80-95	80-91; 93-95
Es13	0 (0)	14 (4)	80-90 ;93-95	82-85	16 (4)			16 (4)		80-95	82-85
Es21	0 (0)	5 (0)	82-86		9 (0)			9		81-87 ;94-95	
Es22	0 (0)	3 (0)	83-85		7 (0)			7		81-87	
Es23	0 (0)	8 (0)	80-87		11(1)			11 (1)		80-88;94-95	83
Es24	0 (0)	2 (0)	83-84		4 (0)			4		82-85	
Es3	0 (0)	16 (12)	80-95	80-88;93-95	16 (13)			14 (13)	2	80-90;93-95/91-92	80-89; 93-95
Es41	0 (0)	16 (11)	80-95	80-88 ;94-95	16 (11)			16 (11)		80-95	80-88; 94-95
Es42	0 (0)	14 (9)	80-90;93-95	80-87;94	16 (9)			16 (9)		80-95	80-87; 94
Es43	0 (0)	16 (16)	80-95	80-95	16 (16)			16 (16)		80-95	80-95
Es51	0 (0)	0 (0)			0 (0)						
Es52	0 (0)	12 (6)	80-88;93-95	81-86	13 (6)			13 (6)		80-89;93-95	81-86
Es53	0 (0)	6 (0)	80-85		10 (0)			10		80-87;94-95	
Es61	0 (0)	16 (16)	80-95	80-95	16 (16)			16 (16)		80-95	80-95
Es62	0 (0)	14 (9)	80-90 ;93-95	80-88	16 (11)			16 (11)		80-95	80-88 ;94-95
Fr1	0 (0)	0 (0)			2 (0)	2				80-81	
Fr21	1 (0)	0 (0)	80		9 (0)	9				80-82 ;85-87 ;93-95	
Fr22	3 (0)	0 (0)	80-81;95		12 (2)	12 (2)				80-83;85-87;93-95	80-81
Fr23	3 (0)	0 (0)	80-82		10 (2)	10 (2)				80-89	80-81
Fr24	3 (0)	0 (0)	80-82		8 (0)	8				80-87	

Appendix 2 (cont.)

	$G_i(10)$ statistics				LISA statistics						
Code	**Signif+**	**Signif-**	**Years 5% s.**	**Years 5% Sidák. s.**	**Signif**	**HH**	**LH**	**LL**	**HL**	**Years 5% s.**	**Years 5% Bonf. s.**
Fr25	4 (0)	0 (0)	80-82 ;85		10 (1)	10 (1)				80-89	81
Fr26	7 (0)	0 (0)	80-86		11 (3)	11(3)				80-90	80-82
Fr3	1 (0)	0 (0)	80		4 (0)	4				80-81;94-95	
Fr41	16 (0)	0 (0)	80-95		16 (7)	16 (7)				80-95	80-81 ;86-87 ; 93-95
Fr42	16 (3)	0 (0)	80-95	93-95	16 (15)	16 (15)				80-95	80-91 ; 93-95
Fr43	14 (0)	0 (0)	80-93		16 (11)	16 (11)				80-95	80-90
Fr51	3 (0)	0 (0)	80-82		8 (0)	8				80-87	
Fr52	3 (0)	0 (0)	80-82		10 (0)	10				80-89	
Fr53	2 (0)	0 (0)	81-82		8 (0)	8				80-87	
Fr61	0 (0)	0 (0)			0 (0)						
Fr62	0 (0)	0 (0)			0 (0)						
Fr63	3 (0)	0 (0)	80-82		9 (0)	9 (0)				80-88	
Fr71	0 (0)	0 (0)			2 (0)	2 (0)				86-87	
Fr72	0 (0)	0 (0)			4 (0)	4 (0)				80-82 ;86	
Fr81	0 (0)	0 (0)			0 (0)	0 (0)					
Fr82	0 (0)	0 (0)			9 (0)	9 (0)				83-91	
Gr11	0 (0)	16 (16)	80-95	80-95	16 (16)			16 (16)		80-95	80-95
Gr12	0 (0)	16 (16)	80-95	80-95	16 (16)			16 (16)		80-95	80-95
Gr13	0 (0)	16 (16)	80-95	80-95	16 (16)			16 (16)		80-95	80-95
Gr14	0 (0)	16 (16)	80-95	80-95	16 (16)			16 (16)		80-95	80-95
Gr21	0 (0)	16 (16)	80-95	80-95	16 (16)			16 (16)		80-95	80-95
Gr22	0 (0)	16 (16)	80-95	80-95	16 (16)			16 (16)		80-95	80-95
Gr23	0 (0)	16 (16)	80-95	80-95	16 (16)			16 (16)		80-95	80-95
Gr24	0 (0)	16 (16)	80-95	80-95	16 (16)			16 (16)		80-95	80-95
Gr25	0 (0)	16 (16)	80-95	80-95	16 (16)			16 (16)		80-95	80-95
Gr3	0 (0)	16 (16)	80-95	80-95	16 (16)			16 (16)		80-95	80-95
Gr41	0 (0)	16 (16)	80-95	80-95	16 (16)			16 (16)		80-95	80-95
Gr42	0 (0)	16 (16)	80-95	80-95	16 (16)			16 (16)		80-95	80-95
Gr43	0 (0)	16 (16)	80-95	80-95	16 (16)			16 (16)		80-95	80-95
It11	10 (0)	0 (0)	83-92		14 (4)	14 (4)				81-94	88-91
It12	13 (0)	0 (0)	80-92		16 (3)	16 (3)				80-95	86 ; 89-90
It13	5 (0)	0 (0)	88-92		10 (3)	10 (3)				83-92	89-91

Appendix 2. (cont.)

	$G_i(10)$ statistics				LISA statistics						
Code	**Signif+**	**Signif-**	**Years 5% s.**	**Years 5% Sidák. s.**	**Signif**	**HH**	**LH**	**LL**	**HL**	**Years 5% s.**	**Years 5% Bonf. s.**
It2	9 (1)	0 (0)	84-92	91	12 (4)	12 (4)				83-94	88-91
It31	13 (3)	0 (0)	83-95	90-92	15 (8)	15 (8)				81-95	84-91
It32	9 (1)	0 (0)	84-92	91	12 (4)	12 (4)				83-94	88-91
It33	11 (1)	0 (0)	84-94	91	14 (5)	14 (5)				82-95	87-91
It4	5 (0)	0 (0)	88-92		9 (3)	9 (3)				84-92	89-91
It51	4 (0)	0 (0)	89-92		7 (0)	7				86-92	
It52	0 (0)	0 (0)			0 (0)						
It53	0 (0)	0 (0)			0 (0)						
It6	0 (0)	0 (0)			0 (0)						
It71	0 (0)	0 (0)			0 (0)						
It72	0 (0)	0 (0)			1 (0)			1		95	
It8	0 (0)	3 (0)	80-81;95		4 (0)			4		80-82;95	
It91	0 (0)	16 (6)	80-95	80-82;93-95	16 (6)			16 (6)		80-95	80-82; 93-95
It92	0 (0)	6 (0)	80-82;93-95		6 (1)			6 (1)		80-82;93-95	95
It93	0 (0)	16 (7)	80-95	80-82;92-95	16 (6)			16 (6)		80-95	80-82; 93-95
Ita	0 (0)	6 (3)	80-82;93-95	80-81;95	8 (3)			8(3)		80-83;85;93-95	80-81; 95
Itb	0 (0)	0 (0)			0 (0)						
Lu	0 (0)	0 (0)			6 (0)	6				80-81;86;93-95	
Nl12	12 (0)	0 (0)	80-88;93-95		16 (6)	10 (6)	6			80-84;86-87;93-95/85;88-92	80-84; 95
Nl13	14 (0)	0 (0)	80-88;90;92-95		16 (0)	15 (10)	1			80-91;93-95 / 92	80-84; 86-87; 93-95
Nl2	5 (0)	0 (0)	80-83;95		12 (0)	12				80-88;93-95	
Nl31	4 (0)	0 (0)	80-82;95		14 (1)	14 (1)				80-90;93-95	80
Nl32	5 (0)	0 (0)	80-83;95		12 (1)	12 (1)				80-88;93-95	80
Nl33	2 (0)	0 (0)	80;95		5 (0)	5				80-81;93-95	
Nl34	4 (0)	0 (0)	80-81;94-95		7 (1)	7(1)				80-82;87;93-95	80
Nl41	4 (0)	0 (0)	80-81;94-95		12 (0)	12				80-83;86-90;93-95	
Nl42	3 (0)	0 (0)	80 ;94-95		5 (0)	5				80-81;87 ;93-95	
Pt11	0 (0)	16 (16)	80-95	80-95	16 (16)			16 (16)		80-95	80-95
Pt12	0 (0)	16 (16)	80-95	80-95	16 (16)			16 (16)		80-95	80-95
Pt13	0 (0)	16 (16)	80-95	80-95	16 (16)			16 (16)		80-95	80-95
Pt14	0 (0)	16 (16)	80-95	80-95	16 (16)			16 (16)		80-95	80-95
Pt15	0 (0)	16 (16)	80-95	80-95	16 (16)			16 (16)		80-95	80-95

Appendix 2 (cont.)

	$G_i(10)$ statistics				LISA statistics						
Code	**Signif+**	**Signif-**	**Years 5% s.**	**Years 5% Sidák. s.**	**Signif**	**HH**	**LH**	**LL**	**HL**	**Years 5% s.**	**Years 5% Bonf. s.**
Uk1	0 (0)	0 (0)			0 (0)						
Uk2	0 (0)	0 (0)			0 (0)						
Uk3	0 (0)	0 (0)			0 (0)						
Uk4	0 (0)	0 (0)			1 (0)	1				81	
Uk5	0 (0)	0 (0)			0 (0)						
Uk6	0 (0)	0 (0)			0 (0)						
Uk7	0 (0)	0 (0)			0 (0)						
Uk8	0 (0)	0 (0)			0 (0)						
Uk9	0 (0)	0 (0)			0 (0)						
Uka	0 (0)	0 (0)			0 (0)						
Ukb	0 (0)	0 (0)			0 (0)						

	$G_i(10)$ statistics			LISA statistics				
	Signif+	**Signif-**	**Total**	**HH**	**LH**	**LL**	**HL**	**Total**
Signif. tot. 5%	686	509	1195	911	15	535	2	1463
% versus total of 2208	31.07%	23.05%	54.12%	41.26%	0.68%	24.23%	0.09%	66.26%
% versus signif. tot. 5%	57.41%	42.59%		62.27%	1.03%	36.57%	0.14%	
Signif. tot. 5% Sidák or Bonf.	228	419	647	492	0	421	0	913
% versus total of 2208	10.33%	18.98%	29.30%	22.28%	0%	19.07%	0%	41.35%
% versus signif. tot. 5%	35.24%	64.76%		53.89%	0%	46.11%	0%	

Notes: *Signif+* and *Signif-*: number of years $G_i(k)$ statistics are positive or negative significant at 5% significance level (in brackets at 5% Sidák pseudo-significance level) based on Normal asymptotic approximation. *Signif*: number of years LISA are significant at 5% pseudo-significance level (in brackets at 5% Bonferroni pseudo-significance level) based on 10000 permutations; *HH, LH, LL* and *HL*: number of years local statistics is significant and is respectively in quadrant HH, LH, LL and HL of Moran's scatterplot for LISA. (1) denotes regions eligible under Objective 1 of Structural Funds over the 1989-1995 period.

3 A Spatial Econometric Analysis of Convergence Across European Regions, 1980-1995

Julie Le Gallo, Cem Ertur and Catherine Baumont

University of Burgundy, France

3.1 Introduction

The convergence of European regions has been largely discussed in the macroeconomic and the regional science literature during the past decade. Two observations are often emphasized. First, the convergence rate among European regions appears to be very slow in the extensive samples considered (Barro and Sala-i-Martin 1991, 1995; Armstrong 1995a; Sala-i-Martin 1996a, 1996b). Second, as shown in Ertur and Le Gallo (see Chap. 2), the geographical distribution of European per capita GDP is highly clustered.

These findings can be linked to several results from economic geography theory, which emphasize the role played by geographic spillovers in growth mechanisms and point out the dominating geographical patterns of Core-Periphery equilibrium and uneven development. These results have two major implications for empirical studies on regional growth. On the one hand, they mean that similar regions tend to cluster. Therefore, econometric estimations based on geographical data have to take into account the fact that economic phenomena may be spatially autocorrelated since spatial dependence between observations leads to unreliable statistical inference based on Ordinary Least Squares (OLS). It is therefore clear-cut that models using geographical data should systematically be tested for spatial autocorrelation like time series models are systematically tested for serial correlation. However, only a few empirical studies in the recent growth theory literature using geographical data apply the appropriate spatial econometric tools (see for example Fingleton 1999; Rey and Montouri 1999). On the other hand, if geographic spillovers do indeed influence growth processes, it is worth estimating these impacts and also examining the way in which the economic performance of each region interacts.

The aim of this chapter is to deal with these empirical issues within the framework of neoclassical growth models applied to the European regions. In this context, what we are interested in is neither regional convergence within countries nor convergence towards the mean European Union income. More precisely, we want to show that some theoretical results about geographic spillovers together with advances in spatial econometric methods can produce an alternative way of analyzing the β–convergence process. In other words, we show that improved results can be obtained when spatial econometric tools are used in the estimation of the regional growth process. First, we avoid bias in statistical inference due to spatial autocorrelation and obtain more reliable estimates of the convergence rate. Second, we estimate the magnitude of geographical spillover effects in regional growth processes and highlight the underlying spatial diffusion process. Third, we show that spatial dependence leads to a *minimal* but unavoidable specification of conditional β–convergence. Therefore, our aim is not to find the determinants of regional differentiation in steady states by including additional explanatory variables in a conditional β–convergence model. We will assume instead that spatial dependence will absorb all these effects in the context of regional information scarcity and unreliability, as suggested by Fingleton (1999).

The empirical study of β–convergence for European regions described in this chapter illustrates these three points. Using a sample of 138 European regions over the 1980-1995 period, we show that the unconditional β–convergence model is misspecified due to spatially autocorrelated errors. Then we estimate different specifications integrating spatial effects explicitly using a distance-based weight matrix. Our results indicate that the most appropriate specification is the spatially autocorrelated error model and that under this model the convergence process is somewhat stronger than would otherwise be the case. We also show that the average growth rate of a region is positively influenced by growth rates in neighboring regions, thus highlighting a spatial spillover effect. Moreover, this specification implies a spatial diffusion process of random shocks that we evaluate by simulation experiments.

In Section 3.2, after having briefly presented the standard β-convergence framework, we explain why spatial dependence should be introduced into this framework and how spatial econometric tools can be applied to the estimation of β-convergence models. Section 3.3 presents the econometric results obtained using different spatial specifications and European regional data. Section 3.4 presents extensive simulation experiments on the spatial diffusion process. Finally, Section 3.5 concludes.

3.2 Convergence Concepts and Spatial Effects

Since the rather informal contribution of Baumol (1986), and the more formal contributions of Barro and Sala-i-Martin (1991, 1992, 1995) and Mankiw et al. (1992) among others, the controversial convergence issue has been extensively debated in the macroeconomic growth and regional science literature and heavily criticized on both theoretical and methodological grounds, and this has led to alternative concepts such as club convergence (Durlauf and Johnson 1995; Quah 1993a, 1993b, 1996a, 1996b) or stochastic convergence (Bernard and Durlauf 1995, 1996; Evans and Karras 1996). Empirical analysis within the convergence framework has also identified econometric problems, such as parameter heterogeneity, omitted variables, model uncertainty, outliers, endogeneity and measurement errors, and alternative techniques like panel data (Islam 1995; Caselli et al. 1996), time series (Bernard and Durlauf 1995, 1996; Carlino and Mills 1993, 1996a, 1996b; Evans and Karras 1996) and probability transition matrices (Quah 1993a, 1996a, 1996b) are often proposed as ways of achieving more compelling empirical analysis. We will not attempt here to discuss this huge literature[1] but we will focus on specific problems implied by the geographical dimension of the data.

Spatial effects have received less attention in the literature although major econometric problems are likely to be encountered if they are present in the standard β–convergence framework, since statistical inference based on OLS will then be flawed. The first study we are aware of that takes up the issue of location and growth explicitly is DeLong and Summers (1991) where the authors explain that: *"Many comparative cross-country regression have assumed there is no dependence across residuals, and that each country provides as informative and independent an observation as any other. Yet it is difficult to believe that Belgian and Dutch economic growth would ever significantly diverge, or that substantial productivity gaps would appear in Scandinavia. The omitted variables that are captured in the regression residuals seem ex ante likely to take on similar values in neighboring countries. This suggests that residuals in nearby nations will be correlated..."* (De Long and Summers 1991, p. 456 and appendix 1, pp. 487-490). However, they are disappointed not to find evidence of spatial correlation in their sample.[2]

Likewise Mankiw (1995) points out that multiple regression in the standard framework treats each country as if it were an independent observation: *"For the reported standard errors to be correct, the residual for Canada must be uncorrelated with the residual for United States. If country residuals are in fact correlated, as is plausible, then the data most likely contain less information then the reported standard errors indicate"* (Mankiw 1995, pp. 304-305). Temple (1999) in his survey on the new growth evidence also draws attention to error correlation and regional spillovers though he interprets these effects as mainly reflecting an omitted variable problem. He notes that: *"Without more evidence that the disturbances are independent, the standard errors in most growth regression should be treated with a certain degree of mistrust."* (Temple 1999, pp. 130-131).

Despite these observations, the appropriate econometric treatment of spatial effects is often neglected in the macroeconomic literature. At best it is handled by the straightforward use of regional dummies or border dummy variables (Chua 1993; Barro and Sala-i-Martin 1995; Ades and Chua 1997; Easterly and Levine 1997).

---

[1] Mankiw (1995), Islam (1998), Durlauf and Quah (1999) and Temple (1999) present outstanding surveys of this debate.

[2] More specifically, their result is based on regressions of normalized products of fitted residuals for all country pairs obtained from a growth equation on different functional forms of the distance between country capitals: "*We are quite surprised at the apparent absence of a significant degree of spatial correlation in our sample...*" (De Long and Summer 1991, p. 489).

It is in fact quite surprising that these effects, although acknowledged, are not studied more fully in the macroeconomic literature. Two kinds of arguments, reinforcing each other, justify the analysis of spatial dependence and spatial spillover effects in the framework of regional convergence models. On the one hand, several theoretical studies (Englmann and Walz 1995; Kubo 1995; Martin and Ottaviano 1999) show that the geographical distribution of economic activities is structured by different economic mechanisms and that geographical patterns interact with growth processes. On the other hand, appropriate statistical techniques and econometric models used for analyzing such spatial processes have been developed (Cliff and Ord 1981; Anselin 1988a; Anselin and Bera 1998; Anselin 2001). They provide relevant tools that are able to identify both "well defined" spatial dependence and also heterogeneity effects related to the regional growth process. Nevertheless, only a few recent empirical studies apply the appropriate spatial econometric tools, namely Moreno and Trehan (1997), Fingleton (1999), Rey and Montouri (1999) and Maurseth (2001).

In this section, we first briefly review the standard β–convergence framework. Next some theoretical arguments, providing a basis for the introduction of spatial effects in the analysis of convergence processes, are given. Finally, we show how spatial dependence is introduced in β–convergence models.

3.2.1 β–Convergence Models

The prediction of the neoclassical growth model (Solow 1956) is that the growth rate of an economy will be positively related to the distance that separates it from its own steady state. If economies have different steady states, this concept is therefore compatible with a persistent high degree of inequality among economies. This is the concept known as conditional β–convergence, which is usually tested on the following cross-sectional model, in matrix form:

$$g_T = \alpha S + \beta y_0 + X\phi + \varepsilon \qquad \varepsilon \sim N(0,\sigma_\varepsilon^2 I) \tag{3.1}$$

where g_T is the $(n\times1)$ vector of average growth rates of per capita GDP between date 0 and T; y_0 is the vector of log per capita GDP levels at date 0; X is a matrix of variables, maintaining constant the steady state of each economy. There is conditional β–convergence if the estimate of β is significantly negative once X is held constant.

If we assume that all the economies are structurally similar, i.e. that they are characterized by the same steady state, and differ only by their initial conditions, we define the concept known as unconditional β–convergence. It is only in that case that the prediction of the basic neoclassical growth model holds true: poor economies grow faster than rich ones and eventually catch them up in the long run. The hypothesis of unconditional β–convergence is usually tested on the following cross-sectional model, in matrix form:

$$g_T = \alpha S + \beta y_0 + \varepsilon \qquad \varepsilon \sim N(0,\sigma_\varepsilon^2 I) \tag{3.2}$$

There is unconditional β–convergence when the estimate of β is significantly negative. This approach is advocated, for example, by Sala-i-Martin (1996a, 1996b) for within country cross-regional analysis, together with an increasing emphasis on the test of the σ–convergence concept, which relates to cross-sectional dispersion.

The convergence process is traditionally characterized by its *convergence speed* and its *half-life*, which can be recovered using the estimated β coefficient in Eqs. 3.1 or 3.2. The convergence speed is $b = -\ln(1+T\beta)/T$. The time necessary for the economies to make up half of the disparity which separates them from their steady state, is called the half-life: $\tau = -\ln(2)/\ln(1+\beta)$.

The β–convergence concepts lead to many controversial debates and we want here to stress those problems which can be tackled by means of a more appropriate econometric approach. Let us underline that in conventional β–convergence tests, the analysis relates to regions observed in cross-sections by usually supposing implicitly that each one of them is a geographically independent entity and by neglecting the possibility of spatial interactions due to geographical spillovers.

From an econometric point of view, the independence hypothesis on the error may be very restrictive and should be tested since, if it is rejected, the statistical inference based on OLS is not reliable. The spatial dimension of the data should then be carefully incorporated as part of the process of estimation.

From an economic point of view, space and location should not be neglected in the analysis of convergence and growth processes. For example, Quah (1996c) using transition probability matrices, finds that: *"[...] physical location and geographical spillover matter more than do national, macro factors"* (Quah 1996c p. 954) and notes that: *"[...] the results highlight the importance of spatial and national spillovers in understanding regional income distribution dynamics. As intended, the paper also carries interest more generally for dynamic geographical and spatial analyses. Methodologically, the paper provides an empirical framework to study the predictions of models such as in Krugman and Venables (1995)."* (Quah 1996c p. 957)

Finally, from a methodological point of view, the hypothesis of conditional β-convergence is often the most appropriate since countries or regions are very likely to have different steady states. This approach is widely used in cross-country analysis, either with specifications formally derived from structural growth models following Mankiw et al. (1992) or with more or less ad hoc specifications to control for the determinants of the steady state (Levine and Renelt 1992). Let us note however that in the latter case, the appropriate choice of these explanatory variables may be problematic because we can't be sure conceptually to include all the variables differentiating steady states. Furthermore, in both cases, data on some of these explanatory variables may not be easily accessible and/or reliable for international comparisons (Fingleton 1999). In addition, some of these variables, including the initial per capita GDP, could be correlated with the error term, thus invalidating estimation by OLS and associated statistical inference (Caselli et al. 1996; Evans 1996). Therefore, we prefer to proceed by modeling the spatial dependence, if it is present in the β–convergence model, so as to catch the effects of omitted explanatory variables resulting from regional information scarcity and unreliability, as suggested by Fingleton (1999).

More generally, economic geography theories can helpfully allow us to gain more insights into these questions, since they explicitly take into account geographical variables in the explanation of economic interaction.

3.2.2 Economic Geography and Regional Growth

Economic geography theories have been developed following Krugman's formalization of inter-regional equilibrium with increasing returns and trade costs (Krugman 1991). These theories aim at explaining the locational behavior of firms and the agglomeration process in a context of regional integration, which is characterized by lower transaction costs, the freer migration of labor and more extensive markets. Economic geography theories give numerous theoretical insights and principles, which help us to understand the uneven spatial distribution of economic activities across regions (see for example, Fujita and Thisse 1997; Fujita et al. 1999 for more details).

More precisely, two principles are clearly apparent. First, driven by dominating agglomeration forces, industrial, high tertiary and R&D activities tend to concentrate in a limited number of places. The geographical distribution of areas characterized by high or low intensities of economic activity is rarely random, and Krugman (1993a) argues that the places where agglomerations take place are identified by what are called first nature or second nature conditions. First nature conditions refers to natural conditions or to random location decisions taken by firms. Second nature conditions means that the attractiveness of a place for a firm is due to the presence of other firms who have previously chosen to locate there. Second, agglomeration processes are strongly cumulative because the agglomeration itself is a component of agglomeration forces. Even if the initial spatial distribution of economic activities is uniform (i.e. there is no agglomeration), an exogenous shock, such as the random decision of a firm to relocate in another place, can lead to agglomeration formation.

More recently, economic geography theories and growth theories have been integrated (Pavilos and Wang 1993; Englmann and Walz 1995; Kubo 1995; Walz 1996; Ottaviano 1998; Martin and Ottaviano 1999, 2001), with some important theoretical results for the analysis of regional growth. It is shown that the spatial concentration of economic activities favors economic growth. As a result, an uneven spatial distribution of economic activities is an efficient geographic equilibrium for economic growth. Moreover, economic growth can be considered to be another agglomerating force, meaning that growth can reinforce polarization processes. Finally, geographic spillovers seem to play an important role in the interaction between geography and growth.

Focusing on geographic spillover effects, these refer to positive knowledge external effects produced by firms at a particular location and affecting the production processes of firms located elsewhere. However, there is a need to distinguish between local and global geographic spillovers. With local spillovers, production processes of the firms located in one region only benefit from the knowledge accumulation in that region. In this case, uneven spatial distributions of economic activities and regional growth divergence are likely to be the outcome. By global geographic spillovers we mean that knowledge accumulation in one region improves productivity of all the firms whatever their region. A global geographic spillover effect doesn't reinforce agglomeration processes and contributes to growth convergence (Englmann and Walz 1995; Martin and Ottaviano 1999, 2001). In practice we may find that the concentration of firms in one region produces both local and global knowledge spillovers of different strengths (Kubo 1995).

Applying these theoretical results to the analysis of an integrated regional space leads to the following observations.

1. Since economic activities are unevenly distributed in space, cumulative agglomeration processes take place and most of the economic activities tend to concentrate in a small number of regions.
2. Since economic growth is stimulated by geographic concentration of economic activities, patterns of uneven development are observed.
3. The shadow effect (Krugman 1993b) contained in the cumulative agglomeration process and the spatial ranges of geographic spillovers may explain why rich and poor regions are or are not regularly distributed. The shadow effect means that cumulative processes of concentration in one region empties its surroundings of economic activities. As a result, rich regions can be close to each other if geographic spillovers are global and regularly distributed among them. On the contrary, the assumption of local spillovers explains a more regular juxtaposition of rich and poor regions.
4. Finally, since history matters through the initial conditions and the cumulative nature of both growth and agglomeration processes, the observed geographic distribution of rich and poor regions will tend to be rather stable through time.

We can easily observe such spatial patterns in the European Union regional area. Rich and attractive regions, i.e. with per capita GDP above the mean of per capita GDP of European regions, remain concentrated in the south of England, in Benelux, in the east of France, in the west of Germany and in the north of Italy, in other words along the London-Munich-Turin axis. In contrast, in Spain, Portugal and south-Italy, poor regions are numerous.

While the foregoing theory is of considerable interest, we should note at this juncture that the empirical research we present in this chapter does not aim to formally test economic geography models per se. Rather, we focus on some issues related to growth models and economic geography models and we put forward an initial evaluation of the importance of the role played by geography in economic processes. More precisely, we investigate the convergence problem for European regions and study the effect of geographic spillovers on growth, using appropriate spatial econometric methods.

3.2.3 Spatial Dependence and Econometric Models

Spatial dependence means that observations are geographically correlated due to some processes, which connect different areas, like for example diffusion or trade processes. Several economic factors, like labor force mobility, capital mobility, technology and knowledge diffusion, transportation or transaction costs may be particularly important because they directly affect regional interactions. However, the problem we want to stress in this chapter is that these various processes induce a particular organization of economic activity in space whatever the underlying explanatory variables. Hence, we focus on the spatial dependence induced by these processes using an exogenously determined weight matrix which is defined below. Therefore we will not try to include these economic factors in this weight matrix nor as explanatory variables in the model. We will assume that our model of spatial dependence will catch the net effects of the omitted economic fcators.

Three kinds of econometric models can be used to deal with spatially dependent observations (Anselin 1988a, 2001; Anselin and Bera 1998): the spatial autoregressive model, the spatial cross-regressive model and the spatial error model. Let us examine now the way they can be applied to the β–convergence model.

3.2.3.1 The Spatial Lag Model

In this model, spatial correlation of observations is handled by the endogenous spatial lag variable Wg_T. The structural model is then written in the following form:

$$g_T = \alpha S + \beta y_0 + \rho W g_T + u \qquad u \sim \mathrm{N}(0, \sigma^2 I) \qquad (3.3)$$

where g_T is the $(n\times 1)$ vector of average growth rates of per capita GDP between date 0 and T; y_0 is the vector of log per capita GDP levels at date 0 and u the $(n\times 1)$ vector of normal i.i.d. error terms; S is the sum vector; α, β and ρ are the unknown parameters to be estimated. ρ is the spatial autoregressive parameter indicating the extent of interaction between the observations according to the spatial pattern exogenously introduced by means of the standardized weight matrix W. The endogenous spatial lag variable Wg_T is then a vector containing the growth rates premultiplied by the weight matrix: for a region i of the vector g_T, the corresponding line of the spatial lag vector contains the spatially weighted average of the growth rates of the neighboring regions.

Estimation of this model by OLS produces inconsistent estimators due to the presence of a stochastic regressor Wg_T, which is always correlated with u, even if the residuals are identically and independently distributed (Anselin 1988a, Chap. 6). Hence, it has to be estimated by the Maximum Likelihood Method (ML) or the Instrumental Variables Method (IV).

This specification can be interpreted in two ways. From the convergence perspective, it yields some information on the nature of convergence through the β parameter once spatial effects are controlled for. From the economic geography perspective, it may help to highlight a spatial spillover effect since it indicates how the growth rate of per capita GDP in a region is affected by those of neighboring regions through the ρ parameter after conditioning on the initial per capita GDP levels.

Moreover, let us stress that this model can be rewritten in reduced form as:

$$(I-\rho W)g_T = \alpha S + \beta y_0 + u \tag{3.4}$$

$$g_T = \alpha(I-\rho W)^{-1}S + \beta(I-\rho W)^{-1}y_0 + (I-\rho W)^{-1}u \tag{3.5}$$

This expression indicates that, in terms of the conditional mean, the average growth rate of a given region is not only affected by its own per capita GDP initial level but also by those of all other regions through the inverse spatial transformation $(I-\rho W)^{-1}$.

It is the so-called spatial multiplier effect or global interaction effect. Concerning the error process this expression means that a random shock in a specific region not only affects the average growth rate of this region, but also has an impact on the growth rates of all other regions through the same inverse spatial transformation.

3.2.3.2 The Spatial Cross-Regressive Model

Another way to deal with spatial dependence is to introduce exogenous spatial lag variables:

$$g_T = \alpha S + \beta y_0 + WZ\gamma + u \qquad u \sim \mathrm{N}(0,\sigma^2 I) \tag{3.6}$$

Here, the influence of h spatially lagged exogenous variables contained in the $(n{\times}h)$ matrix Z is reflected by the parameter vector γ. This general specification allows handling of spatial spillover effects explicitly and can be interpreted like a conditional convergence model integrating spatial environment variables possibly affecting growth rates. The set of explanatory variables in Z can include or not include y_0. This model can be estimated by OLS.

An interesting special case appears when Z includes only y_0. We have then:

$$g_T = \alpha S + \beta y_0 + \gamma W y_0 + u \qquad u \sim \mathrm{N}(0,\sigma^2 I) \tag{3.7}$$

In this specification, only the spatially lagged initial per capita GDP levels affect growth rates. This model can be interpreted as the minimal specification allowing a spatially lagged exogenous effect in a conditional β–convergence model. It gives estimates of both a direct effect and a spatially lagged effect of initial per capita GDP levels on the growth rates.

3.2.3.3 The Spatial Error Model

This specification is useful when we think that spatial dependence works through omitted variables. This misspecification is then handled by the error process with errors from different regions displaying spatial covariance. When the errors follow a first order spatial autoregressive process, the model is:

$$g_T = \alpha S + \beta y_0 + \varepsilon \qquad \varepsilon = \lambda W \varepsilon + u \qquad u \sim \mathrm{N}(0, \sigma^2 I) \tag{3.8}$$

where λ is the scalar parameter expressing the intensity of spatial correlation between regression residuals. Generally, the use of OLS in the presence of non-spherical errors yields unbiased but inefficient estimators. However, failure to capture the effect of omitted variables will produce bias. In addition inference based on OLS may be misleading due to biased estimates of the parameters variances. Therefore this model should be estimated by ML or Generalized Methods of Moments (GMM).

This model has two interesting properties.

The first one relates to the spatial diffusion effect of random shocks we have highlighted in equation (3.5), which appears here also since $\varepsilon = \lambda W\varepsilon + u$, so that $\varepsilon = (I - \lambda W)^{-1} u$ and the model (3.8) can be rewritten as follows:

$$g_T = \alpha S + \beta y_0 + (I - \lambda W)^{-1} u \tag{3.9}$$

Second, this model can be rewritten in another form, which can be interpreted like a minimal model of conditional β–convergence integrating two spatial environment variables. Indeed, let us note that premultiplying equation (3.8) by $(I - \lambda W)$ we get:

$$(I - \lambda W) g_T = \alpha (I - \lambda W) S + \beta (I - \lambda W) y_0 + u \tag{3.10}$$

Hence :

$$g_T - \lambda W g_T = \alpha (I - \lambda W) S + \beta y_0 - \lambda \beta W y_0 + u \tag{3.11}$$

$$g_T = \alpha (I - \lambda W) S + \beta y_0 + \lambda W g_T + \gamma W y_0 + u \tag{3.12}$$

with the restriction $\gamma = -\lambda\beta$.

The model (3.12) is the *spatial Durbin model* and can be estimated by ML. The restriction $\gamma + \lambda\beta = 0$ can be tested by the common factor test (Burridge 1981). If it cannot be rejected then model (3.12) reduces to model (3.8).

It should be stressed that model (3.12) encompasses model (3.3) and (3.7) in the sense that it incorporates the spatially lagged (endogenous and exogenous) variables Wg_T and Wy_0. It reveals two types of spatial spillover effects. Indeed, the average growth rate of a region i may be influenced by the average growth rate of neighboring regions, by means of the endogenous spatial lag variable. It may also be influenced by the initial per capita GDP of neighboring regions, due to the presence of the exogenous spatial lag.

Other specifications for the error process are sometimes used: the moving average specification (Haining 1978; Cliff and Ord 1981) and the Kelejian-Robinson specification (Kelejian and Robinson 1993, 1995). These specifications imply *local* spatial spillovers, instead of *global* spatial spillovers implied by the autoregressive specification used in this chapter. However, they are rarely used in the literature. For the moving average specification, this can be explained by the form of the log-likelihood, which includes the inverse of a matrix of dimension $(n \times n)$.

This is computationally very demanding when the number observations is large (Sneek and Rietveld 1998).

Spatial econometric models appear to be useful as a way to highlight spatial spillover effects. The application of decision rules based on different spatial autocorrelation tests helps us to choose between the alternative specifications.

3.3 Econometric Results

Using Exploratory Spatial Data Analysis, Ertur and Le Gallo (see Chap. 2) show that a strong positive spatial autocorrelation characterizes both per capita GDP levels and average growth rates in a sample of 138 European Union regions for 1980-1995. Therefore, it is likely that spatial autocorrelation will also be present in the unconditional β–convergence model. Hence, we use spatial econometrics techniques briefly described above to detect and to treat spatial autocorrelation in this model. We first estimate the unconditional β–convergence model by OLS and carry out various tests aimed at detecting the presence of spatial dependence using the spatial weight matrices specified below. We then consider the specifications that explicitly incorporate these spatial effects.

3.3.1 Data

Data limitations remain a serious problem in the European regional context. Harmonized and reliable data allowing consistent regional comparisons are scarce, in particular for the beginning of the time period under study. There is clearly a lack of appropriate or easily accessible data, to allow us to include a wide array of control and environmental variables and to estimate a regional conditional β–convergence model which is as complex as those using international series (typically the Summers and Heston data set, also called the Penn World Table)[3] as for example in Barro and Sala-i-Martin (1995) or Mankiw et al. (1992). Consequently, we use per capita GDP in logarithms measured in ECUs (former European Currency Unit) in an unconditional β–convergence framework. The sample we use is the same as in Ertur and Le Gallo (see Appendix 1, Chap. 2). More precisely, we consider 138 European regions belonging to 11 countries over the 1980-1995 period at NUTS1 level for United Kingdom (11) and at NUTS2 level for Belgium (11), Denmark (1), France (21), Germany (30), Greece (13), Italy (20), Luxembourg (1), Netherlands (9), Portugal (5) and Spain (16).

All our data are extracted from the same database and are based on the ESA 79 (European System of Accounts) prevailing until 1995. It is worth mentioning that our sample is far more consistent and encompasses many more regions than the one initially used by Barro and Sala-i-Martin (1991 with a sample of 73 regions;

[3] Levine and Renelt (1992) discuss the wide range of variables (over 50) used in various studies.

1995 with a sample of 91 regions[4]) and Sala-i-Martin (1996a with a sample of 73 regions; 1996b with a sample of 90 regions). In fact, their samples mix different sources and different regional breakdowns as noted also by Button and Pentecost (1995). Moreover the smaller 73 regions data set is largely confined to prosperous European regions belonging to Western Germany, France, United-Kingdom, Belgium, Denmark, Netherlands and Italy, excluding Spanish, Portuguese and Greek regions, which are indeed less prosperous. This may result in a selection bias problem raised by DeLong (1988)[5].

3.3.2 The Spatial Weight Matrix

The spatial weight matrix is the fundamental tool used to model the spatial interdependence between regions. More precisely, each region is connected to a set of neighboring regions by means of a *spatial pattern* introduced *exogenously* as a spatial weight matrix W. As mentioned in Chap. 2, various matrices can be considered: a simple binary contiguity matrix, a binary spatial weight matrix with a distance-based critical cut-off, above which spatial interactions are assumed negligible, more sophisticated generalized distance-based spatial weight matrices with or without a critical cut-off. The notion of distance is quite general and different functional forms based on distance decay can be used (for example inverse distance, inverse squared distance, negative exponential etc.). It is important to stress that the weights should be exogenous to the model to avoid the identification problems raised by Manski (1993). This is the reason why we consider pure geographical distance, more precisely great circle distance between regional centroid, which is indeed strictly exogenous; the functional form we use is the inverse of squared distance, which can be interpreted as reflecting a gravity function. The critical cut-off can be the same for all regions or can be defined to be specific to each region leading in the latter case, for example, to k-nearest neighbors weight matrices when the critical cut-off for each region is determined so that each region has the same number of neighbors.

The general form of the distance weight matrix $W(h)$ we use is defined as follows for $h = 1,2,3,4$:

$$\begin{cases} w_{ij}^{*}(h) = 0 \text{ if } i = j \\ w_{ij}^{*}(h) = 1/d_{ij}^{2} \text{ if } d_{ij} \leq D(h) \quad \text{and} \quad w_{ij}(h) = w_{ij}^{*}(h) \Big/ \sum_{j} w_{ij}^{*}(h) \\ w_{ij}^{*}(h) = 0 \text{ if } d_{ij} > D(h) \end{cases} \tag{3.13}$$

---

[4] For example, for the sample of 91 regions used by Barro and Sala-i-Martin (1995): GDP data are collected by Molle (1980) for the pre-1970 period whereas they are extracted from Eurostat data for the recent period and referred to personal income data from Banco de Bilbao for Spanish regions.

[5] Armstrong (1995a, 1995b) tries to overcome these problems by expanding the original Barro and Sala-i-Martin (1991) 73 regions data set to southern less prosperous regions using a more consistent sample of 85 regions.

where d_{ij} is the great circle distance between centroids of regions i and j. $D(h)$ is the cutoff parameter for h = 1,2,3 above which interactions are assumed negligible: $D(1) = Q_1$, $D(2) = Me$ and $D(3) = Q_3$, where Q_1, Me and Q_3 are respectively the lower quartile (321 miles), the median (592 miles) and the upper quartile (933 miles) and the maximum (2093 miles) of the great circle distance distribution. For $h = 4$, $D(4) = Max$, which is the maximum (2093 miles) of the great circle distance distribution and the corresponding distance matrix is full without cutoff. We consider therefore 4 different spatial weight matrices $W(h)$ for h = 1,2,3,4. Each matrix is row-standardized so that it is relative and not absolute distance, which matters.

The general form of the k-nearest neighbors weight matrix $W(k)$ we also use is defined as follows for k = 10,15,20,25:

$$\begin{cases} w^*_{ij}(k) = 0 \text{ if } i = j \\ w^*_{ij}(k) = 1 \text{ if } d_{ij} \le d_i(k) \quad \text{and} \quad w_{ij}(k) = w^*_{ij}(k) \Big/ \sum_j w^*_{ij}(k) \\ w^*_{ij}(k) = 0 \text{ if } d_{ij} > d_i(k) \end{cases} \tag{3.14}$$

where $d_i(k)$ is a critical cut-off distance defined for each region i. More precisely, $d_i(k)$ is the k-th order smallest distance between regions i and j so that each region i has exactly k neighbors. In our analysis, we consider k = 10,15,20,25 and therefore we can compute our results with four additional matrices $W(k)$.

In fact, it is important to keep in mind that all subsequent analyses are conditional upon the choice of the spatial weight matrix: the results of statistical inference depend on spatial weights. Consequently, we use h = 1,2,3,4 for the distance-based weight matrix and k = 10,15,20,25 for the k-nearest neighbors weight matrix to check for robustness of our results.

Let us also note that, first, even when using the lowest cut-off: $h = 1$ or $k = 10$, some islands such as Sicilia, Sardegna, and Baleares are connected to continental Europe so that we avoid rows and columns in W with only zero values. Second, United-Kingdom is also connected to continental Europe. Third, we note that connections between southern European regions are assured so that eastern Spanish regions are connected to Baleares, which are connected to Sardegna, which is in turn connected to Italian regions, which are finally connected to Western Greek regions. Finally, for all the W matrices used in this chapter, the block-diagonal structure of the simple contiguity matrix when ordered by country is thus avoided and the spatial connections between regions belonging to different countries are guaranteed. In our opinion, these matrices have therefore more appealing features than the simple but less appropriate contiguity matrix when working on a sample of European regions, which are less closely connected than US states.

3.3.3 Estimation : the Unconditional β–Convergence Model and Tests

Let us take as a starting point the following model of unconditional β–convergence:

$$g_T = S\alpha + \beta y_0 + \varepsilon \qquad \varepsilon \sim \mathrm{N}(0,\sigma_\varepsilon^2 I) \tag{3.15}$$

where g_T is the vector of dimension $n = 138$ of the average per capita GDP growth rates for each region i between 1995 and 1980; $T = 15$; y_0 is the vector containing the observations of log per capita GDP for all the regions in 1980; α and β are the unknown parameters to be estimated; S is the unit vector and ε is the vector of errors with the usual properties.[6]

The results of the estimation by OLS of this model are given in Table 3.1.[7] The estimated β coefficient is significant and negative (0.0079), which confirms the convergence hypothesis for the European regions. The speed of convergence associated with this estimation is 0.84%, far below 2% usually found in the convergence literature, but closer to about 1% found by Armstrong (1995a). The half-life is 88 years. These results indicate that the process of convergence is indeed very weak.

The Jarque-Bera test (1987) does not reject Normality (p-value of 0.011): the reliability of all subsequent testing procedures and the use of Maximum Likelihood estimation method are therefore conserved. We note that the White (1980) test rejects homoskedasticity and that the Breusch-Pagan test (1979) rejects it *vs.* y_0. Nevertheless, further consideration of spatial heterogeneity per se is beyond the scope of this chapter. We only take into account spatial dependence and consider, as a first approximation, that heteroskedasticity is implied by spatial autocorrelation (Anselin 1988a; Anselin and Griffith 1988).

Five spatial autocorrelation tests using the distance-based spatial weight matrix $W(1)$ are then carried out: Moran's I test adapted to the regression residuals by Cliff and Ord (1981) is very powerful against all forms of spatial dependence but it does not allow discriminating between them (Anselin and Florax 1995). In this purpose, we use two Lagrange Multiplier tests (Anselin 1988b) as well as their robust counterparts (Anselin et al. 1996), which allow testing the presence of the two possible forms of autocorrelation: LMLAG for an autoregressive spatial lag variable and LMERR for a spatial autocorrelation of errors. The two robust tests RLMLAG and RLMERR have good power against their specific alternatives.

---

[6] All computations are carried out by means of the SpaceStat 1.90 (Anselin 1999) and Arcview 3.2 (Esri) softwares.

[7] All our empirical results are quite similar either when this regression and the following ones are estimated with per capita GDP measured in ECUs or in Purchasing Power Standards (PPS). Therefore, we present only the results for data measured in ECUs and results in PPS are available from the authors upon request.

Table 3.1 Estimation results using $W(1)$ with GDP per capita (ECUs)

Model	β-convergence	Spatial error	Spatial error	Spatial lag	Cross-regressive
Estimation	OLS-White	ML	GMM	ML	OLS-White
$\hat{\alpha}$	0.129	0.158	0.159	0.053	0.123
	(0.000)	(0.000)	(0.000)	(0.001)	(0.000)
$\hat{\beta}$	-0.0079	-0.0112	-0.0113	-0.0044	-0.0109
	(0.002)	(0.000)	(0.000)	(0.005)	(0.006)
conv. speed	0.84%	1.23%	1.24%	0.46%	1.19%
	(0.000)	(0.000)	(0.000)	(0.002)	(0.021)
Half-life	88	62	61	157	64
$\hat{\lambda}$	-	0.783	0.822	-	-
		(0.000)			
$\hat{\rho}$	-	-	-	0.769	-
				(0.000)	
$\hat{\gamma}$	-	-	-	-	-0.0037
					(0.437)
R^2 or Sq. Corr*	0.13	0.13*	0.13*	0.54*	0.14
LIK	456.14	494.32	-	491.78	456.44
AIC	-908.27	-984.65	-	-977.57	-906.90
BIC	-902.42	-978.80	-	-968.79	-898.11
$\hat{\sigma}^2$	$7.996.10^{-5}$	$4.088.10^{-5}$	-	$4.263.10^{-5}$	$8.019.10^{-5}$
JB	8.976	-	-	-	9.821
	(0.011)				(0.007)
BP or BP-S*	14.786	3.324*	-	6.972*	8.849**
or KB** vs. y_0	(0.000)	(0.068)		(0.008)	(0.003)
White	29.903	-	-	-	37.41
	(0.000)				(0.000)
MORAN	13.425	-	-	-	13.310
	(0.000)				(0.000)
LMERR	151.683	-	-	-	149.247
	(0.000)				(0.000)
R-LMERR	18.102	-	-	-	1.448
	(0.000)				(0.228)
LMERR*	-	-	-	2.198	-
				(0.138)	
LMLAG	134.199	-	-	-	150.828
	(0.000)				(0.000)
R-LMLAG	0.618	-	-	-	3.029
	(0.432)				(0.0818)
LMLAG*	-	0.705	-	-	-
		(0.401)			
LR-com-fac	-	0.151	-	-	-
		(0.697)			
$\gamma = -\hat{\lambda}\hat{\beta}$	-	0.01	-	-	-
		(0.869)			

Notes for Table 3.1. and Table 3.2: *p*-values are in parentheses. *OLS-White* indicates the use of the heteroskedasticity consistent covariance matrix estimator of White (1980) in the OLS estimation. *ML* indicates Maximum Likelihood estimation. *GMM* indicates iterated Generalized Moments Estimation (Kelejian and Prucha 1999). *Sq. Corr.* Is the squared correlation between predicted and actual values. *LIK* is the value of the maximum likelihood function. *AIC* is the Akaike (1974) information criterion. *BIC* is the Schwarz information criterion (1978). *JB* is the Jarque-Bera (1987) estimated residuals Normality test. *BP* is the Breusch-Pagan (1979) test for heteroskedasticity, *BP-S* is the spatially adjusted version of this test (Anselin, 1988a, 1988b) and *KB* is the Koenker-Bassett (1982) test for heteroskedasticity in presence of non-normality. *White* is the White (1980) test for heteroskedasticity. *MORAN* is the Moran's *I* test adapted to estimated residuals (Cliff and Ord 1981). *LMERR* is the Lagrange multiplier test for residual spatial autocorrelation and *R-LMERR* is its robust version. *LMLAG* is the Lagrange multiplier test for spatially lagged endogenous variable and *R-LMLAG* is its robust version (Anselin and Florax 1995; Anselin et al. 1996). *LMERR** is the Lagrange multiplier test for an additional residual spatial autocorrelation in the spatial autoregressive model; *LMLAG** is the Lagrange multiplier test for an additional spatially lagged endogenous variable in the spatial error model (Anselin 1988a). *LR-com-fac* is the likelihood ratio common factor test (Burridge 1981).

The decision rule suggested by Anselin and Florax (1995) is then used to decide which specification is the more appropriate. If LMLAG is more significant than LMERR and RLMLAG is significant but RLMERR is not, then the appropriate model is the spatial autoregressive model. Conversely, if LMERR is more significant than LMLAG and RLMERR is significant but RLMLAG is not, then the appropriate specification is the spatial error model.

Here, this decision rule indicates the presence of spatial autocorrelation rather than a spatial lag variable: the spatial error model appears to be the appropriate specification. Furthermore, this result is robust to the choice of the spatial weight matrix, since the same conclusion is reached using the 10-nearest neighbors weight matrix $W(10)$ (see Table 3.2).

Therefore, the unconditional β–convergence model is misspecified due to the omission of spatial autocorrelation of the errors. Actually, each region is not independent of the others, as it is frequently supposed in the previous studies at the regional level. Statistical inference based on OLS estimators is not reliable. Consequently, the model of absolute β–convergence must be modified to integrate this form of spatial dependence explicitly.

3.3.4 Spatial Dependence

The estimation results by ML for the spatial error model using the distance-based matrix $W(1)$ are given in Table 3.1. The coefficients are all strongly significant. The estimated β coefficient is higher than in the unconditional β–convergence model estimated by OLS: the convergence speed raises to 1.23% and the half-life reduces to 62 years once the spatial effects are controlled for. The convergence process appears then to be a little stronger but it remains actually weak. Moreover, estimation of this model by GMM (Kelejian and Prucha 1999) leads sensibly to the same results on the convergence parameter in both cases. This first implication may seem qualitatively negligible, but we must underline that this is the only proper way of estimating a β–convergence model once spatially autocorrelated errors are detected and the only proper way of drawing reliable statistical inferences.

Table 3.2 Estimation results using $W(10)$ with GDP per capita (ECUs)

Model	β-convergence	Spatial error	Spatial error	Spatial lag	Cross-regressive
Estimation	OLS-White	ML	GMM	ML	OLS-White
$\hat{\alpha}$	0.129 (0.000)	0.186 (0.000)	0.187 (0.000)	0.060 (0.001)	0.107 (0.000)
$\hat{\beta}$	-0.0079 (0.002)	-0.0144 (0.000)	-0.0145 (0.000)	-0.0053 (0.001)	-0.017 (0.008)
conv. speed	0.84% (0.000)	1.62% (0.000)	1.63% (0.000)	0.55% (0.000)	1.90% (0.000)
Half-life	88	47.80	47.49	131	41.5
$\hat{\lambda}$	-	0.802 (0.000)	0.813	-	-
$\hat{\rho}$	-	-	-	0.772 (0.000)	-
$\hat{\gamma}$	-	-	-	-	0.011 (0.085)
R^2 or Sq. Corr*	0.13	0.13*	0.13*	0.49*	0.17
LIK	456.14	497.42	-	485.60	458.81
AIC	-908.27	-990.85	-	-965.21	-911.61
BIC	-902.42	-984.99	-	-956.43	-902.83
$\hat{\sigma}^2$	$7.996.10^{-5}$	$4.332.10^{-5}$	-	$4.748.10^{-5}$	$7.750.10^{-5}$
JB	8.976 (0.011)	-	-	-	9.245 (0.010)
BP or BP-S* or KB** vs. y_0	14.786 (0.000)	3.315* (0.068)	-	14.124* (0.000)	5.237** (0.022)
White	29.903 (0.000)	-	-	-	34.364 (0.000)
MORAN	13.937 (0.000)	-	-	-	13.449 (0.000)
LMERR	155.704 (0.000)	-	-	-	144.230 (0.000)
R-LMERR	36.484 (0.000)	-	-	-	2.295 (0.130)
LMERR*	-	-	-	4.343 (0.037)	-
LMLAG	124.457 (0.000)	-	-	-	150.769 (0.000)
R-LMLAG	5.237 (0.022)	-	-	-	8.833 (0.003)
LMLAG*	-	0.0274 (0.868)	-	-	-
LR-com-fac	-	0.189* (0.663)	-	-	-
$\hat{\gamma} = -\hat{\lambda}\hat{\beta}$	-	0.01 (0.823)	-	-	-

Notes : see notes Table 3.1

There is also strong evidence of a significant positive spatial autocorrelation of the errors since the estimated λ coefficient is equal to 0.783. The common factor test indicates that the restriction $\gamma + \lambda\beta = 0$ cannot be rejected so the spatial error model can be rewritten as the constrained spatial Durbin model:

$$g_T = \alpha(I - \lambda W)S + \beta y_0 + \lambda W g_T + \gamma W y_0 + u \tag{3.16}$$

with $\gamma = -\lambda\beta$, but using the asymptotic delta method it appears that this coefficient is not significant.

From the convergence perspective, this expression can be interpreted as *a minimal conditional β–convergence model* integrating two spatial environment variables. This reformulation has also an interesting interpretation from the economic geography perspective: the average growth rate of a region *i* is positively influenced by the average growth rate of neighboring regions, through the endogenous spatial lag variable Wg_T. However, it doesn't seem to be influenced by the initial per capita GDP of neighboring regions, through the exogenous spatial lag variable Wy_0. This spillover effect indicates that the spatial association patterns are not neutral for the economic performance of European regions. The more a region is surrounded by dynamic regions with high growth rates, the higher will be its growth rate. In other words, the geographical environment has an influence on growth processes.

Let us note that the LMLAG* test does not reject the null hypothesis of the absence of an additional autoregressive lag variable. The spatially adjusted Breusch-Pagan test is no longer significant, indicating absence of heteroskedasticity *vs.* y_0. Therefore heteroskedasticity found in the unconditional β–convergence model estimated by OLS can be interpreted as entirely due to spatial autocorrelation and is no more a problem.

According to information criteria this model seems to perform better than the previous one (Akaike 1974; Schwarz 1978). It is worth mentioning that all of our results are quite robust to the choice of the spatial weight matrix and are qualitatively similar when either different distance-based matrices or nearest-neighbors matrices are used.[8]

In order to check for the decision rule applied, we also estimate the spatial lag model including the endogenous lag variable and the special case of the spatial cross-regressive model with only the spatially lagged initial per capita GDP level. The estimation results are given in Table 3.1 for the distance-based matrix $W(1)$ and in Table 3.2 for the 10-nearest neighbors matrix $W(10)$. In the spatial lag model, we note that the convergence process appears to be even weaker and that a spatial spillover effect is still found: the average growth rate of per capita GDP in a given region is influenced by those of neighboring regions. However, the spatially adjusted Breusch-Pagan test is significant indicating heteroskedasticity. This model does not perform better than the previous one in terms of information criteria. Finally, estimation of the spatial cross-regressive model shows that there is no spatial spillover effect associated with the exogenous lag variable. Moreover, there are some problems with normality and heteroskedasticity. There is also a problem

---

[8] The complete results using $W(h)$ for h = 2, 3, 4 for distance-based weight matrices and k = 15, 20, 25 for k-nearest neighbors weight matrices are all available from the authors upon request.

with the spatial autocorrelation tests, which still indicate the presence of spatial error autocorrelation. This latter model seems therefore to be strongly misspecified and is also the worst in terms of information criteria. Indeed, the spatial error model appears as the most appropriate specification as the use of the decision rule suggested by Anselin and Florax (1995) showed previously.

3.4 The Spatial Diffusion Process

Let us reconsider the reduced form of the spatial error model:

$$g_T = \alpha S + \beta y_0 + (I - \lambda W)^{-1} u \qquad u \sim N(0, \sigma^2 I) \tag{3.17}$$

We saw that this spatial error specification has an interesting property concerning the diffusion of a random shock. This property can be used to illustrate a spatial diffusion process among all the regions of the sample since a random shock affecting one region propagates to all the other regions through the inverse spatial transformation $(I - \lambda W)^{-1}$. In this framework, it is interesting to analyze how a random shock affecting a given region diffuses to all the regions of the sample. More generally, the question we address now is the following: which regions emit the biggest impulses (respectively the smallest) and which receive the largest (respectively the least)? The magnitude of these diffusion effects are influenced by the location of each region inside the global European regional system through the spatial weight matrix W and by the spillover effects through the estimated λ parameter.

More formally, the simulation experiment can be described as follows. We assume a random shock set to two times the residual standard-error of the estimated model.

Let $\hat{u}^i$ be the $(n \times 1)$ vector containing the estimated error of the model (3.17) with a shock affecting region i:

$$\hat{u}^{i\prime} = \begin{pmatrix} \hat{u}_1 & \ldots & \hat{u}_i + 2\hat{\sigma} & \ldots & \hat{u}_n \end{pmatrix}' \tag{3.18}$$

Therefore, the $(n \times 1)$ vector y^{i*} containing the observations on the simulated average growth rates of per capital GDP after a shock in region i, can be computed in the following way:

$$y^{i*} = X\hat{\gamma} + (I - \hat{\lambda} W)^{-1} \hat{u}^i \tag{3.19}$$

where $X = [S \quad y_0]$; $\hat{\gamma} = [\hat{\alpha} \quad \hat{\beta}]'$; $\hat{\alpha}$, $\hat{\beta}$ and $\hat{\lambda}$ are the ML estimations of α, β and λ in the spatial error model.

The relative impact of a shock affecting the region i is computed by dividing each term of the vector $y^{i*} - y$, which represents the difference between the vector of simulated average growth rates and the vector of actual average growth rates, by each corresponding term of the vector y.

To illustrate this property[9], we consider a shock affecting Ile-de-France. This region is basically chosen as an illustration because it has many direct neighbors and is centrally located in Europe. Therefore, we expect that the shock will have a large impact, which will spread over many regions. The spatial diffusion process, using the distance-based matrix $W(1)$ is displayed in Fig. 3.1. We note that this shock has the largest relative impact on Ile-de-France where the simulated average growth rate is 22.19% higher than the actual average growth rate without the shock. Nevertheless, we observe a clear spatial diffusion pattern of this shock to all other regions of the sample. The magnitude of the impact of this shock is between 1.6% and 3.7% for the regions neighboring Ile-de-France and gradually decreases when we move to peripheral regions (Fig. 3.1). The median impact is 0.231% excluding Ile-de-France.

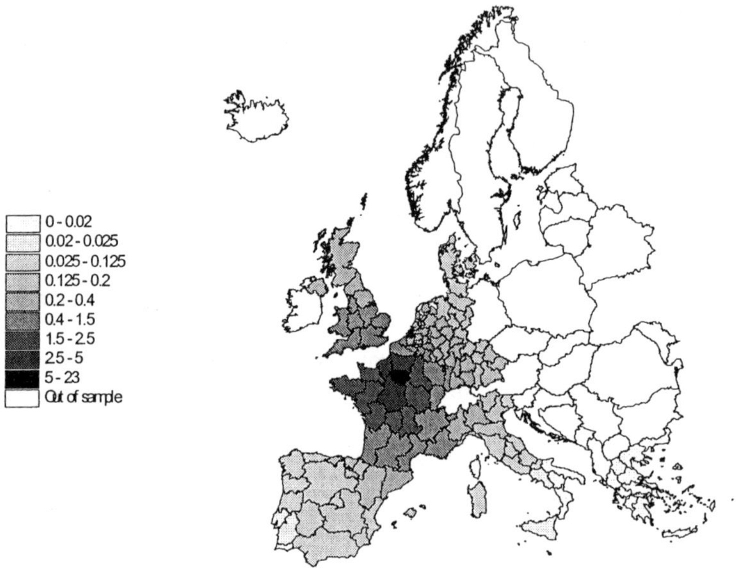

Fig. 3.1 Spatial diffusion effect measured in percent variation of average growth rates due to a shock in Ile-de-France using $W(1)$ (median: 0.231%; mean: 0.414%, excluding Ile-de-France)

The picture is similar with the 10-nearest neighbors weight matrix $W(10)$ (Fig. 3.2). The median impact is 0.0625% again excluding Ile-de-France. Therefore, the spatial error specification underlines that the geographical diffusion of shocks is at least as important as the dynamic diffusion of these shocks in the analysis of convergence processes.

---

[9] These simulations have been carried out using GAUSS 3.2.26.

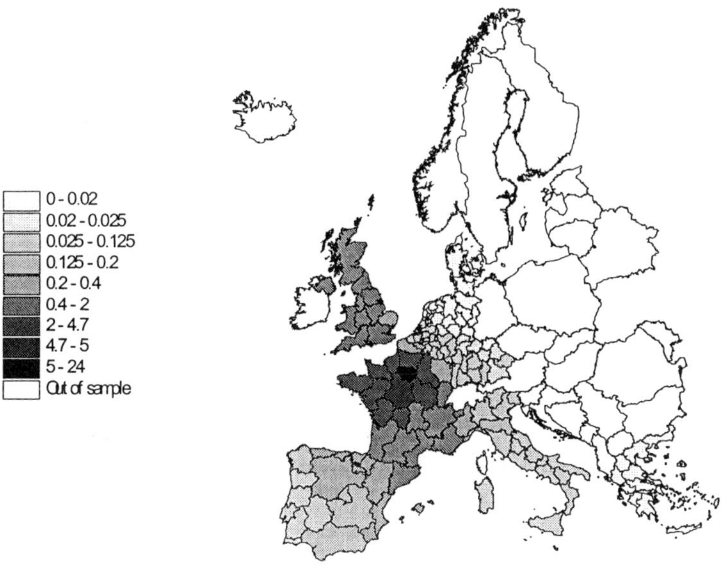

Fig. 3.2 Spatial diffusion effect measured in percent variation of average growth rates due to a shock in Ile-de-France using *W*(10) (median: 0.062%; mean: 0.488%, excluding Ile-de-France)

Let us now study the relative impacts of shocks affecting all the regions of our sample. Indeed, this would allow us to detect which regions are the most influential, i.e. the regions for which the spatial diffusion of a shock is the most important. Conversely, this would also indicate which regions are the most affected by a shock in another region.

If there is a shock in each region of the sample, then Y^* the matrix of dimension $(n \times n)$ containing the observations on the simulated average growth rates of per capital GDP is written as follows:

$$Y^* = \left[y^{1*} \quad \dots \quad y^{n*} \right] = \left[X\hat{\gamma} \quad \dots \quad X\hat{\gamma} \right] + A^{-1} \left[\hat{u}^1 \quad \dots \quad \hat{u}^n \right] \tag{3.20}$$

with $A = I - \hat{\lambda} W$.

Equation (3.20) can be rewritten as follows:

$$Y^* = S' \otimes X\hat{\gamma} + A^{-1}\hat{U} \tag{3.21}$$

where $\otimes$ is the Kronecker product; $\hat{U}$ is the matrix of dimension $(n \times n)$ defined as: $\hat{U} = \left[\hat{u}^1 \quad \dots \quad \hat{u}^n \right]$.

Since each element $\hat{u}^i$ is defined following Eq. (3.18), this matrix $\hat{U}$ can also be written in the following way:

$$\hat{U} = \begin{pmatrix} \hat{u}_1 + 2\hat{\sigma} & \hat{u}_1 & \cdots & \hat{u}_1 \\ \hat{u}_2 & \hat{u}_2 + 2\hat{\sigma} & \cdots & \hat{u}_2 \\ \vdots & \vdots & \ddots & \vdots \\ \hat{u}_n & \hat{u}_n & \cdots & \hat{u}_n + 2\hat{\sigma} \end{pmatrix} \tag{3.22}$$

where $\hat{\sigma}$ is the estimated residual standard-error of the spatial error model.

Equivalently, (3.22) can be rewritten as:

$$\hat{U} = S' \otimes \hat{u} + 2\hat{\sigma} I \tag{3.23}$$

Combining (3.21) and (3.23), we finally obtain:

$$Y^* = S' \otimes X\hat{\gamma} + A^{-1}(S' \otimes \hat{u} + 2\hat{\sigma} I) \tag{3.24}$$

This expression yields a matrix of dimension $(n \times n)$ where the column i indicates the simulated average growth rates of capita GDP for all regions in the sample after a shock in region i. Each column i gives information on the *emission side* of the spatial diffusion process from a specific region i to all the other regions. Moreover, each row j indicates how the average growth rate of a specific region j is influenced by a shock affecting all the other regions: each row j gives information on the *reception side* of the spatial diffusion process.

To assess the impact of these shocks on the average growth rates of the regions in our sample, we first compute the difference D between the matrix of simulated average growth rates Y^* and the matrix of actual average growth rates Y.

Let us note:

$$D = Y^* - Y \tag{3.25}$$

Since:

$$Y = S' \otimes y \qquad \text{where} \qquad y = X\hat{\gamma} + A^{-1}\hat{u} \tag{3.26}$$

Then:

$$D = 2\hat{\sigma} A^{-1} \qquad \text{with} \qquad A = I - \hat{\lambda} W \tag{3.27}$$

We therefore highlight once more the role played by the matrix W through the inverse spatial transformation $(I - \lambda W)^{-1}$.

Let us consider now the matrix V, containing the variation in percentage between the simulated and the actual average growth rates. V is obtained by dividing each term of the D matrix by each corresponding term of the Y matrix. The column i of the matrix V indicates how the region i affects the other regions of the sample when there is a shock in this region (emission). On the other hand, the row j of matrix V indicates how the region j is affected by a shock in each of all the other regions of the sample (reception). Note that the first diagonal of this matrix has been set to zero, so that the impact of a shock in a region on itself is not taken into account. Therefore, the columns of V allow detection of the regions which are the most influential, while the rows of V allow detection of the regions which are the most influenced. The emission and reception characteristics of each region are described by the box-plots associated to the row and column distributions of the V matrix.[10]

Figures 3.3. and 3.4. display the simulation results obtained with $W(1)$ when we focus on the way each region experiencing a shock affects all other regions in the sample, e.g. the emission side of the spatial diffusion process. The quartiles of the distribution of median impact of emission are displayed in Fig. 3.3 and the box-plot is presented in decreasing order relative to the median of each region in Fig. 3.4. It appears that the most influential regions are rich northern European regions mainly belonging to Belgium, Germany, Netherlands and France; Luxembourg is also highly influential. Let us note that all these regions belong to the so-called Core of Europe. On the contrary, poor southern Objective 1 regions are the less influential.

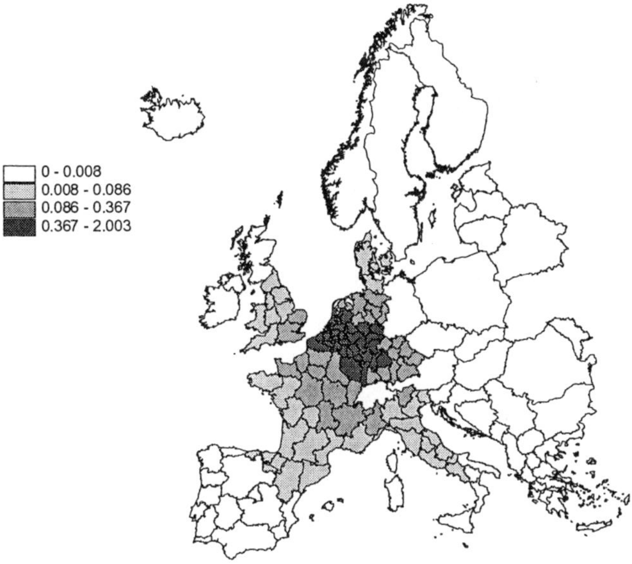

Fig. 3.3 Quartiles of the distribution of the median impact of emission using $W(1)$ in the spatial diffusion process of random shocks affecting all the European regions of the sample

[10] All the results presented below are quite similar when log per capita GDP is measured in PPS, instead of ECUs, although all the magnitudes are lower in the former case.

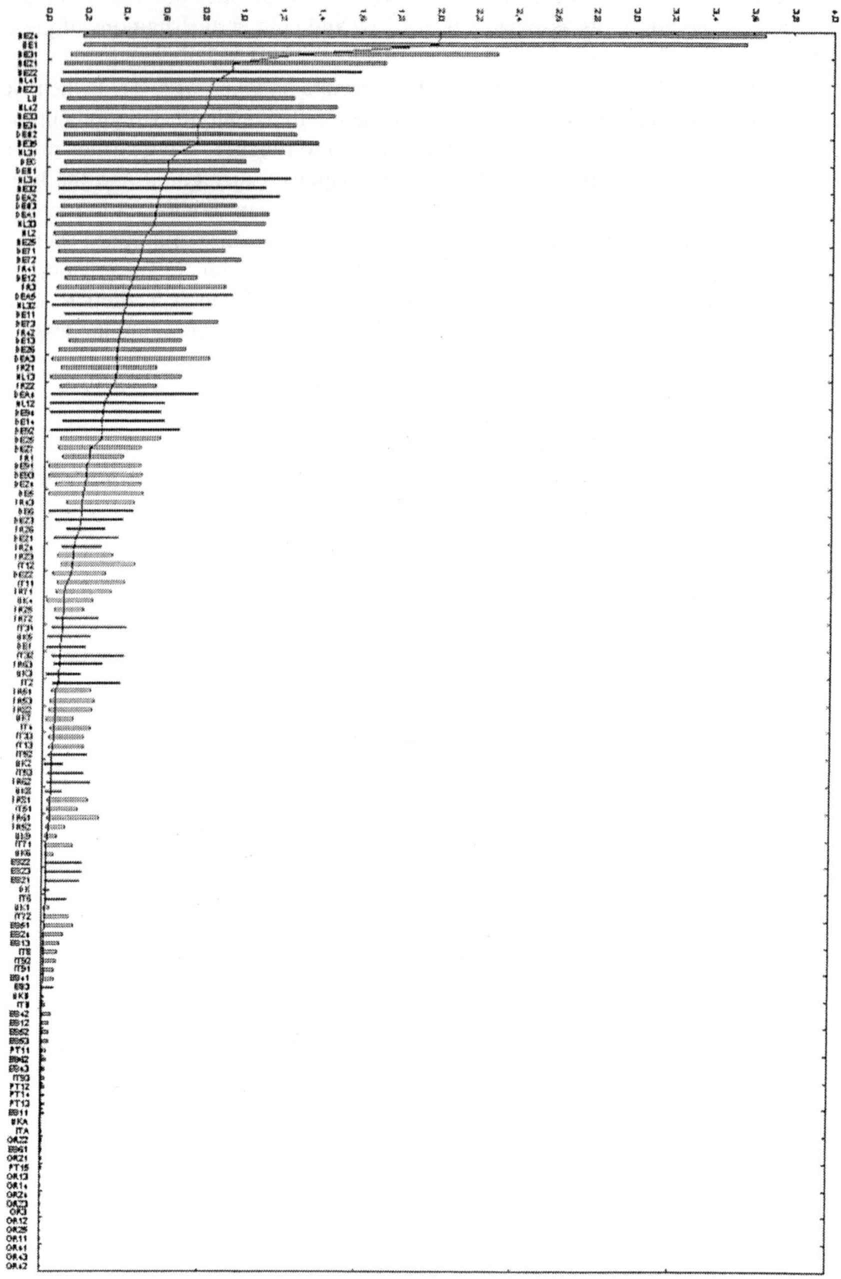

Fig. 3.4 Box plot and median of impact in emission using $W(1)$ (Q_1 -median- Q_3 ; list of region codes in Chap. 2)

The impacts of the spatial diffusion process between these two European regional groups appear therefore to be significantly different. This result is confirmed by non-parametric tests on the equality of the medians and the general form of the distributions between Objective 1 regions and non Objective 1 regions[11] (Table 3.3). All these tests strongly reject the null hypothesis.

Table 3.3 Non-parametric tests for the emission impact of the spatial diffusion process using $W(1)$

Tests	Kruskal-Wallis	U Mann-Withney	Wald-Wolfowitz
Standardized value	73.633	8.581	-9.330
p-value	0.000	0.000	0.000

Turning now to the reception side of the spatial diffusion process, we can underline again that the rich northern regions are the most influenced while poor southern Objective 1 regions are the less influenced (see Figs. 3.5 and 3.6). The impact differences between these two groups are significantly different from zero since non-parametric tests on the equality of the medians and the general form of the distributions between Objective 1 regions and non Objective 1 regions again strongly reject the null hypothesis (Table 3.4).

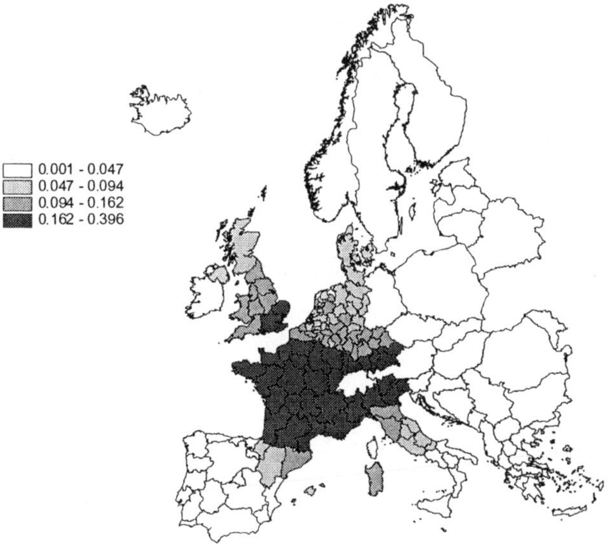

Fig. 3.5 Quartiles of the distribution of the median impact of reception using $W(1)$ in the spatial diffusion process of random shocks affecting all the European regions of the sample

[11] All these tests have to be taken with caution due to the presence of spatial autocorrelation (Rey 2001)

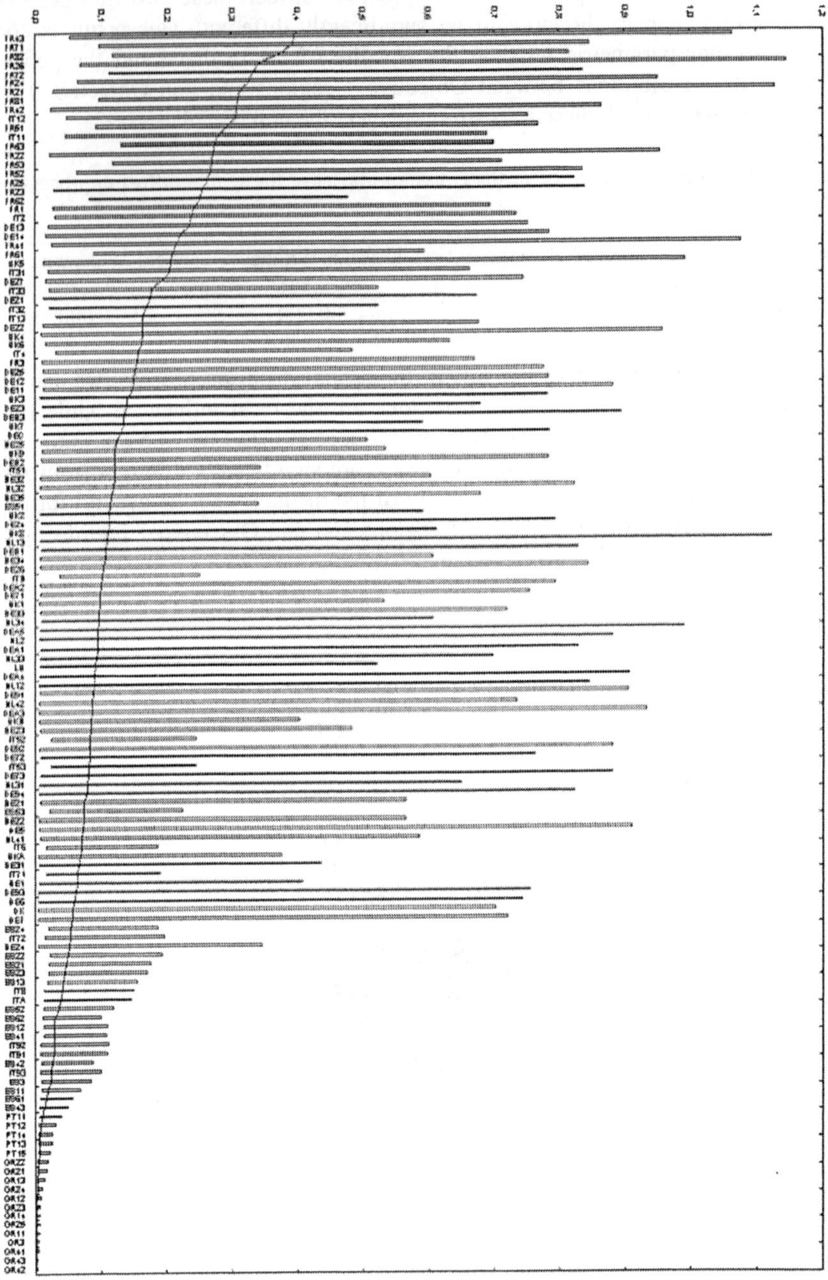

Fig. 3.6 Box plot and median of impact in reception using $W(1)$ (Q_1-median-Q_3; list of region codes in Chap. 2)

These simulation results highlight the fact that the spatial diffusion process is therefore dominated by the well-known Core-Periphery pattern and suggest potential heterogeneity in the growth process.

Table 3.4 Non-parametric tests for the reception impact of the spatial diffusion process using $W(1)$

Tests	Kruskal-Wallis	U Mann-Withney	Wald-Wolfowitz
Standardized value	69.494	8.336	-9.330
p-value	0.000	0.000	0.000

Finally, comparing the quartiles of the distributions of the median impact of emission and the median impact of reception (Figs 3.3 and 3.5), it is possible to note that the composition of the lower quartile remains globally unchanged and involves the European Objective 1 regions, while the composition of the highest quartiles differs. Interestingly, the most influential regions, i.e. regions in the last quartile (Fig. 3.3), are mainly located in Belgium, Netherlands and north-western Germany, while the most influenced regions (Fig. 3.5) are located in France, south Germany and northern Italy. These findings are also illustrated by the non-parametric rank correlation tests results between median emission and median reception displayed in Table 3.5. These tests strongly reject the null hypothesis of no correlation between rankings, although the positive correlation is relatively weak.

Table 3.5 Rank correlation tests for the emission effects *vs.* reception effects of the spatial diffusion processes using $W(1)$

Tests	R-Spearman	Tau-Kendall	Gamma-Kendall
Values	0.512	0.353	0.353
Standardized value	6.955	6.150	6.150
p-value	0.000	0.000	0.000

These results suggest that the geographical dimension of the growth process is not neutral since regions that are centrally located receive more benefits from all others regions than the peripheral regions. Moreover, if central regions are also more dynamic than peripheral regions then they also diffuse more benefits. Finally, in the case of the European regional context that we have studied here, the spatial diffusion process suggests a kind of geographical lock-in for southern poor regions.

3.5 Conclusion

In this chapter, we analyze the consequences of spatial dependence on regional growth and convergence processes. Indeed, spatial error autocorrelation is detected in the unconditional β–convergence model estimated for a sample of 138 European regions over the 1980-1995 period leading to unreliable statistical inference. Therefore, we have to find the most appropriate specification taking into account the spatial autocorrelation detected and estimate it with the appropriate econometric method. Among all the specifications integrating spatial autocorrelation, the spatial error model is the best one according to the decision rule suggested by Anselin and Florax (1995) and according to information criteria. We interpret this specification as the minimal conditional β–convergence model in the sense that it captures the effects of all other variables that could explain differentiated steady states along the lines of Fingleton (1999). This specification reveals that the convergence process is somewhat stronger. Moreover, it reveals spatial spillover effects in that the average growth rate of per capita GDP of a region is affected by the average growth rate of neighboring regions. The spatial diffusion process implied by this model is also highlighted by simulation experiments. Let us finally note that spatial heterogeneity, which may be present in the sample as indicated in these extensive simulation experiments and that we discarded here to focus on spatial autocorrelation, is studied in Baumont et al. (see Chap. 4).

References

Ades A, Chua HB (1997) Thy Neighbor's Curse: Regional Instability and Economic Growth. *Journal of Economic Growth* 2: 279-304

Akaike H (1974) A New Look at the Statistical Model Identification. *IEEE Transactions on Automatic Control* AC-19: 716-723

Anselin L (1988a) *Spatial Econometrics: Methods and Model.* Kluwer Academic Publishers, Dordrecht

Anselin L (1988b) Lagrange Multiplier Test Diagnostics for Spatial Dependence and Spatial Heterogeneity. *Geographical Analysis* 20: 1-17

Anselin L (1999) *SpaceStat, a Software Package for the Analysis of Spatial Data. Version 1.90.* Ann Arbor, BioMedware

Anselin L (2001) Spatial Econometrics. In Baltagi B (ed) *Companion to Econometrics.* Basil Blackwell, Oxford

Anselin L, Bera A (1998) Spatial Dependence in Linear Regression Models with an Introduction to Spatial Econometrics. In Ullah A, Giles DEA (eds) *Handbook of Applied Economic Statistics.* Springer-Verlag, Berlin

Anselin L, Florax RJGM (1995) Small Sample Properties of Tests for Spatial Dependence in Regression Models. In Anselin L, Florax RJGM (eds) *New Directions in Spatial Econometrics.* Springer, Berlin

Anselin L, Griffith D (1988) Do Spatial Effects really matter in Regression Analysis? *Papers in Regional Science* 65: 11-34

Anselin L, Bera AK, Florax R, Yoon MJ (1996) Simple Diagnostic Tests for Spatial Dependence. *Regional Science and Urban Economics* 26: 77-104

Armstrong H (1995a) Convergence among the Regions of the European Union. *Papers in Regional Science* 74: 143-152

Armstrong H (1995b) An Appraisal of the Evidence from Cross-Sectional Analysis of the Regional Growth Process within the European Union. In Armstrong H, Vickerman R (eds) *Convergence and Divergence among European Regions.* Pion, London

Barro RJ, Sala-i-Martin X (1991) Convergence across States and Regions. *Brookings Papers on Economic Activity* 1991: 107-182
Barro RJ, Sala-i-Martin X (1992) Convergence. *Journal of Political Economy* 100: 223-251
Barro RJ, Sala-i-Martin X (1995) *Economic Growth Theory*. McGraw-Hill, Boston
Baumol WJ (1986) Productivity Growth, Convergence and Welfare: What the Long Run Data Show. *American Economic Review* 76: 1072-1085
Bernard AB, Durlauf SN (1995) Convergence in International Output. *Journal of Applied Econometrics* 10: 97-108
Bernard AB, Durlauf SN (1996) Interpreting Tests of the Convergence Hypothesis. *Journal of Econometrics* 71: 161-173
Breusch T, Pagan A (1979) A Simple Test for Heterosckedasticity and Random Coefficient Variation. *Econometrica* 47: 1287-1294
Burridge P (1981) Testing for a Common Factor in a Spatial Autoregressive Model. *Environment and Planning Series A* 13: 795-800
Button KJ, Pentecost EJ (1995) Testing for Convergence of the EU Regional Economies. *Economic Inquiry* 33: 664-671
Carlino G, Mills L (1993) Are US Regional Incomes Converging? *Journal of Monetary Economics* 32: 335-346
Carlino G, Mills L (1996a) Convergence and the US States: A Time-Series Analysis. *Journal of Regional Science* 36: 597-616
Carlino G, Mills L (1996b) Testing Neoclassical Convergence in Regional Incomes and Earnings. *Regional Science and Urban Economics* 26: 565-590
Caselli F, Esquivel G, Lefort F (1996) Reopening the Convergence Debate: A New Look at Cross-Country Growth Empirics. *Journal of Economic Growth* 1: 363-390
Chua HB (1993) *On Spillovers and Convergence*. Ph.D. Thesis. Harvard University
Cliff AD, Ord JK (1981) *Spatial Processes: Models and Applications*. Pion, London
DeLong JB (1988) Productivity Growth, Convergence and Welfare: Comment. *American Economic Review* 78: 1138-1154
DeLong JB, Summers LH (1991) Equipment Investment and Economic Growth. *Quarterly Journal of Economics* 106: 445-502
Durlauf SN, Johnson PA (1995) Multiple Regimes and Cross-Country Growth Behaviour. *Journal of Applied Econometrics* 10: 365-384
Durlauf SN, Quah D (1999) The New Empirics of Economic Growth. In Taylor J and Woodford M (eds), *Handbook of Macroeconomics*. North-Holland Elsevier Science
Easterly W, Levine R (1997) Africa's Growth Tragedy: Policies and Ethnic Divisions. *Quarterly Journal of Economics* 112:1203-1250
Englmann FC, Walz U (1995) Industrial Centers and Regional Growth in the Presence of Local Inputs. *Journal of Regional Science* 35: 3-27
Ertur C, Le Gallo J (2002) An Exploratory Spatial Data Analysis of European Regional Disparities, 1980-1995. In: Fingleton B (ed) *European regional growth*. Springer, Berlin
European Commission (1999) *The European Regions: Sixth Periodic Report on the Socio-Economic Situation in the Regions of the European Union*. Official Publication Office, Luxembourg
Eurostat (1999) *Regio Database, User's Guide, Methods and Nomenclatures*. Official Publication Office, Luxembourg
Evans P (1996) Using Cross-Country Variances to Evaluate Growth Theories. *Journal of Economic Dynamics and Control* 20: 1027-1049

Evans P, Karras G (1996) Convergence Revisited. *Journal of Monetary Economics* 37: 249-265

Fingleton B (1999) Estimates of Time to Economic Convergence: An Analysis of Regions of the European Union. *International Regional Science Review* 22: 5-34

Fujita M, Krugman P, Venables AJ (1999) *The Spatial Economy.* The MIT Press, Cambridge

Fujita M, Thisse J (1997) Economie géographique. Problèmes anciens et nouvelles perspectives. *Annales d'Economie et Statistique* 45/46: 37-87

Haining R (1978) The Moving Average Model for Spatial Interaction. *Transactions and Papers of the Institute of British Geographers* 3: 202-225

Islam N (1995) Growth Empirics: A Panel Data Approach. *Quarterly Journal of Economics* 110: 1127-1170

Islam N (1998) Convergence: Variation in Concept and Empirical Results. (mimeo) Department of Economics, Emory University

Jarque CM, Bera AK (1987) A test for Normality of Observations and Regression Residuals. *International Statistical Review* 55: 163-172

Kelejian HH, Prucha IR (1999) A Generalized Moments Estimator for the Autoregressive Parameter in a Spatial Model. *International Economic Review* 40: 509-534

Kelejian HH, Robinson DP (1993) A Suggested Method of Estimation for Spatial Interdependent Models with Autocorrelated Errors, and an Application to a County Expenditure Country Police Expenditure. *Papers in Regional Science* 72: 297-312

Kelejian HH, Robinson DP (1995) Spatial Correlation: a Suggested Alternative to the Autoregressive Model. In Anselin L, Florax R (eds), *New Directions in Spatial Econometrics.* Springer, Berlin

Koenker R, Basset G (1982) Robust Tests for Heteroskedasticity Based on Regression Quantiles. *Econometrica* 50: 43-61

Krugman P (1991) Increasing Returns and Economic Geography. *Journal of Political Economy* 99: 483-499

Krugman P (1993a) First Nature, Second Nature and Metropolitan Location. *Journal of Regional Science* 33: 129-144

Krugman P (1993b) On the Number and Location of Cities. *European Economic Review* 37: 293-293

Krugman P, Venables AJ (1995) The Seamless World: A Spatial Model of International Specialization, *National Bureau of Economic Research Working Paper*, n° 5220

Kubo Y (1995) Scale Economies, Regional Externalities, and the Possibility of Uneven Development. *Journal of Regional Science* 35: 29-42

Levine R, Renelt D (1992) A Sensitivity Analysis of Cross-Country Growth Regressions. *American Economic Review* 82: 942-963

López-Bazo E, Vayá E, Mora A, Suriñach J (1999) Regional Economic Dynamics and Convergence in the European Union. *Annals of Regional Science* 33: 343-370

Mankiw NG (1995) The Growth of Nations. *Brooking Papers on Economic Activity* 1: 275-310

Mankiw NG, Romer D, Weil DN (1992) A Contribution to the Empirics of Economic Growth. *Quarterly Journal of Economics* 107: 407-437

Manski CF (1993) Identification of Endogenous Social Effects: The Reflection Problem. *Review of Economic Studies* 60: 531-542

Martin P, Ottaviano GIP (1999) Growing Locations: Industry Location in a Model of Endogenous Growth. *European Economic Review* 43: 281-302

Martin P and GIP Ottaviano (2001) Growth and Location. *International Economic Review* 42: 947-968
Maurseth B (2001) Convergence, Geography and Technology, *Structural Change and Economic Dynamics.* 12: 247-276
Molle W (1980) *Regional Disparity and Economic Development in the European Community.* Saxon House, Farnborough, Hantes
Moreno R, Trehan B (1997) Location and Growth of Nations. *Journal of Economic Growth* 2: 399-418
Ottaviano GIP (1998) *Dynamic and Strategic Considerations in International and Interregional Trade.* Ph.D. Thesis. CORE, Louvain-la-Neuve
Palivos T and P Wang (1993) Spatial Agglomeration and Endogenous Growth. *Regional Science and Urban Economics* 26: 645-669
Quah D (1993a) Empirical Cross-section Dynamics in Economic Growth. *European Economic Review* 37: 426-434
Quah D (1993b) Galton's Fallacy and Tests of the Convergence Hypothesis. *Scandinavian Journal of Economics* 95: 427-443
Quah D (1996a) Empirics for Economic Growth and Convergence. *European Economic Review* 40: 1353-1375
Quah D (1996b) Twin Peaks: Growth and Convergence in Models of Distribution Dynamics. *Economic Journal* 106: 1045-1055
Quah D (1996c) Regional Convergence Clusters across Europe. *European Economic Review.* 40: 951-958
Rey SJ (2001) Spatial Dependence in the Evolution of Regional Income Distributions. Working Paper, REAL 01-T-03. University of Illinois at Urbana-Champaign
Rey SJ, Montouri BD (1999) U.S. Regional Income Convergence: a Spatial Econometric Perspective. *Regional Studies* 33: 145-156
Sala-i-Martin X (1996a) Regional Cohesion: Evidence and Theories of Regional Growth and Convergence. *European Economic Review* 40: 1325-1352
Sala-i-Martin X (1996b) The Classical Approach to Convergence Analysis. *Economic Journal* 106: 1019-1036
Schwarz G (1978) Estimating the Dimension of a Model. *The Annals of Statistics* 6: 461-464
Sneek JM, Rietveld P (1998) On the Estimation of the Spatial Moving Average Model. Working Paper. Tinbergen Institute, Rotterdam
Solow RM (1956) A Contribution to the Theory of Economic Growth. *Quarterly Journal of Economics* 70: 65-94
Summers R, Heston A (1988) A New Set of International Comparisons of Real Product and Price Levels Estimates for 130 Countries, 1950-1985. *Revue of Income and Wealth* 34: 1-25
Temple J (1999) The New Growth Evidence. *Journal of Economic Literature* 37: 112-156.
Walz U (1996) Transport Costs, Intermediate Goods and Localized Growth. *Regional Science and Urban Economics* 26: 671-695
White H (1980) A Heteroskedastic-Consistent Covariance Matrix Estimator and a Direct Test for Heteroskedasticity. *Econometrica* 48: 817-838

4 Spatial Convergence Clubs and the European Regional Growth Process,1980-1995

Catherine Baumont, Cem Ertur and Julie Le Gallo

University of Burgundy, France

4.1 Introduction

In international cross-country studies, evidence for club convergence has often been found using different methodologies (Baumol 1986; Durlauf and Johnson 1995; Quah 1996a, 1997). In the case of the European regions, Ertur and Le Gallo (see Chap. 2) and Le Gallo et al. (see Chap. 3) have shown that the convergence rate among European regions is slow and that GDP disparities seem to be persistent despite the European economic integration process and higher growth rates of some poorer regions, as highlighted as well in the European Commission reports (1996, 1999). Moreover, over the 1980-1995 period, Ertur and Le Gallo (see Chap. 2) found that the geographical distribution of European regions exhibits a persistent polarization pattern between rich regions in the north and poor regions in the south.

These evidences can in fact be linked to several results of new economic geography theories (Krugman 1991), which show that locations of economic activities are spatially structured by some agglomerative and cumulative processes. Therefore, we may say that the geographical distribution of areas characterized by high or low economic activities is spatially dependent and tends to exhibit persistence. Moreover, the economic surrounding of a region seems to influence the economic development perspectives for this region: a poor (respectively rich) region surrounded by poor (respectively rich) regions will stay in this state of economic development whereas a poor region surrounded by richer regions has more probability to reach a higher state of economic development. These results are highlighted for European regions by Le Gallo (2001) who analyses the transitional dynamics of per capita GDP over the 1980-1995 period by means of a spatial Markov chains approach: the results emphasizes a poverty trap since the cluster of the poorest European regions in Southern Europe creates a great disadvantage for these regions.

All these observations may indicate the existence of different groupings of regions in the European Union as found in cross-country studies using international data sets. The aim of this chapter is therefore to analyze the convergence and growth processes among European regions over the period 1980-1995 in both a more disaggregated and comprehensive way. Indeed, both economic and geographic disparities embodied in the European regional polarization pattern should be taken into account. Actually, the purpose of this chapter is to show that the introduction of the spatial effects as defined by Anselin (1988a) in the estimation of the β–convergence model allows one to do just that.

Spatial effects refer to both spatial autocorrelation and spatial heterogeneity. On the one hand, we emphasize the link between the detection of a positive spatial autocorrelation of regional GDPs and the regional polarization of the economies in Europe. Moreover, we show that modeling spatial autocorrelation in the β–convergence model allows one to estimate geographic spillover effects. On the other hand, spatial heterogeneity means that economic behavior is not stable over space. Such spatial heterogeneity in patterns of economic development can be modeled via spatial regimes and/or groupwise heteroskedasticity: a cluster of rich regions (the core) being distinguished from a cluster of poor regions (the periphery).

From an econometric point of view, the presence of spatial dependence and/or spatial heterogeneity leads to unreliable statistical inference based on Ordinary Least Squares (OLS) estimations. The spatial dependence issue has been studied in Le Gallo et al. (see Chap. 3) by using the appropriate spatial econometric tools to test for its presence and to estimate the appropriate spatial specification. Concerning the spatial heterogeneity problem, we now define spatial regimes, which are interpreted as spatial convergence clubs, using Exploratory Spatial Data Analysis (ESDA), in order to capture the North-South polarization pattern observed in European regions. Taking into account both these effects, we show two results. First, the convergence process is different within each of the regimes. Actually there is no such convergence process for northern regions, whereas it is weak for southern regions. Second, a significant geographic spillover effect appears in the growth process: the average growth rate for a given region is positively influenced by the average growth rates of neighboring regions.

In Section 4.2 the club convergence concept used in this chapter is presented, as well as the way spatial effects can be interpreted in such a framework. In Section 4.3, the empirical methodology and the econometric results are presented. As a first step, we define *spatial* convergence clubs using Exploratory Spatial Data Analysis (ESDA). The second step is to show that the global and a-spatial unconditional β–convergence model is misspecified and that a spatial regimes model with spatially autocorrelated errors is more appropriate. Section 4.4 concludes.

4.2 Convergence Clubs and Spatial Heterogeneity

As pointed out by Le Gallo et al. (see Chap. 3), the standard β–convergence framework has been forcefully criticized both on theoretical and methodological grounds. In this chapter, we focus more precisely on the parameter heterogeneity problem by considering the possibility of convergence clubs based on groupings of regions within the European Union. We first define more precisely this concept, which can be based both on endogenous and neoclassical growth models. Finally, we show how the concept of club convergence can be tested in a spatial framework, by using the appropriate spatial econometric specifications.

4.2.1 Club Convergence

The concept of club convergence, used for example by Durlauf and Johnson (1995), is consistent with economic polarization, persistent poverty and clustering. In case of unconditional convergence, there is only one equilibrium level to which all economies approach. In case of conditional convergence, equilibrium differs by economy, and each economy approaches its own unique, globally stable, steady state equilibrium. In contrast, the concept of club convergence, is based on endogenous growth models characterized by the possibility of multiple, locally stable, steady state equilibria as in Azariadis and Drazen (1990). Which of these different equilibria an economy will be converging on depends on the range to which its initial conditions belong. In other words, economies converge to one another if their initial conditions are in the "basin of attraction" of the same steady state equilibrium. Galor (1996) shows that multiplicity of steady state equilibria and thus club convergence is also consistent with standard neoclassical growth models that exhibit diminishing marginal productivity of capital and constant return to scale if heterogeneity across individuals is permitted. The problem is then to distinguish evidence of club convergence from that of conditional convergence.

In such a framework, as noted by Durlauf and Johnson (1995), it is difficult for standard convergence tests to discriminate between these multiple steady state models and the standard Solow model. Moreover, Bernard and Durlauf (1996) show that a linear regression applied to data generated by economies converging to multiple steady states can produce a negative initial per capita GDP coefficient. The standard global β–convergence result appears then to be an artifact. Durlauf and Johnson (1995), using the Summers and Heston data set over the 1960-1985 period and the Mankiw et al. (1992) framework, show that convergence is indeed stronger within groups of countries once they arbitrarily split the whole sample based on the initial per capita GDP level and the adult literacy rate at the beginning of the period. Moreover, estimated parameter values associated to conditioning variables differ significantly across the groups. They endogenize the splitting using the regression tree method and note the geographic homogeneity within each group. However, they fail to find evidence of convergence among the high-output economies, i.e. North-American and European countries. This result is furthermore qualitatively similar to that obtained by De Long (1988). They interpret the overall parameter instability as indicative of countries belonging to different regimes.

The standard β–convergence concept and test are also, more deeply, criticized by Friedman (1992) and Quah (1993b) who raise the Galton's fallacy problem. Moreover, Quah (1993a, 1996a, 1996b, 1997) argues that convergence should be studied by taking into account the shape of the entire distribution of per capita

GDP and its intra-distribution dynamics over time and not by estimating the cross-section correlation between growth rates and per capita GDP levels or computing first or higher moments. Using an alternative empirical methodology based on Markov chains and probability transition matrices, Quah (1993a, 1996a, 1996b, 1997) finds evidence on the formation of convergence clubs, the international income distribution polarizing into "twin-peaks" of rich and poor countries. Quite surprisingly Quah (1996c) does not find evidence supporting "twin-peakedness" in the European regional income distribution for a sample of 82 regions, excluding southern poor Portuguese and Greek regions, over the 1980-1989 period. Yet Le Gallo (2001), using the same empirical approach, finds such evidence for an extended sample of 138 European regions over the 1980-1995 period.

4.2.2 Economic Geography and Spatial Clubs

From a different theoretical perspective, some economic geography models, are also characterized by the possibility of multiple spatial equilibria[1] (symmetric and center-periphery). For example, in a two regions economy with an agricultural sector and an industrial sector, two type of equilibrium may arise: we obtain a symmetric equilibrium if firms are equally distributed between regions, while the center-periphery equilibrium is characterized by the concentration of the firms in one region.[2] Whereas "spatial convergence" is linked to the symmetric equilibrium, "spatial divergence" refers to the center-periphery equilibrium. The type of equilibrium reached depends on initial conditions on the geographic division of firms between regions and the characteristics of the economy concerning the level of interactions costs, the degree of factor mobility, the market size, the existence of vertical linkages between firms and of geographic spillovers between regions. An important prediction of economic geography theories is that tighter regional integration favors the relevance of the center-periphery equilibrium or economic polarization since lower transaction costs and higher labor mobility contributes to the agglomeration process and hence to uneven regional development (Krugman and Venables 1995, 1996; Montfort and Nicolini 2000; Puga 1999).

The role played by knowledge spillovers can be highlighted for two reasons (Englmann and Walz 1995; Kubo 1995; Martin and Ottaviano 1999; Baldwin and Forslid 2000). On the one hand, the concentration in one region of many economic activities creates a traditional external scale effect of knowledge spillover, which improves the production process of each firms located in this region. Such external scale economies could incite firms to locate in tertiary and industrially developed regions. On the other hand, knowledge accumulation produced in one region

---

[1] A spatial equilibrium is a set of prices and wages associated with a particular geographic distribution of the industry.

[2] Other equilibrium may arise in a multi-locational framework (Krugman, 1993 or Brakman and al. 1996) or in models where the production of intermediate goods or innovations is considered (Krugman and Venables 1995b, 1996; Englmann and Walz 1995; Walz 1996).

improves the productivity of all the firms whatever the regions they are located in. Such regional externalities or geographic spillovers are based on an implicit spatial process of knowledge diffusion. For example, if knowledge spillovers decrease with distance between regions then rich regions tend to cluster. An important theoretical result based on these different and complex mechanisms demonstrates a circular causation process between growth and agglomeration: economic growth favors firms' concentration (i.e. uneven spatial distribution of firms inside an integrated economic space), which in turn has an effect on the growth rate. Uneven distribution of economic activities between regions and uneven regional development, illustrated for example by the well-known core-periphery pattern, naturally emerge from this circular causality. We can observe such spatial orders in the European Union where rich and attractive regions are geographically concentrated in Northern Europe while poor regions mainly cluster in Southern Europe and form therefore spatial clubs. Let us now consider what kinds of spatial effects are involved in this polarization pattern.

4.2.3 Spatial Effects and Polarization Patterns

Spatial effects refer to both spatial dependence and spatial heterogeneity. Applied to the study of income disparities, spatial autocorrelation means that rich regions tend to be geographically clustered as well as poor regions. Therefore, there is a link between the detection of a positive spatial autocorrelation among regional GDPs and the theoretical mechanisms previously discussed leading to regional polarization of the economies.

Spatial heterogeneity means in turn that economic behavior is not stable over space. In a regression model, spatial heterogeneity can be reflected by varying coefficients, i.e. structural instability, or by varying error variances across observations, i.e. heteroskedasticity. These variations follow for example specific geographical patterns such as East and West, or North and South. Such a spatial heterogeneity probably characterizes patterns of economic development in the form of spatial regimes and/or groupwise heteroskedasticity: a cluster of rich regions (the core) being distinguished from a cluster of poor regions (the periphery).

The links between spatial autocorrelation and spatial heterogeneity are quite complex. First, as pointed out by Anselin (2001), spatial heterogeneity often occurs jointly with spatial autocorrelation in applied econometric studies. Moreover, in cross-section, spatial autocorrelation and spatial heterogeneity may be observationally equivalent. For example, in polarization phenomena, a spatial cluster of extreme residuals in the center may be interpreted as heterogeneity between the center and the periphery or as spatial autocorrelation implied by a spatial stochastic process yielding clustered values in the center. Finally, spatial autocorrelation of residuals may be implied by some spatial heterogeneity that is not correctly modeled in the regression (Brunsdon et al. 1999 provide such an example). In other words, in a regression, spatial error autocorrelation may indicate that the regression is misspecified.

Three kinds of issues arise from these complex links between spatial dependence and spatial heterogeneity.

First, we must identify spatial clusters of regional wealth upon which a spatial regimes convergence model could be based. Each spatial cluster contains all regions connected by a spatial association criterion whereas the type of spatial association differs between clusters. Then both spatial dependence and heterogeneity effects are associated in the construction of our spatial clubs.

Second, statistical inference based on OLS when heterogeneity or spatial dependence is present is not reliable. For example, if we try to estimate a model characterized by a specific form of structural instability, we cannot rely on standard tests of structural instability in the presence of spatial autocorrelation and/or heteroskedasticity. It is therefore necessary to test if both effects are present. Furthermore, when spatial autocorrelation and spatial heterogeneity occur jointly in a regression, the properties of the White (1980) and Breusch-Pagan (1979) tests for heteroskedasticity may be flawed (Anselin and Griffith 1988). Therefore, it is necessary to adjust structural instability and heteroskedasticity tests for spatial autocorrelation and to use appropriate econometric methods as proposed by Anselin (1988b, 1990a, 1990b).

Third, the role played by geographic spillovers in the convergence of European regions has to be considered. Le Gallo et al. (see Chap. 3) have shown that if spatial autocorrelation is detected in the unconditional β–convergence model, then it leads to specifications integrating potential geographic spillovers in the convergence process. However, since spatial heterogeneity is now integrated in the estimation of the β–convergence model, we must use appropriate specifications and tests if we want to obtain reliable estimates of the effect of geographic spillovers on regional growth in Europe.

In the following section, we will define more precisely our empirical methodology, which extends the approach developed by Durlauf and Johnson (1995) by explicitly taking into account the possible spatial effects that we have outlined above. This is applied within the framework of the standard β–convergence process.

4.3 Econometric Results

In the first stage of our analysis, we use ESDA to look for possible spatial autocorrelation in conjunction with spatial structural instability in the distribution of (log) per capita GDP measured in ECUs across the European regions. We use the same data set extracted from the EUROSTAT-REGIO database (see Ertur and Le Gallo, Chap. 2) and the distance-based spatial weight matrices $W(h)$ as well as the k-nearest neighbors weight matrices $W(k)$ defined in Le Gallo et al. (see Chap. 3 for more details). ESDA is a set of techniques aimed at describing and visualizing spatial distributions, at detecting patterns of global and local spatial association and at suggesting spatial regimes or other forms of spatial heterogeneity (see Ertur and Le Gallo, Chap. 2). Here we only use Moran's I statistic to test for global spatial autocorrelation and Moran scatterplot to visualize patterns of local spatial association and spatial instability. In the second step, we estimate an unconditional β–convergence model by OLS and carry out various tests aimed at detecting the presence of spatial dependence and spatial heterogeneity. We finally suggest the most appropriate specification.

4.3.1 Exploratory Spatial Data Analysis: Detection of Spatial Clubs

We first test for global spatial autocorrelation in log per capita GDP (in ECUs) using Moran's I statistic and the distance-based spatial weight matrix $W(1)$. It appears that it is positively spatially autocorrelated since the statistics are significant with p approximately equal to 0.0001 for every year (Table 4.1).

Table 4.1 Moran's I statistics for log per capita GDP over 1980-1995 using $W(1)$

Year	Moran's I	Standard deviation	Standardized value
1980	0.812	0.040333	20.310
1981	0.788	0.040367	19.689
1982	0.774	0.040351	19.359
1983	0.803	0.040453	20.030
1984	0.789	0.040452	19.668
1985	0.795	0.040419	19.851
1986	0.813	0.040418	20.270
1987	0.819	0.040430	20.396
1988	0.804	0.040494	20.041
1989	0.786	0.040432	19.612
1990	0.790	0.040359	19.755
1991	0.782	0.040321	19.564
1992	0.791	0.040280	19.823
1993	0.809	0.040344	20.226
1994	0.816	0.040367	20.390
1995	0.817	0.040346	20.422
Growth 80-95	0.509	0.039441	13.085

Note: The expected value for Moran's I statistic is constant for each year: $E(I) = -0.007$. *Growth* is the average growth rate of per capita GDP over the 1980-1995 period.

These results suggest that the null hypothesis of no spatial autocorrelation should be rejected and that the distribution of per capita regional GDP is clustered throughout the period under study.[3] In other words, the regions with relatively high per capita GDP (resp. low) are located close to other regions with relatively high per capita GDP (resp. low) more often than if their locations were purely random. A similar result holds for the average growth rates of regional per capita GDP over the whole period. Moreover, these results are extremely robust in respect to the choice of the spatial weight matrix $W(h)$, $h = 1,\ldots,4$. In addition the

[3] All computations are carried out by means of SpaceStat 1.90 (Anselin 1999a).

results are also robust to the use of a *k*-nearest neighbors spatial weight matrices, for $k = 10, 15, 20, 25$ as presented in Ertur and Le Gallo (see Chap. 2), where the distance-based spatial weight matrix was not used since it was not possible to obtain statistical inference using the permutation approach for LISA statistics.

Spatial instability in the form of spatial regimes is then investigated by means of a Moran scatterplot (Anselin 1996). Given our context of β–convergence analysis, we choose to define such local spatial association using the logarithm of the *initial* level of per capita GDP. As noted by Durlauf and Johnson (1995), the use of split variables which are known at the beginning of the period are necessary to avoid the sample selection bias problem raised by De Long (1988).

The Moran scatterplot displays the spatial lag Wy_0 against y_0 both in standardized values, where y_0 is the vector of log per capita GDP levels in 1980. A four-way split of the sample based on the two control variables, initial per capita GDP and initial spatially lagged per capita GDP, allowing for interactions between them, is based on this Moran scatterplot.

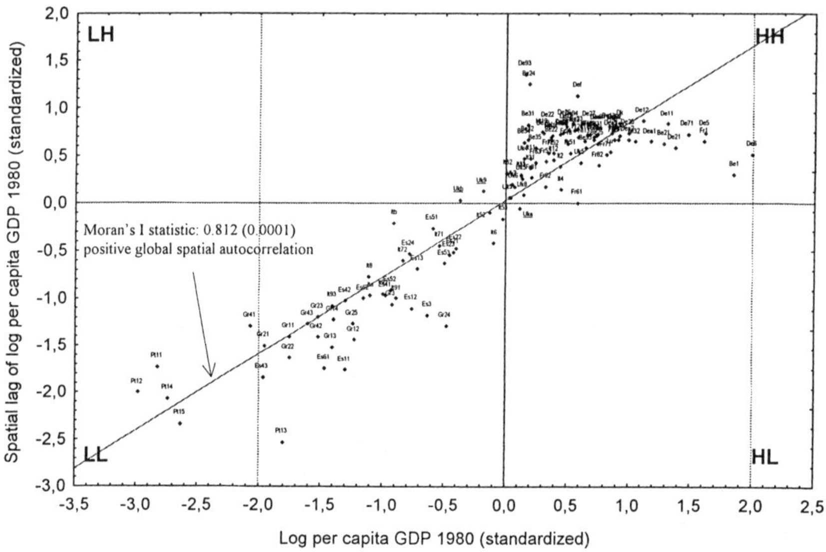

Fig. 4.1 Moran scatterplot for log per capita GDP in 1980 using $W(1)$ (list of region codes in Ertur and Le Gallo, see Chap. 2)

Fig. 4.1 displays this Moran scatterplot computed with $W(1)$ for log per capita GDP in 1980. It reveals the predominance of high-high and low-low clustering: almost all the European regions are characterized by positive spatial association since 90 regions are of type HH and 45 regions of type LL. The Moran scatterplot confirms the clear North-South polarization of the European regions: northern regions are located in the HH quadrant while southern regions are located in the LL quadrant. Only three regions show a spatial association of dissimilar values: Wales, and Northern Ireland are located in quadrant LH, which indicates poor regions, surrounded on average by rich regions, conversely Scotland is located in quadrant HL.

This suggests some kind of spatial heterogeneity in the European regional economies: the convergence process, if it exists, could be different across regimes. We consider therefore two spatial clubs constituted by HH and LL regions, which we call "North" and "South". Since Wales, Scotland and Northern Ireland are deleted[4], our new sample contains 135 regions, which belong to North and South as follows:

- *North* = {France, Germany, Netherlands, Belgium, Denmark, Luxembourg, United Kingdom (except Wales, Scotland) and northern Italy (Piemonte, Valle d'Aosta, Liguria, Lombardia, Trentino-Alto Adige, Veneto, Friuli-Venezia Guilia, Emilia-Romagna and Toscana)}.
- *South* = {Portugal, Spain, Greece and Southern Italy (Umbria, Marche, Lazio, Abruzzo, Molise, Campania, Puglia, Basilacata, Calabria, Sicilia and Sardegna)}.

Not surprisingly, regions belonging to the South regime correspond to the Objective 1 regions and mainly belong to the so-called "cohesion countries" defined by the European Commission.

The Moran scatterplots computed with the other spatial weight matrices $W(2)$, $W(3)$ and $W(4)$ sensibly lead to the same clubs: the only difference is the presence of Scotland in the North regime. This highlights again the robustness of our results with regard to the choice of the spatial weight matrix.[5] Moreover, the observed polarization seems to be persistent over the whole period since the composition of the clubs defined by the Moran scatterplots computed for each year remains globally unchanged.

The Moran scatterplot is illustrative of the complex interrelations between global spatial autocorrelation and spatial heterogeneity in the form of spatial regimes. Global spatial autocorrelation is reflected by the slope of the regression line of Wy_0 against y_0, which is formally equivalent to Moran's I statistic for a row-standardized weight matrix and which seems to be inherent to the layout of the spatial regimes corresponding to a clear North-South polarization pattern.

These exploratory results suggest that great care must be taken in the second step of our analysis, which is concerned with the estimation of the standard β–convergence model, due to the presence of spatial autocorrelation and spatial heterogeneity. Standard estimation by OLS and statistical inference based on it are

---

[4] The spatial clubs LH and HL containing only two regions and one region respectively are omitted due to the small number of observations in each and lack of degrees of freedom for the second step of our analysis.

[5] Using the nearest neighbors spatial weight matrix $W(10)$, we obtained practically the same North-South polarization: Aquitaine (France) replacing Scotland (United Kingdom) in the HL quadrant. The complete results are presented in Ertur and Le Gallo (see Chap. 2) and Fig. 4.2 in the Appendix. Furthermore, this result is fairly robust to the use of k-nearest neighbors spatial weight matrices, for k = 10, 15, 20, 25.

therefore likely to be misleading. Moreover, with respect to the simulation results presented by Anselin (1990a) on the size and power of traditional tests of structural instability in presence of spatially autocorrelated errors, we are potentially in the worst case: positive global spatial autocorrelation and two regimes corresponding to closely connected or compact observations. Therefore, these standard tests are likely to be highly misleading. Concerning the methodological approach to be taken in empirical studies we will follow Anselin's suggestion: "*...it is prudent to always carry out a test for the presence of spatial error autocorrelation... If there is a strong indication of spatial autocorrelation, and particularly when it is positive and/or the regimes correspond to compact contiguous observations, the standard techniques are likely to be unreliable and a maximum-likelihood approach should be taken*" (Anselin 1990a p. 205). We are aware that this empirical approach raises the well known pretest problem invalidating the use of the usual asymptotic distribution of the tests, but the simulation results presented by Anselin (1990a) indicate that this problem may not be so harmful in this case.

Finally, the determination of the different regimes or clubs should, ideally, be endogenous as in, for example, Durlauf and Johnson (1995) within an a-spatial framework. However, to our knowledge, such an attempt has still not yet been made in a setting that also takes into account spatial dependence[6] and remains beyond the scope of this chapter.

4.3.2 Estimation Results

In the second step of our empirical approach, we are interested in convergence among the European Union's regions within the context of European economic integration, although our approach implies neither within-country regional convergence nor convergence towards the mean European Union per capita GDP growth rate. Therefore, we first estimate the unconditional β–convergence model by OLS and carry out various tests aimed at detecting the presence of spatial dependence using the distance-based spatial weight matrix $W(1)$. We also test for spatial heterogeneity in the form of groupwise heteroskedasticity and/or structural instability across the spatial regimes which were defined earlier. However, testing for one effect in presence of the other requires some caution (Anselin and Griffith 1988; Anselin 1990a, 1990b). We then estimate the appropriate specifications integrating these spatial effects separately. Two kinds of econometric specifications can be used to deal with the problem of spatial dependence (Anselin 1988a; Anselin and Bera 1998; Anselin 2001): the spatial error model (spatial autoregressive error or SAR model) and the spatial lag model (mixed regressive, spatial autoregressive model). The way these models are estimated and interpreted in the context of β–convergence models is presented in detail in Le Gallo et al. (see Chap. 3). We

[6] This matter of fact is also noted by Anselin and Cho (2000). This issue is much more complex than in the standard non-spatial framework due to the spatial weight matrix and the spatial ordering of the observations.

integrate spatial heterogeneity in the conventional way: we estimate a groupwise heteroskedastic model by Feasible Generalized Least Squares (FGLS) and a two-regimes model by OLS. However, taking into account all effects jointly and estimating an appropriate econometric specification appears to be less straightforward: we overcome the problem by estimating a spatial regimes model with spatially autocorrelated errors. We present all the estimation results using the distance-based spatial weight matrix $W(1)$, the results obtained with the nearest neighbors weight matrix $W(10)$ are displayed in the Appendix.

4.3.2.1 OLS Estimation of the Unconditional β–Convergence Model and Tests

Let us take as a starting point the following model of unconditional β–convergence:

$$g_T = \alpha S + \beta y_0 + \varepsilon \qquad \varepsilon \sim \mathrm{N}(0, \sigma_\varepsilon^2 I) \qquad (4.1)$$

where g_T is the vector of dimension $n = 135$ of the average per capita GDP growth rates for each region i between 1995 and 1980; $T = 15$, y_0 is the vector containing the observations of per capita GDP in logarithms for all the regions in 1980; α and β are the unknown parameters to be estimated; S is the unit vector and ε is the vector of errors with the usual properties.

In this context, the choice of the cutoff for the distance-based spatial weight matrix W can be based on the OLS residual correlogram with ranges defined by minimum, lower quartile, median, upper quartile and maximum great circle distances, as suggested for example by Fingleton (1999). With the sample of 135 regions we consider now, $Q1$, Me, $Q3$ and Max are modified as following: $Q1 = 312$ miles, $Me = 582$ miles, $Q3 = 928$ miles and $Max = 1997$ miles (Table 4.2).

Table 4.2 Residual correlogram

Range (miles)	[*Min*;*Q*1[[8; 312[	[*Q*1;*Me*[[312;582[	[*Me*;*Q*3[[582;928[	[*Q*3;*Max*[[928;1997[
Moran's *I*	15.54	-3.35	-12.41	10.99
p-value	0.000	0.001	0.000	0.000
LMERR	157.38	10.45	91.74	29.93
p-value	0.000	0.001	0.000	0.000
R-LMERR	44.97	0.0097	34.92	0.0138
p-value	0.000	0.922	0.000	0.907

Note: $Q1$, Me, $Q3$ and Max are respectively the lower quartile (312 miles), the median (582 miles), the upper quartile (928 miles) and the maximum (1997 miles) of the great circle distance distribution between centroids of each region. For each range, we estimate the absolute β-convergence model and we perform the Moran's I test, the Lagrange multiplier test and its robust version for residual spatial autocorrelation based on the binary matrix computed for that range.

The determination of the cutoff that maximizes the absolute value of the significant Moran's I test statistic adapted to regression residuals (Cliff and Ord

1981) or Lagrange Multiplier test statistic for spatial autocorrelation of the errors (Anselin 1988a, 1988b) leads to $Q1$: we retain a cutoff of 312 miles for the distance-based weight matrix.

The results of the estimation by OLS of this model are then given in Table 4.3. The coefficient associated with the initial per capita GDP is significant and negative, which confirms the hypothesis of convergence for the European regions. The speed of convergence associated with this estimation is 0.85% (the half-life is 87 years), far below 2% usually found in the convergence literature, but closer to about 1% found by Armstrong (1995). These results indicate that the process of convergence is indeed very weak.

Table 4.3 Estimation results for the unconditional β-convergence model

OLS-White		Tests	
$\hat{\alpha}$	0.130 (0.000)	JB	8.50 (0.014)
$\hat{\beta}$	-0.00797 (0.002)	MORAN	12.94 (0.000)
Convergence speed	0.85% (0.000)	LMERR	140.68 (0.000)
Half-life	87	R-LMERR	16.61 (0.000)
R^2-adj	0.14	LMLAG	124.58 (0.000)
LIK	446.35	R-LMLAG	0.509 (0.475)
AIC	-888.69	BP / y_0	14.57 (0.000)
BIC	-882.88	BP / D1	5.85 (0.015)
$\hat{\sigma}_{\varepsilon}^2$	$7.984.10^{-5}$	White test	28.39 (0.000)
JLM1	155.25 (0.000)	JLM2	146.53 (0.000)

Note: p-values are in parentheses. *OLS-White* indicates the use of the White (1980) heteroskedasticity consistent covariance matrix estimator for statistical inference in the OLS estimation. *LIK* is the value of the maximum likelihood function. *AIC* is the Akaike (1974) information criterion. *BIC* is the Schwarz (1978) information criterion. *JB* is the Jarque-Bera (1987) estimated residuals Normality test. *MORAN* is the Moran's *I* test adapted to OLS residuals (Cliff and Ord 1981). *LMERR* is the Lagrange multiplier test for residual spatial autocorrelation and *R-LMERR* is its robust version. *LMLAG* is the Lagrange multiplier test for spatially lagged endogenous variable and *R-LMLAG* is its robust version (Anselin and Florax 1995; Anselin et al. 1996). *BP* is the Breusch-Pagan (1979) test for heteroskedasticity. *White* is the White (1980) test of heteroskedasticity. *JLM1* is the LM test of the joint null hypothesis of absence of heteroskedasticity linked to y_{1980} and residual spatial autocorrelation, *JLM2* is the LM test of the joint null hypothesis of absence of heteroskedasticity linked to D_1 and residual spatial autocorrelation (Anselin 1988a, 1988b).

Evidence in favor of normality is rather weak according to the Jarque-Bera test (1987) with a p-value of 0.014. We also note that the White (1980) test clearly rejects homoskedasticity as does the Breusch-Pagan (1979) test versus the explanatory variable y_0. Versus D_1, which is the dummy variable for the northern regime, the rejection is slightly weaker with a p-value of 0.015. Further consideration of spatial heterogeneity is therefore needed: we could think of some general form of heteroskedasticity or a more specific heteroskedasticity linked to the explanatory variable y_0 in the regression, or groupwise heteroskedasticity possibly associated to structural instability across regimes.

Five tests of spatial autocorrelation are then carried out: Moran's I test adapted to regression residuals (Cliff and Ord 1981) indicates the presence of spatial dependence. To discriminate between the two forms of spatial dependence - spatial autocorrelation of errors or endogenous spatial lag - we perform the Lagrange Multiplier tests: respectively LMERR and LMLAG and their robust versions (Anselin 1988b; Anselin et al. 1996). Applying the decision rule suggested by Anselin and Florax (1995), these tests indicate the presence of spatial autocorrelation rather than a spatial lag variable: the spatial error model appears to be the appropriate specification. The LM test of the joint null hypothesis of absence of heteroskedasticity and residual spatial autocorrelation is highly significant whatever the form of the heteroskedasticity assumed (Anselin 1988a, 1988b).

In addition to the apparent non-normality of the residuals, we are faced with two interconnected problems, which we have to deal with: spatial heterogeneity and spatial autocorrelation. A direct implication of these results is that the OLS estimator is inefficient and that all the statistical inference based on it is unreliable. In addition, as pointed out earlier, we must keep in mind that in presence of heteroskedasticity, results of the spatial autocorrelation tests may be misleading and conversely results of the heteroskedasticity tests may also be misleading in presence of spatial autocorrelation (Anselin 1988a; Anselin and Griffith 1988; Anselin 1990a,b). Therefore they must be interpreted with caution. More precisely, although the tests indicate heteroskedasticity, this may not be a problem because it can be due to the presence of spatial dependence (McMillen 1992).

The unconditional β-convergence model is strongly misspecified due to the spatial autocorrelation and heteroskedasticity of the errors. Actually, each region cannot be considered as independent of the others. The model must be modified to integrate this spatial dependence explicitly and to take into account spatial heterogeneity. Moreover, these two aspects may be linked.

4.3.2.2 Spatial Dependence

We first deal with the spatial dependence issue. We saw that the decision rule suggested by Anselin and Florax (1995) indicates a clear preference for the spatial error model over the spatial lag model. We then estimate the following SAR model:

$$g_T = \alpha S + \beta y_0 + \varepsilon \qquad \varepsilon = \lambda W \varepsilon + u \qquad u \sim \mathrm{N}(0, \sigma_u^2 I) \tag{4.2}$$

Estimation results by Maximum Likelihood (ML) are presented in Table 4.4. The coefficients are all strongly significant. From the convergence perspective, the estimated β coefficient is higher than in the unconditional β–convergence model estimated by OLS: the convergence speed is 1.2 % and the half-life reduces to 63 years once the spatial autocorrelation is controlled for. The convergence process appears then to be a little stronger but it remains actually quite weak.

Table 4.4 Estimation results for the spatial error model

Estimation results	ML	GMM	Tests	
$\hat{\alpha}$	0.156	0.157	LR-SED	74.15
	(0.000)	(0.000)		(0.000)
$\hat{\beta}$	-0.0110	-0.0110	LMLAG*	0.808
	(0.000)	(0.000)		(0.369)
$\hat{\lambda}$	0.788	0.828	LR-com-fac	0.177
	(0.000)			(0.674)
Convergence speed	1.2%	1.2%	Wald-com-fac	0.185
	(0.000)	(0.000)		(0.667)
Half-life	63	63	$\hat{\gamma}$	0.0084
				(0.871)
Sq. Corr.	0.14	0.14	S-BP / y_0	3.06
LIK	483.42			(0.080)
AIC	-962.85		S-BP / D1	4.27
BIC	-957.03			(0.039)
$\hat{\sigma}_u^2$	$4.078.10^{-5}$			

Note: *p*-values are in parentheses. *ML* indicates maximum likelihood estimation. *GMM* indicates iterated generalized moments estimation (Kelejian and Prucha 1999). *Sq. Corr.* is the squared correlation between predicted values and actual values. *LIK* is value of the maximum likelihood function. *AIC* is the Akaike (1974) information criterion. *BIC* is the Schwarz information criterion (1978). *LR-SED* is the likelihood ratio test for spatial error autocorrelation. *LMLAG** is the Lagrange multiplier test for an additional spatially lagged endogenous variable in the spatial error model (Anselin 1988a). *LR-com-fac* is the likelihood ratio common factor test. *Wald-com-fac* is the Wald common factor test (Burridge 1981). *S-BP* is the spatially adjusted Breusch-Pagan test for heteroskedasticity (Anselin 1988a, 1988b). γ is not estimated but computed using the restriction; its significance is assessed using the asymptotic delta method.

It is also important to note the presence of significant positive error spatial autocorrelation ($\hat{\lambda} = 0.788$). The LR and Wald common factor tests (Burridge, 1981) indicate that the restriction $\gamma + \lambda\beta = 0$ cannot be rejected so the spatial error model can be rewritten as the constrained spatial Durbin model (see Le Gallo et al. Chap. 3 for the derivation):

$$g_T = \alpha(I - \lambda W)S + \beta y_0 + \lambda W g_T + \gamma W y_0 + u \tag{4.3}$$

with $\gamma = -\lambda\beta$, but this coefficient is not significant. From the convergence perspective, this expression can be interpreted as a *minimal conditional β-convergence model* integrating two spatial environment variables (see Le Gallo et al. Chap. 3). This reformulation has also an interesting interpretation from the economic perspective: the average growth rate of a region i is positively influenced by the average growth rate of neighboring regions, through the endogenous spatial lag variable Wg_T. However, it doesn't seem to be influenced by the initial per capita GDP of neighboring regions, through the exogenous spatial lag variable Wy_0.

This spillover effect indicates that the spatial association patterns are not neutral for the economic performances of European regions. The more a region is surrounded by dynamic regions with high growth rates, the higher will be its growth rate. In other words, the geographical environment has an influence on growth processes.

The LMLAG* test does not reject the null hypothesis of the absence of an additional autoregressive lag variable in the spatial error model. According to information criteria this model seems to perform better than the preceding one (Akaike 1974; Schwarz 1978). Moreover, estimation of this model by the iterated Generalized Moments Method (GMM), as suggested by Kelejian and Prucha (1999), leads to practically the same results on the parameters of interest. This estimation method avoids the normality hypothesis of the error term and the problems linked to the accurate computation of the eigenvalues of W required by the ML estimator. However, it does not provide additional inference for the spatial autoregressive parameter, which is considered as a nuisance parameter.

The spatially adjusted Breusch-Pagan test (Anselin 1988a, 1988b) is no longer significant (p-value of 0.08), indicating absence of heteroskedasticity versus y_0. If this test was the only one carried out to detect heteroskedasticity in the spatial error model, we could say that heteroskedasticity found in the previous model is not a problem and was due to the presence of spatial dependence as in Chap. 3. However, the spatially adjusted Breusch-Pagan test remains significant versus D_1 (p-value of 0.04). We can deduce from these results that only a part of the heteroskedasticity found in the previous model is due to the spatial autocorrelation of the error term and that groupwise heteroskedasticity remains a problem that must be taken into account.

4.3.2.3 Spatial Heterogeneity: Groupwise Heteroskedasticity and/or Structural Instability

Let us turn now to the spatial heterogeneity issue, which can be considered from two points of view. The first one relates to the heteroskedasticity problem in the form of groupwise heteroskedasticity across the regimes previously defined. The second one relates to the structural instability problem across the two regimes and furthermore may be associated to groupwise heteroskedasticity.

We estimate the following model to take account of groupwise heteroskedasticity:

$$g_T = \alpha S + \beta y_0 + \varepsilon \qquad \varepsilon \sim N\left(0, \begin{bmatrix} \sigma^2_{\varepsilon,1} I_{90} & 0 \\ 0 & \sigma^2_{\varepsilon,2} I_{45} \end{bmatrix}\right) \tag{4.4}$$

Estimation results by FGLS are displayed in Table 4.5. The coefficients are all strongly significant. The estimated β coefficient is smaller than in all the preceding models leading to a convergence speed of 0.71 %. The half-life raises to 102 years indicating a very weak convergence process. The difference between regimes variances doesn't seem to be significant (p-value of 0.052) as assessed by the Wald test. However, this result should be interpreted with caution due to the presence of spatial dependence detected by the LMERR and LMLAG tests with a slight preference for spatially autocorrelated errors. Taking into account groupwise heteroskedasticity doesn't seem to eliminate the spatial dependence and globally leads to unreliable results.

Table 4.5 Estimation results for the groupwise heteroskedastic model

FGLS		Tests	
$\hat{\alpha}$	0.120 (0.000)	$\hat{\sigma}^2_{\varepsilon,1}$	$6.228.10^{-5}$ (0.000)
$\hat{\beta}$	-0.00677 (0.000)	$\hat{\sigma}^2_{\varepsilon,2}$	$11.141.10^{-5}$ (0.000)
Convergence speed	0.71% (0.000)	Wald het. test	3.78 (0.052)
Half-life	102	LMERR	129.59 (0.000)
Sq. Corr.	0.14	LMLAG	119.20 (0.000)

Note: p-values are in parentheses. *FGLS* indicates feasible generalized least square estimation. *Sq. Corr.* is the squared correlation between predicted values and actual values. $\hat{\sigma}^2_{\varepsilon,1}$ and $\hat{\sigma}^2_{\varepsilon,2}$ are respectively the estimated variances for the north and south regimes. *Wald het. test* is the Wald test for different variances across regimes. *LMERR* and *LMLAG* are respectively the Lagrange multiplier tests for residual spatial autocorrelation and endogenous spatial lag.

Let us consider more closely the possibility of structural instability. We estimate a spatial regimes model of unconditional β-convergence, which can be specified as follows:

$$g_T = \alpha_1 D_1 + \alpha_2 D_2 + \beta_1 D_1 y_0 + \beta_2 D_2 y_0 + \varepsilon \qquad \varepsilon \sim N(0, \sigma^2_\varepsilon I) \tag{4.5}$$

where D_1 and D_2 are dummy variables qualifying the two spatial regimes previously defined. More precisely, $D_{1,i}$ equals to 1 if region i belongs to the North and 0 if region i belongs to the South; $D_{2,i}$ equals to 0 if region i belongs to the North and 1 if region i belongs to the South. This model can also be formulated in matrix form as follows:

$$\begin{bmatrix} g_{T,1} \\ g_{T,2} \end{bmatrix} = \begin{bmatrix} S_1 & y_{0,1} & 0 & 0 \\ 0 & 0 & S_2 & y_{0,2} \end{bmatrix} \begin{bmatrix} \alpha_1 \\ \beta_1 \\ \alpha_2 \\ \beta_2 \end{bmatrix} + \begin{bmatrix} \varepsilon_1 \\ \varepsilon_2 \end{bmatrix} \Rightarrow Z = X\delta + \varepsilon \tag{4.6}$$

with $\varepsilon' = [\varepsilon_1' \quad \varepsilon_2']$ and $\varepsilon \sim N(0, \sigma_\varepsilon^2 I)$, the subscribe 1 standing for the north regime and the subscribe 2 for the south regime.

This type of specification takes into account the fact that the convergence process, if it exists, could be different within each of the regimes. Actually, this approach can be interpreted as a spatial convergence clubs approach, where the clubs are identified using a spatial criterion with the Moran scatterplot as described above. Our approach extends the empirical methodology elaborated by Durlauf and Johnson (1995) to take into account explicitly the spatial dimension of data.

The estimation results by OLS are displayed in Table 4.6. We see that the estimated β_1 coefficient does not have the expected sign and is not significant for the North. However, the estimated β_2 coefficient has the expected sign and is significant for southern regions leading to a convergence speed of 2.8% and a half-life of 30 years.

Table 4.6 Estimation results for the spatial regimes model

OLS -White	North 1	South 2	Tests	
$\hat{\alpha}$	-0.000825	0.252	Ind. stability test	18.64
	(0.981)	(0.000)		(0.000)
$\hat{\beta}$	0.00663	-0.0228	Ind. stability test	18.39
	(0.093)	(0.000)		(0.000)
Conv. speed	-	2.80%	Chow - White test	18.86
		(0.000)	overall stability	(0.000)
Half-life	-	30		
R^2-adj	0.25		MORAN	11.95
LIK	457.81			(0.000)
AIC	-907.62		LMERR	109.57
BIC	-896.00			(0.000)
$\hat{\sigma}_\varepsilon^2$	$6.840.10^{-5}$		R-LMERR	12.59
				(0.000)
JB	0.395		LMLAG	97.48
	(0.821)			(0.000)
BP / D1	4.015		R-LMLAG	0.512
	(0.045)			(0.474)

Note: *p*-values are in parentheses. *OLS-White* indicates the use of the White (1980) heteroskedasticity consistent covariance matrix estimator for statistical inference in the OLS estimation. *LIK* is the value of the maximum likelihood function. *AIC* is the Akaike (1974) information criterion. *BIC* is the Schwarz information criterion (1978). *JB* is the Jarque-Bera (1987) estimated residuals Normality test. *BP* is the Breusch-Pagan (1979) test for heteroskedasticity. The individual coefficient stability tests are based on asymptotic Wald statistics using adjusted White (1980) covariance matrix, distributed as χ^2 with 1 degree of freedom. The *Chow - White test of overall stability* is also based on an asymptotic Wald statistic using adjusted White (1980) covariance matrix, distributed as χ^2 with 2 degrees of freedom. *MORAN* is the Moran's *I* test adapted to OLS residuals (Cliff and Ord 1981). *LMERR* is the Lagrange multiplier test for residual spatial autocorrelation and *R-LMERR* is its robust version. *LMLAG* is the Lagrange multiplier test for spatially lagged endogenous variable and *R-LMLAG* is its robust version (Anselin and Florax 1995; Anselin et al. 1996).

The convergence process among southern regions seems to be stronger than the one in the initial model. This result is consistent with that obtained in an a-spatial framework by Beine and Jean-Pierre (2000) using a sample of 62 NUTS1 regions over the 1980-1995 period with an endogenous determination of convergence clubs. It is also consistent with that obtained by Durlauf and Johnson (1995) for rich versus poor countries using an international data set.

The Chow test of overall stability strongly rejects the joint null hypothesis. The individual coefficient stability tests reject the corresponding null hypotheses. The convergence process seems therefore to be quite different across regimes.

It is worth mentioning that the Jarque-Bera test (1987) doesn't reject normality (p-value of 0.82) in clear contrast to the result on the initial model: the reliability of all subsequent testing procedures and the use of Maximum Likelihood estimation method are then strengthened. Concerning the Breusch-Pagan test versus D_1, we note that the rejection of groupwise heteroskedasticity is weaker than in the initial model with a p-value of 0.045. The diagnostic tests for spatial dependence still indicate a preference for spatially autocorrelated errors as in the preceding model. However, all these tests should be interpreted with caution due to the potential presence of spatially autocorrelated errors and of groupwise heteroskedasticity.

4.3.2.4 Spatial Dependence and Spatial Heterogeneity

To take into account spatial error autocorrelation in conjunction with structural instability, we estimate the following spatial regimes model, in which we assume that the same spatial autoregressive process affects all the errors:

$$g_T = \alpha_1 D_1 + \alpha_2 D_2 + \beta_1 D_1 y_0 + \beta_2 D_2 y_0 + \varepsilon \qquad (4.7)$$

with $\varepsilon = \lambda W \varepsilon + u$ and $u \sim \mathrm{N}(0, \sigma_u^2 I)$. Or equivalently in matrix form:

$$\begin{bmatrix} g_{T,1} \\ g_{T,2} \end{bmatrix} = \begin{bmatrix} S_1 & y_{0,1} & 0 & 0 \\ 0 & 0 & S_2 & y_{0,2} \end{bmatrix} \begin{bmatrix} \alpha_1 \\ \beta_1 \\ \alpha_2 \\ \beta_2 \end{bmatrix} + \begin{bmatrix} \varepsilon_1 \\ \varepsilon_2 \end{bmatrix} \Rightarrow Z = X\delta + \varepsilon \qquad (4.8)$$

with $\varepsilon' = \begin{bmatrix} \varepsilon_1' & \varepsilon_2' \end{bmatrix}$; $\varepsilon = \lambda W \varepsilon + u$ and $u \sim \mathrm{N}(0, \sigma_u^2 I)$.

The subscribe 1 stands for the north regime and the subscribe 2 for the south regime. This specification allows the convergence process to be different within each of the regimes and in the same time deals with spatially autocorrelated errors previously detected. However, spatial effects are assumed to be identical in northern regions and southern regions but all the regions are still interacting spatially through the spatial weight matrix W. In addition, it seems meaningless to estimate separately the two regressions allowing for different spatial effects possibly based on different spatial weight matrices across regimes. This would imply that northern and southern regions do not interact spatially and are independent. There is also no obvious reason to consider different spatial weight matrices across regimes; the weight matrix contains the pure distance based spatial pattern, which is completely exogenous. This assumption would appear to be even more unlikely.

The ML estimates are presented in Table 4.7 First we note that $\hat{\beta}_1$ and $\hat{\beta}_2$ now have both the expected sign but $\hat{\beta}_1$ is still not significant for the North. For southern regions, $\hat{\beta}_2$ is strongly significant and negative. The convergence speed and the half-life are slightly different, compared to the preceding OLS model, once the spatial effects are controlled for (respectively 2.94% and 29 years).

Table 4.7 Estimation results for the spatial regimes spatial error model

Estimation results	North 1 ML	North 1 GMM	South 2 ML	South 2 GMM	Tests	
$\hat{\alpha}_r$	0.0798	0.0837	0.263	0.280	Ind. stability test	12.88
	(0.014)	(0.009)	(0.000)	(0.000)		(0.000)
$\hat{\beta}_r$	-0.0026	-0.0030	-0.0238	-0.0261	Ind. stability test	12.57
	(0.438)	(0.405)	(0.000)	(0.000)		(0.000)
$\hat{\lambda}$	0.788 (ML)			0.793 (GMM)	Chow-Wald test	13.06
	(0.000)				Overall stability	(0.001)
Conv.speed	-		2.94%	3.24%	LR-SED	63.68
			(0.000)	(0.000)		(0.000)
Half-life	-		29	26	LMLAG*	0.032
						(0.857)
Sq. corr.	0.22 (ML)			0.25 (GMM)	LR com. fac.	5.38
LIK	489.65					(0.068)
AIC	-971.31 (k=4)		-969.31 (k=5)		S-BP / D1	3.396
BIC	-959.68 (k=4)		-954.78 (k=5)			(0.065)
$\hat{\sigma}_\varepsilon^2$	$3.719.10^{-5}$					
$\hat{\gamma}_r$	0.002		0.0187			
	(0.970)		(0.729)			

Note: p-values are in parentheses. *ML* indicates maximum likelihood estimation. *GMM* indicates iterated generalized moments estimation (Kelejian and Prucha 1999). *Sq. Corr.* is the squared correlation between predicted values and actual values. *LIK* is value of the maximum likelihood function. *AIC* is the Akaike (1974) information criterion. *BIC* is the Schwarz information criterion (1978). The information criteria are computed both for 4 and 5 parameters, as λ may be considered as a nuisance parameter. The individual coefficient stability tests are based on spatially adjusted asymptotic Wald statistics, distributed as χ^2 with 1 degree of freedom. The Chow – White test of overall stability is also based on a spatially adjusted asymptotic Wald statistic, distributed as χ^2 with 2 degrees of freedom (Anselin 1988a). *LR-SED* is the likelihood ratio test for spatial error autocorrelation, *LMLAG** is the Lagrange multiplier test for an additional spatially lagged endogenous variable in the spatial error model (Anselin 1988a, 1990a). *LR-com-fac* is the likelihood ratio common factor test. *S-BP* is the spatially adjusted Breusch-Pagan test for heteroskedasticity (Anselin 1988a, 1988b). The γ_r parameters are not estimated but computed using the accepted restrictions; their significance is assessed using the asymptotic delta method.

The spatially adjusted Chow test (Anselin 1988a, 1990a) strongly rejects the joint null hypothesis of structural stability and the individual coefficient stability tests reject the corresponding null hypotheses. These results clearly indicate that the convergence process differs across regimes. Furthermore, if there is a convergence process among European regions, it mainly involves the southern regions rather than the northern regions.

The second aspect of these results we want to stress in this chapter relates to spatial spillover effects. We first note that a significant positive spatial autocorrelation is found under this assumption ($\hat{\lambda} = 0.788$). Recall that the spatial error model can also be expressed as the constrained spatial Durbin model, which can be formulated here as:

$$g_T = \alpha_1(I - \lambda W)D_1 + \alpha_2(I - \lambda W)D_2 + \beta_1 D_1 y_{1980} + \beta_2 D_2 y_{1980} + \lambda W g_T + \gamma_1 W D_1 y_{1980} + \gamma_2 W D_2 y_{1980} + u \tag{4.9}$$

with $u \sim N(0, \sigma_u^2 I)$ and the two nonlinear restrictions: $\gamma_1 = -\lambda\beta_1$ and $\gamma_2 = -\lambda\beta_2$. The LR common factor tests (Burridge 1981) indicate that these restrictions cannot be rejected. Nevertheless, these two coefficients do not seem to be significant. We saw previously that this reformulation of the spatial error model has an interesting interpretation from the spatial spillover perspective. It appears therefore that, whatever the regime, the average growth rate of a region *i* is positively influenced by the average growth rates of neighboring regions, through the endogenous spatial lag variable Wg_T. However, it doesn't seem to be influenced by the initial per capita GDP of neighboring regions, through the exogenous spatial lag variable Wy_0.

The LMLAG* test does not reject the null hypothesis of the absence of an additional autoregressive lag variable in the spatial error model. The spatially adjusted Breusch-Pagan heteroskedasticity test versus D_1 is not significant (*p*-value of 0.065) indicating that there is no need to further allow for groupwise heteroskedasticity in the model. According to information criteria (Akaike 1974; Schwarz 1978) this model seems to perform better than all the preceding ones. Moreover, estimation of this model by GMM (Kelejian and Prucha 1999) leads to almost the same results on the parameters of interest.

4.4 Conclusion

The aim of this chapter was to assess whether spatial dependence and spatial heterogeneity really matter in the estimation of β–convergence processes. Based on a sample of 138 European regions over the period 1980-1995, we showed that they do matter. Faced with the well-known theoretical inadequacy and the econometric problems associated with the standard β–convergence model, we have improved the model in both aspects.

First, from the econometric point of view, the unreliability of statistical inference based on OLS estimation in presence of non-spherical errors is well known.

Using the appropriate econometric tools, we detected spatial autocorrelation and overcame the problem by estimating the appropriate spatial error model that can be interpreted as a minimal conditional β–convergence model. Concerning spatial heterogeneity, it appeared that the problem was essentially due to structural instability in the form of spatial regimes. These spatial regimes, interpreted as spatial convergence clubs, were defined using Exploratory Spatial Data Analysis (ESDA), or more precisely by using a Moran scatterplot. We therefore took into account spatial autocorrelation in conjunction with structural instability. The estimation of the appropriate spatial regimes spatial error model showed that indeed the convergence process is different across regimes. Furthermore, it appeared that actually there is no such a process for northern regions, but only a weak one for southern regions. This non convergence result, which is consistent with that obtained for rich countries by De Long (1988) and Durlauf and Johnson (1995) using international data sets, might be due to residual intra-regime heterogeneity not being taken into account. Inclusion of additional variables in a conditional β–convergence framework might lead to a convergence result for the North regime using the Mankiw et al (1992) framework for example. Unfortunately, data for doing this are not available in the EUROSTAT-REGIO database. The global weak convergence found in the estimation of the standard β–convergence model appears then as an artifact.

Second, from the economic point of view, use of spatial econometric tools have helped us to integrate predictions coming from 'new' economic geography theory, leading to the estimation of a spatial spillover effect in the framework of spatial convergence clubs. This effect appeared to be strongly significant indicating that the average growth rate of per capita GDP of a given region is positively affected by the average growth rate of neighboring regions. The geographic environment plays therefore an important role in the study of economic growth processes.

References

Akaike H (1974) A New Look at the Statistical Model Identification. *IEEE Transactions on Automatic Control* AC-19:716-723

Anselin L (1988a) *Spatial Econometrics: Methods and Models*. Kluwer Academic Publishers, Dordrecht

Anselin L (1988b) Lagrange Multiplier Test Diagnostics for Spatial Dependence and Spatial Heterogeneity. *Geographical Analysis* 20:1-17

Anselin L (1990a) Spatial Dependence and Spatial Structural Instability in Applied Regression Analysis. *Journal of Regional Science* 30:185-207

Anselin L (1990b) Some Robust Approach to Testing and Estimating in Spatial Econometrics. *Regional Science and Urban Economics* 20:141-163

Anselin L (1996) The Moran Scatterplot as an ESDA Tool to Assess Local Instability in Spatial Association. In: Fisher M, Scholten HJ, Unwin D (eds) *Spatial Analytical Perspectives on GIS*. Taylor & Francis, London

Anselin L (1999) *SpaceStat, a Software Package for the Analysis of Spatial Data*, Version 1.90. BioMedware, Ann Arbor

Anselin L (2001) Spatial Econometrics. In: Baltagi B (ed) *Companion to Econometrics*. Basil Blackwell, Oxford, Chap. 14, pp 310-330

Anselin L, Bera A (1998) Spatial Dependence in Linear Regression Models with an Introduction to Spatial Econometrics. In: Ullah A, Giles DEA (eds) *Handbook of Applied Economic Statistics*. Springer-Verlag, Berlin, Chap. 7, pp 237-289

Anselin L, Cho WKT (2000) Spatial Effects and Ecological Inference. *Working Paper*. University of Illinois, Urbana-Champaign

Anselin L, Florax R (1995) Small Sample Properties of Tests for Spatial Dependence in Regression Models. In: Anselin L, Florax R (eds) *New Directions in Spatial Econometrics*. Springer-Verlag, Berlin

Anselin L, Griffith D (1988) Do Spatial Effects Really Matter in Regression Analysis? *Papers in Regional Science* 65:11-34

Anselin L, Bera AK, Florax R, Yoon MJ (1996) Simple Diagnostic Tests for Spatial Dependence. *Regional Science and Urban Economics* 26:77-104

Armstrong H (1995) Convergence among the Regions of the European Union. *Papers in Regional Science* 74:143-152

Azariadis C, Drazen A (1990) Threshold Externalities in Economic Development. *Quarterly Journal of Economics* 105:501-526

Baldwin RD, Forslid R (2000) The Core-Periphery Model and Endogenous Growth: Stabilizing and Destabilizing Integration. *Economica* 67:307-324

Baumol W.J., 1986, Productivity Growth, Convergence and Welfare: What the Long Run Data Show. *American Economic Review* 76:1072-1085

Beine M, Jean-Pierre P (2000) L'apport des tests de racine unitaire en panel à l'identification des club de convergence. In: Beine M, Docquier F (eds.), *Croissance et Convergence Economiques des Régions: Théorie, Faits et Déterminants*, De Boeck Université, Brussels

Bernard A.B., Durlauf S.N. (1996) Interpreting Tests of the Convergence Hypothesis. *Journal of Econometrics* 71:161-173

Brakman S., Garresten H., Gigengack, van Marrewijk C. Wagenvoort R (1996) Negative Feedbacks in the Economy and Industrial Location. *Journal of Regional Science* 36:631-651

Breusch T, Pagan A (1979) A Simple Test for Heterosckedasticity and Random Coefficient Variation. *Econometrica* 47:1287-1294

Brunsdon C, Fotheringham AS, Charlton M (1999) Some Notes on Parametric Significance Tests for Geographically Weighted Regression. *Journal of Regional Science* 39:497-524

Burridge P (1981) Testing for a Common Factor in a Spatial Autoregression Model. *Environment and Planning Series A* 13:795-800

Cliff AD, Ord JK (1981) *Spatial Processes: Models and Applications*. Pion, London

DeLong JB (1988) Productivity Growth, Convergence and Welfare: Comment. *American Economic Review* 78:1138-1154

Durlauf SN, Johnson PA (1995) Multiple Regimes and Cross-Country Growth Behaviour. *Journal of Applied Econometrics* 10:365-384

Englmann FC, Walz U (1995) Industrial Centers and Regional Growth in the Presence of Local Inputs. *Journal of Regional Science* 35:3-27

European Commission (1996) *First report on economic and social cohesion*. Official Publication Office, Luxembourg

European Commission (1999) *The European Regions: Sixth Periodic Report on the Socio-Economic Situation in the Regions of the European Union.* Official Publication Office Luxembourg

Fingleton B (1999) Estimates of Time to Economic Convergence: An Analysis of Regions of the European Union. *International Regional Science Review* 22:5-34

Friedman M (1992) Do Old Fallacies Ever Die? *Journal of Economic Literature* 30:2129-2132

Galor O (1996) Convergence? Inference from Theoretical Models. *Economic Journal* 106:1056-1069

Jarque CM, Bera AK (1987) A test for Normality of Observations and Regression Residuals. *International Statistical Review* 55:163-172

Kelejian HH, Prucha IR (1999) A Generalized Moments Estimator for the Autoregressive Parameter in a Spatial Model. *International Economic Review* 40:509-534

Krugman P (1991) Increasing Returns and Economic Geography. *Journal of Political Economy* 99:483-499

Krugman P., 1993, On the Number and Location of Cities, *European Economic Review*, 37:293-298

Krugman P, Venables AJ (1995) Globalization and the Inequality of Nations. *Quarterly Journal of Economics* 110:857-880

Krugman P, Venables AJ (1996) Integration, Specialization, and Adjustment. *European Economic Review* 40:959-967

Kubo Y (1995) Scale Economies, Regional Externalities, and the Possibility of Uneven Development. *Journal of Regional Science* 35:29-42

Le Gallo J (2001) Space-Time Analysis of GDP Disparities among European regions: A Markov chains approach. *Working Paper n°2001-05*, LATEC, University of Burgundy, Dijon

Mankiw NG, Romer D, Weil DN (1992) A Contribution to the Empirics of Economic Growth. *Quarterly Journal of Economics* 107:407-437

Martin P, Ottaviano GIP (1999) Growing Locations: Industry Location in a Model of Endogenous Growth. *European Economic Review*, 43:281-302

McMillen DP (1992) Probit with Spatial Autocorrelation. *Journal of Regional Science*, 32:335-348

Monfort P, Nicolini R (2000) Regional Convergence and International Integration. *Journal of Urban Economics* 48:286-306

Puga D (1999) The Rise and Fall of Regional Inequalities. *European Economic Review*, 43:303-334

Quah D (1993a) Empirical Cross-section Dynamics in Economic Growth. *European Economic Review* 37:426-434

Quah D (1993b) Galton's Fallacy and Tests of the Convergence Hypothesis. *Scandinavian Journal of Economics* 95:427-443

Quah D (1996a) Empirics for Economic Growth and Convergence. *European Economic Review* 40:1353-1375

Quah D (1996b) Twin Peaks: Growth and Convergence in Models of Distribution Dynamics. *Economic Journal* 106:1045-1055

Quah D (1996c) Regional Convergence Clusters across Europe. *European Economic Review* 40:951-958

Quah D (1997) Empirics for Growth and Distribution: Stratification, Polarization and Convergence Clubs. *Journal of Economic Growth* 2:27-59

Schwarz G (1978) Estimating the Dimension of a Model. *The Annals of Statistics*, 6:461-464

Walz U (1996) Transport Costs, Intermediate Goods and Localized Growth. *Regional Science and Urban Economics* 26:671-695

White H (1980) A Heteroskedastic-Consistent Covariance Matrix Estimator and a Direct Test for Heteroskedasticity. *Econometrica* 48:817-838

Appendix

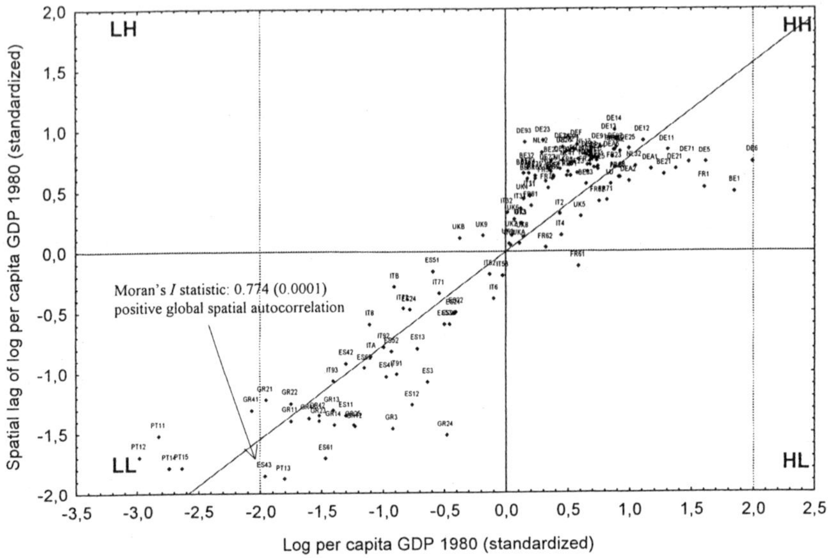

Fig. 4.2 Moran scatterplot for log per capita GDP 1980 in ECUs using W(10). 90 HH regions (North), 45 LL regions (South), 2 LH regions (Wales and Northern Ireland), 1 HL region (Aquitaine). Spatial heterogeneity ⇒ two spatial regimes: North (D1) and South (D2) with a new sample of 135 regions (list of region codes in Le Gallo et al., see Chap. 2)

Estimation results with W(10)

OLS estimation results of the unconditional β-convergence model obtained with the nearest neighbors weight matrix $W(10)$ are quite similar to those obtained using the distance-based weight matrix $W(1)$ and point to the spatial error model (Table 4.8). We also note that the White (1980) test again clearly rejects homoskedasticity as does the Breusch-Pagan (1979) test versus the explanatory variable y_0. Versus D_1 the rejection is slightly weaker with the same p-value of 0.015.

Table 4.8 Estimation results for the unconditional β-convergence model

OLS-White		Tests	
$\hat{\alpha}$	0.130 (0.000)	JB	8.63 (0.0134)
$\hat{\beta}$	-0.00792 (0.0025)	MORAN	13.32 (0.000)
Convergence speed	0.84% (0.000)	LMERR	140.88 (0.000)
Half-life	87	R-LMERR	31.14 (0.000)
R^2-adj	0.13	LMLAG	113.57 (0.000)
LIK	446.32	R-LMLAG	3.83 (0.050)
AIC	-888.64	BP / y_0	14.68 (0.000)
BIC	-882.83	BP / D_1	5.88 (0.015)
$\hat{\sigma}_{\varepsilon}^2$	$7.987.10^{-5}$	White test	28.606 (0.000)
JLM1	155.56 (0.000)	JLM2	146.76 (0.000)

Note: *p*-values are in parentheses. *OLS-White* indicates the use of the White (1980) heteroskedasticity consistent covariance matrix estimator for statistical inference in the OLS estimation. *LIK* is the value of the maximum likelihood function. *AIC* is the Akaike (1974) information criterion. *BIC* is the Schwarz information criterion (1978). JB is the Jarque-Bera (1987) estimated residuals Normality test. *MORAN* is the Moran's *I* test adapted to OLS residuals (Cliff and Ord 1981). *LMERR* is the Lagrange multiplier test for residual spatial autocorrelation and *R-LMERR* is its robust version. *LMLAG* is the Lagrange multiplier test for spatially lagged endogenous variable and *R-LMLAG* is its robust version (Anselin and Florax 1995; Anselin et al. 1996). *BP* is the Breusch-Pagan (1979) test for heteroskedasticity. *White* is the White (1980) test of heteroskedasticity. *JLM1* is the LM test of the joint null hypothesis of absence of heteroskedasticity linked to y_{1980} and residual spatial autocorrelation, *JLM2* is the LM test of the joint null hypothesis of absence of heteroskedasticity linked to *D1* and residual spatial autocorrelation (Anselin 1988a, 1988b).

The estimation by ML of the spatial error model reveals that the LR common factor test can not be computed (Table 4.9). Indeed, the maximized log-likelihood value of the unconstrained spatial Durbin model appears to be smaller than the log-likelihood value of the spatial error model, which is the constrained model. This is a surprising result which may be due to a multicolinearity problem in the unconstrained spatial Durbin model or to computational problems involved in the maximum likelihood estimation procedure, even in small samples, when the spatial weight matrix is not symmetric as it is the case with *W*(10) (Kelejian and

Prucha 1999, p. 510 and p. 514). Therefore, all statistical inference based on ML estimators may not be reliable. As suggested by Kelejian and Prucha (1999) the GMM estimation procedure is then the only valid alternative, even though the spatial autoregressive parameter is in that case considered as a nuisance parameter for which no further statistical inference can be made. Therefore, only the GMM estimation results should be interpreted. We note that the convergence speed is practically the same as the one obtained with $W(1)$.

Table 4.9 Estimation results for the spatial error model

Estimation results	ML	GMM	Tests	
$\hat{\alpha}$	0.182	0.183	LR-SED	76.31
	(0.000)	(0.000)		(0.000)
$\hat{\beta}$	-0.0139	-0.0140	LMLAG*	0.240
	(0.000)	(0.000)		(0.624)
$\hat{\lambda}$	0.798	0.815	LR-com-fac	-11.66**
	(0.000)			
Convergence speed	1.56%	1.58%	Wald-com-fac	0.046
	(0.000)	(0.000)		(0.830)
Half-life	49	49	$\hat{\gamma}$	0.011
				(0.834)
Sq. Corr.	0.14	0.14	S-BP / y_0	3.74
LIK	484.48			(0.053)
AIC	-964.96		S-BP / D_1	5.36
BIC	-959.15			(0.020)
$\hat{\sigma}_u^2$	$4.471.10^{-5}$			

Note: *p*-values are in parentheses. *ML* indicates maximum likelihood estimation. *GMM* indicates iterated generalized moments estimation (Kelejian and Prucha, 1999). *Sq. Corr.* is the squared correlation between predicted values and actual values. *LIK* is value of the maximum likelihood function. *AIC* is the Akaike (1974) information criterion. *BIC* is the Schwarz information criterion (1978). *LR-SED* is the likelihood ratio test for spatial error autocorrelation. *LMLAG** is the Lagrange multiplier test for an additional spatially lagged endogenous variable in the spatial error model (Anselin 1988a). *LR-com-fac* is the likelihood ratio common factor test, note that there is problem since its value is negative indicating that the log-likelihood of the unconstrained spatial Durbin model is lower then the log-likelihood of the spatial error model, which is the constrained model. *Wald-com-fac* is the Wald common factor test (Burridge, 1981). *S-BP* is the spatially adjusted Breusch-Pagan test for heteroskedasticity (Anselin 1988a, 1988b). γ is not estimated but computed using the accepted restriction; its significance is assessed using the asymptotic delta method.

Estimation results for the groupwise heteroskedastic model are quite similar to those obtained with $W(1)$ (Table 4.10). This is also the case of the OLS estimation results of the spatial regimes model (Table 4.11) pointing again to the spatial regimes spatial error model. In this latter model estimated by GMM (Table 4.12), the convergence speed is slightly enhanced in the South (4.14% with a half-life reduced to 22 years), there is still no convergence in the North. All these results globally confirm those obtained with $W(1)$ indicating their robustness with respect to the choice of the weight matrix.

Table 4.10 Estimation results for the groupwise heteroskedastic model

FGLS		Tests	
$\hat{\alpha}$	0.119 (0.000)	$\hat{\sigma}^2_{\varepsilon,1}$	6.226.10-5 (0.000)
$\hat{\beta}$	-0.00669 (0.000)	$\hat{\sigma}^2_{\varepsilon,2}$	11.153.10-5 (0.000)
Convergence speed	0.70% (0.000)	Wald het. test	3.80 (0.051)
Half-life	103	LMERR	142.86 (0.000)
Sq. Corr.	0.14	LMLAG	125.65 (0.000)

Note: p-values are in parentheses. *FGLS* indicates feasible generalized least square estimation. *Sq. Corr.* is the squared correlation between predicted values and actual values. $\hat{\sigma}^2_{\varepsilon,1}$ and $\hat{\sigma}^2_{\varepsilon,2}$ are respectively the estimated variances for the north and south regimes. *Wald het. test* is the Wald test for different variances across regimes. *LMERR* and *LMLAG* are respectively the Lagrange multiplier tests for residual spatial autocorrelation and endogenous spatial lag.

Table 4.11 Estimation results for the spatial regimes model

OLS -White	North 1	South 2	Tests	
$\hat{\alpha}$	-0.0033 (0.938)	0.252 (0.000)	Ind. stability test	23.17 (0.000)
$\hat{\beta}$	0.0069 (0.093)	-0.0228 (0.000)	Ind. stability test	23.78 (0.000)
Convergence speed	-	2.80% (0.000)	Chow - White test overall stability	12.43 (0.000)
Half-life	-	30		
R^2-adj	0.26		MORAN	13.744 (0.000)
LIK	458.05			
AIC	-909.00		LMERR	138.324 (0.000)
BIC	-896.48			
$\hat{\sigma}_{\varepsilon}^2$	6.815.10-5		R-LMERR	38.68 (0.000)
JB	0.442 (0.802)		LMLAG	101.44 (0.000)
BP / D_1	4.13 (0.042)		R-LMLAG	1.80 (0.180)

Note: p-values are in parentheses. *OLS-White* indicates the use of the White (1980) heteroskedasticity consistent covariance matrix estimator for statistical inference in the OLS estimation. *LIK* is the value of the maximum likelihood function. *AIC* is the Akaike (1974) information criterion. *BIC* is the Schwarz information criterion (1978). *JB* is the Jarque-Bera (1987) estimated residuals Normality test. *BP* is the Breusch-Pagan (1979) test for heteroskedasticity. The individual coefficient stability tests are based on asymptotic Wald statistics using adjusted White (1980) covariance matrix, distributed as χ^2 with 1 degree of freedom. The Chow – White test of overall stability is also based on an asymptotic Wald statistic using adjusted White (1980) covariance matrix, distributed as χ^2 with 2 degrees of freedom. *MORAN* is the Moran's *I* test adapted to OLS residuals (Cliff and Ord 1981). *LMERR* is the Lagrange multiplier test for residual spatial autocorrelation and *R-LMERR* is its robust version. *LMLAG* is the Lagrange multiplier test for spatially lagged endogenous variable and *R-LMLAG* is its robust version (Anselin and Florax 1995; Anselin et al. 1996).

Table 4.12 Estimation results for the spatial regimes spatial error model

Estimation results	North 1 ML	North 1 GMM	South 2 ML	South 2 GMM	Tests	
$\hat{\alpha}$	0.0629 (0.069)	0.0651 (0.063)	0.298 (0.000)	0.316 (0.000)	Ind. stability test	24.05 (0.000)
$\hat{\beta}$	-0.00078 (0.841)	-0.00085 (0.827)	-0.0281 (0.000)	-0.0308 (0.000)	Ind. stability test	24.15 (0.000)
$\hat{\lambda}$	0.811 (ML) (0.000)			0.793 (GMM)	Chow-Wald test Overall stability	24.15 (0.000)
Conv.speed	-		3.64% (0.000)	4.14% (0.000)	LR-SED	77.64 (0.000)
Half-life	-		24	22	LMLAG*	41.20 (0.000)
Sq. corr.	0.26 (ML)			0.26 (GMM)	Wald com. fac.	2.81 (0.245)
LIK	489.72					
AIC	-971.43 (k=4)		-969.44 (k=5)		S-BP / D1	7.14 (0.007)
BIC	-959.81 (k=4)		-954.91 (k=5)			
$\hat{\sigma}_{\varepsilon}^2$	3.778.10-5					
$\hat{\gamma}_r$	0.0006 (0.991)		0.0228 (0.685)			

Note: *p*-values are in parentheses. *ML* indicates maximum likelihood estimation. *GMM* indicates iterated generalized moments estimation (Kelejian and Prucha 1999). *Sq. Corr.* is the squared correlation between predicted values and actual values. *LIK* is value of the maximum likelihood function. *AIC* is the Akaike (1974) information criterion. *BIC* is the Schwarz information criterion (1978). The information criteria are computed both for 4 and 5 parameters, as λ may be considered as a nuisance parameter. The individual coefficient stability tests are based on spatially adjusted asymptotic Wald statistics, distributed as χ^2 with 1 degree of freedom. The Chow – White test of overall stability is also based on a spatially adjusted asymptotic Wald statistic, distributed as χ^2 with 2 degrees of freedom (Anselin 1988a). *LR-SED* is the likelihood ratio test for spatial error autocorrelation, *LMLAG** is the Lagrange multiplier test for an additional spatially lagged endogenous variable in the spatial error model (Anselin 1988a, 1990a). *LR-com-fac* is the likelihood ratio common factor test; *Wald-com-fac* is the Wald common factor test (Burridge 1981). *S-BP* is the spatially adjusted Breusch-Pagan test for heteroskedasticity (Anselin 1988a, 1988b). The γ_r parameters are not estimated but computed using the accepted restrictions; their significance is assessed using the asymptotic delta method.

5 Spatial Distance in a Technology Gap Model

Marjolein Caniëls[1] and Bart Verspagen[2]

[1] *Open University of the Netherlands, The Netherlands*

[2] *Eindhoven University of Technology, The Netherlands*

5.1 Introduction

The issue of convergence of GDP per capita is the topic of a large and growing literature in economics (e.g., Barro and Sala-i-Martin 1995; Fagerberg 1994). The general conclusion from this literature is that convergence, as opposed to divergence, is a special outcome that may prevail between a set of countries that is relatively homogenous in terms of variables such as knowledge generation (R&D), infrastructure, educational systems, etc.

This relationship between homogeneity of 'structural' characteristics and convergence arises from at least two different theoretical perspectives. On the one hand, there is the notion of 'conditional convergence' that arises from new growth models in the neoclassical tradition (e.g., Barro and Sala-i-Martin 1995). Here, the basic idea is that long-run economic growth can be described by a steady state balanced path. In the short run, countries that have not yet reached their steady state show higher (or lower, depending on which side they are approaching the steady state from) growth than countries closer to the steady state. If steady states are similar between countries, convergence is unconditional, but when, as in the new models, steady states differ, convergence becomes conditional upon variables such as R&D and infrastructure. In the latter case of heterogeneity, observed empirical patterns of growth may show little or no convergence. The more homogenous the group of countries under consideration becomes, however, the more convergence becomes likely.

The other theoretical perspective on convergence emerges from so-called technology gap theory (e.g., Fagerberg 1994). Here convergence results from the international diffusion of knowledge. Initially backward countries benefit from the inflow of knowledge that was developed in the more advanced countries. However, the extent to which they are able to benefit from this depends on so-called social capability and technological congruence (Abramovitz 1994). Social capability refers to variables that are largely similar to the variables defining the steady state in neoclassical models, such as education and infrastructure. Technological congruence refers, for example, to the match between the sectoral structure in the backward and advanced countries. When technological congruence is large, and social capabilities do not differ too much between countries, convergence will become more likely.

One aspect that remains relatively obscured in both theoretical perspectives on convergence, is the impact of spatial proximity on the diffusion of technological knowledge. The models in the technology gap tradition (e.g., Fagerberg 1988; Verspagen 1991; Amable 1993) do not take into account spatial proximity as a determinant of knowledge diffusion. Nevertheless, there is a large literature in economic geography that underlines the importance of proximity for knowledge spillovers. The concept of interest in this literature (for an overview see Baptista 1998) is the existence of agglomeration economies and its effects on growth. Agglomeration economics involve the positive effects on a firm or a region generated by a spatial concentration of economic activity. Agglomeration economies are induced, among others, by a large opportunity for communication of ideas and experience, which is enhanced by spatial proximity. In this Chapter we focus on knowledge spillovers as the prime form of agglomeration economies. Several studies (e.g., Acs, Audretsch and Feldman 1992; Jaffe, Trajtenberg and Henderson 1993) have confirmed such a positive relation between geographic proximity and knowledge spillovers.

Theoretical reasons for the localized nature of knowledge spillovers are as follows. Technological knowledge is often informal, tacit and uncodified in its nature (e.g., Pavitt 1987). This implies that there are differences between knowledge and information, where the former concept is more far-reaching than the latter. Audretsch and Feldman (1996) argue that although the cost of transmitting *information* may be invariant to distance, presumably the cost of transmitting *knowledge* rises with distance. Possibilities for learning-by-doing and learning-by-using, important for the transmission of knowledge, to a large extent come from direct contacts with competitors, customers, suppliers and providers of services (Von Hippel 1988, 1994) and are therefore highly dependent on proximity.

Uncertainty is another characteristic of the innovative process. Interaction between innovators, e.g. in regional networks, helps to reduce this uncertainty. This kind of interaction is highly dependent on geographical proximity. In this respect, Freeman (1991) points out that networks frequently tend to be localised. Another reason why proximity has an effect on the innovative process lies in the fact that innovation relies heavily upon sources of basic scientific knowledge. Jaffe (1989) and Acs, Audretsch and Feldman (1992) have empirically shown that knowledge spillovers from university research to private firms are facilitated by geographic proximity. Furthermore, innovative activity is cumulative, meaning that new innovations build upon scientific knowledge generated by previous innovations. Breschi (1995) and Malerba and Orsenigo (1995) point out that the accumulation of innovative activity in a geographic area facilitates the generation of new innovations in this area.

The aim of this Chapter is to integrate space into a model of technology gaps, and to investigate whether or not the conclusion of a relationship between homogeneity of countries and convergence remains in tact. In order to do so, we incorporate spatial proximity into a technology gap growth model developed earlier by

Verspagen (1991)[1]. The resulting model is one in which a multitude of geographic units (which will be called regions) interact with each other in terms of knowledge diffusion. These regions may differ with respect to their R&D efforts and their social capability to assimilate knowledge from other regions. Ceteris paribus, knowledge from regions close by diffuses more easily than knowledge from regions far away.

The rest of this Chapter is organised as follows. In Section 2, the part of the model that describes technological spillovers is first presented for the case of two regions. Later on, this is extended to a multi-region model. Section 3 explores the outcomes of our model in terms of patterns of convergence or divergence by means of simulation techniques. Finally, Section 4 summarises the main conclusions from this Chapter.

5.2 Description of the Spillover System

For simplicity, we disregard any sources of output growth other than the growth of technological knowledge. Specifically, it is assumed that output growth is a linear function of the growth of the knowledge stock:

$$\dot{Q}_i / Q_i = \beta(\dot{K}_i / K_i) \tag{5.1}$$

in which Q_i denotes the level of output of region i and K_i points to the level of the knowledge stock of region i. β is a parameter, indicating the proportion of the knowledge stock growth that results in output growth. Dots above variables denote time derivatives.

New knowledge is assumed to stem from three sources: learning-by-doing (modelled as a Verdoorn effect[2]), spillovers received from surrounding (not necessarily contingent) regions (S_i), and an exogenous rate of growth (ρ_i), which can be thought of as reflecting the impact of exogenous R&D activities in the region. This yields the following equation:

$$\dot{K}_i / K_i = \alpha(\lambda(\dot{Q}_i / Q_i) + S_i + \rho_i) \tag{5.2}$$

in which α and λ are parameters. α points out the extent to which the knowledge stock growth is influenced by the above factors, and λ reflects the intensity of the Verdoorn effect.

For the explanation of the spillover term S, it is convenient to first consider two regions, later on this framework will be extended, and a multi-region model will be constructed. In the two-region setting, it is assumed that there is one techno-

[1] The model is further analyzed in Caniëls and Verspagen (2001).

[2] The Verdoorn-Kaldor law states that a positive relation exists between the growth of productivity and the growth of output.

logically advanced region and one backward region. Spillovers depend on the size of the knowledge gap, as well as three different parameters reflecting distinct effects related to the realisation of potential spillovers. We use the following equation to model spillovers:

$$S_j = \frac{\delta_j}{\gamma_{ij}} e^{-(\frac{1}{\delta_j} G_{ij} - \mu_j)^2} \qquad with \qquad i \neq j \tag{5.3}$$

$$G_{ij} = \ln \frac{K_i}{K_j} \tag{5.4}$$

in which S_j denotes the spillovers generated by region i and received by region j[3]. G_{ij} denotes the technology gap of region i towards region j, and is defined as the log of the ratio of the knowledge stocks of two regions. The realisation of the potential spillover level depends on the three parameters γ, δ and μ, which we will now discuss in turn.

γ_{ij} is the geographical distance between two regions. If γ_{ij} increases, the spillover is reduced. This assumption stems from the geographical literature. As was discussed in the introduction of this Chapter, this is based on the assumption that spatial proximity eases spillovers (agglomeration economies), because interaction between the receiver and generator of the spillovers is easier when distance is small. μ_j and δ_j are two parameters that are related to the intrinsic learning capability of region j. These parameters thus reflect the broad concept of 'social capability' to assimilate spillovers (e.g., Abramovitz 1994). Regions that have a high social capability to learn (e.g., a highly educated workforce, good infrastructure, an efficient financial system, etc.), can implement the knowledge from other regions more easily. μ_j and δ_j reflect different parts of the learning capability that will be explained further by means of graphical analysis.

Fig. 5.1 presents the spillover functions for two regions, assuming all parameters to be equal between the regions. The horizontal axis displays the size of the technology gap. Note that a positive value of G_{ij} by definition implies that region j is the backward region. A first characteristic of our model is that, in contrast with most of the catch-up literature, we allow spillovers to occur in two directions, i.e., from the technological leader to backward region(s), and vice versa. Fig. 5.1 shows that, for equal parameter values between the regions, the spillover stream from the advanced to the backward region will be larger than the reverse stream

---

[3] Note that the lower the initial stock of knowledge a region is endowed with, the more spillovers it will receive. This is similar to the concept of β-convergence (Barro 1984; Baumol 1986; De Long 1988; Barro and Sala-i-Martin 1991, 1992a, 1992b) in which a backward economy (an economy with a low initial level of GDP per capita) will grow faster than a rich economy and therefore catch up.

(S_i is always below S_j to the right of the vertical axis). This reflects the notion that the backward region can learn more than the advanced region. However, spillovers from a backward region to the leader region also take place because it could well be possible that the backward region has (developed) complementary knowledge, knowledge that was not yet in the hands of the leader. So there always is a small flow of knowledge from laggard to leader, although this quantity quickly goes to zero for large gaps.

The *net* spillover will be equal to zero when the gap between the two regions is zero (i.e., they have equal knowledge stocks). In this situation there are still spillovers, but these are of equal size in both directions. This only holds, however, when the parameters (ρ, λ, μ, δ) are equal between the two regions. In the more general case of unequal parameters between regions, net spillovers may be positive or negative for a gap of value zero.

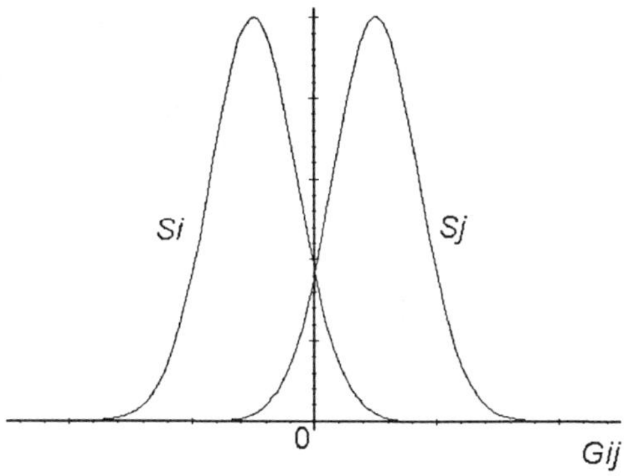

Fig. 5.1 Spillover curve for two regions

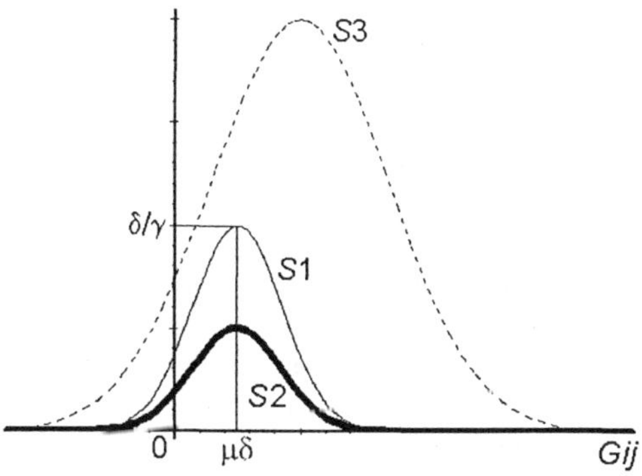

Fig. 5.2 The influence of geographic distance ($S2$) and the learning capability ($S3$) on the spillover curve of a region

Fig. 5.2 displays the spillovers received by one region for this two-region model. Note that the top of each spillover curve lies at a technology gap equal to $\mu_j\delta_j$. The maximal spillover corresponding to this is equal to δ_j/γ_{ij}. We take the curve labeled *S*1 as the starting point, and we consider what happens to the spillover function under certain conditions. First, an enlargement of the geographical distance between two regions (higher γ) will lead to lower spillovers received by each region, depicted by the thick line *S*2. Note that an increase in distance shifts the curve down, but leaves the value of the gap for which spillovers are maximal unchanged.

Second, an increase in the learning capability parameter δ of the lagging region will cause the spillover function to shift up, and the maximum of the curve to shift to the right (dotted line *S*3).[4] Thus, with higher δ, the laggard is able to learn more (magnitude of the spillover function) and more easily, or earlier (at a larger technological distance).

As will be explained below, the value of *G* at which the spillover curve peaks ($\mu\delta$) is important for the result in terms of convergence or divergence. We therefore want to allow for the possibility that maximum of the spillover curve shifts left or right, without affecting the value of the maximum itself. Fig. 5.3 shows how the parameter μ does exactly this.

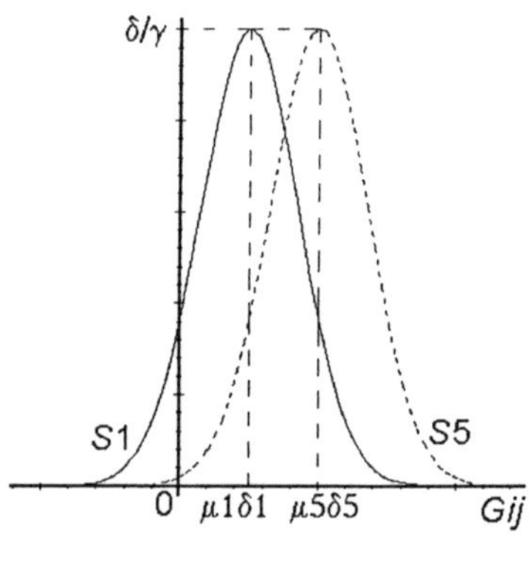

Fig. 5.3 The influence of geographic distance (*S*2) and the learning capability (*S*3) on the spillover curve of a region

If μ_j is increased, all other things being equal, the curve will shift to the right (*S*5). This has several effects. First, the level of spillovers in the case of equal knowledge stocks across regions ($G = 0$), is smaller. This indicates that for relatively large μ, the model resembles a regular catch-up model, which is characterized by zero spillovers for zero technological distance. Second, because the top of the curve moves to the right, catch-up becomes easier. At a larger technological

[4] To achieve this reaction of the spillover curve, the learning capability had to appear in two places in the spillover function (Eq. (5.3)).

distance, it is still possible to catch up. How the distinction between catching-up of falling behind works exactly will become clearer after we discuss the net spill-over function.

Thus, the difference between the parameters μ and δ is mainly a technical matter. In practice, they can hardly be disentangled in terms of the variables that make up social capability to assimilate spillovers. We mainly use the parameter μ to calibrate the model (i.e., to generate a setup that implies a reasonable borderline between catching-up and falling-behind), while δ is used more actively in the simulation experiments below as an indicator of the learning capability of a region.

In order to be able to analyze the dynamics of convergence and divergence, we take the time derivative of the technology gap in Eq. (5.4) and substitute Eqs. (5.1), (5.2) and (5.3). For a two-region model this yields:

$$\dot{G}_{ij} = \frac{d}{dt} \ln \frac{K_i}{K_j} = \frac{\dot{K}_i}{K_i} - \frac{\dot{K}_j}{K_j} \tag{5.5}$$

$$= \frac{\alpha}{1 - \alpha\beta\lambda}((\rho_i - \rho_j) - (S_j - S_i)), \text{with } 0 < \alpha\beta\lambda < 1$$

in which α, β and λ are assumed to have the same value in each region. This expression can be analyzed using Fig. 5.4.

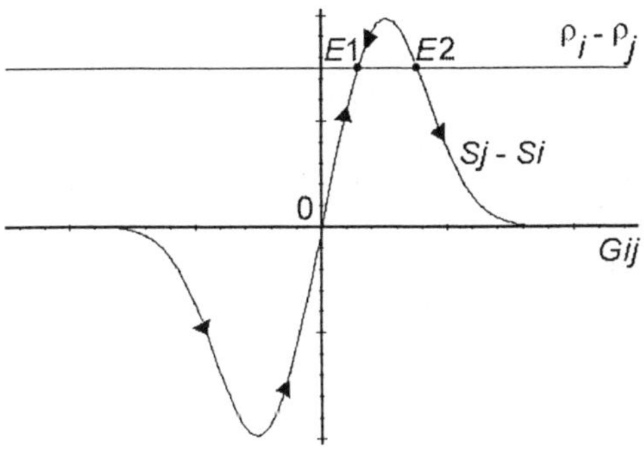

Fig. 5.4 The dynamics of the model

We will restrict ourselves to describing only one case, namely the one in which region i is the leader, i.e., where the initial gap is positive. We also assume that leadership implies larger R&D efforts, such that $\rho_i > \rho_j$.[5] In Fig. 5.4, $S_j - S_i$ represents the difference in received spillovers between the two regions. The lagging region receives positive net spillovers, as discussed above. Note that we have again assumed $\delta_i = \delta_j$ and $\mu_i = \mu_j$. In the more general case where these assumptions do not hold, the net spillover curve will not intersect with the origin, but this does not change the dynamics in a major way. The horizontal line $\rho_i - \rho_j$ displays the difference in the exogenous rate of growth of the knowledge stock between the two regions.

It is straightforward from Eq. (5.5) that when the curve in Fig. 5.4 intersects with the horizontal line $\rho_i - \rho_j$, the time derivative of the technology gap is equal to zero. In other words, the intersection points correspond to equilibrium points. The (leftmost) intersection point at which the S-curve has a positive slope is stable, whereas the other intersection point is unstable. Thus, what happens to the knowledge gap in the long run depends on where the process starts. Starting points to the left of *E*2 will yield convergence to a stable technology gap (corresponding to *E*1). Starting values to the right of *E*2 will yield falling behind, with an ever-growing knowledge gap.[6]

Now consider what happens with changing parameter values. We will first consider a variation in the difference in the exogenous rate of growth of the knowledge stock between the two regions, $\rho_i - \rho_j$. If the difference is enlarged in favor of the leader, the $\rho_i - \rho_j$ line in Fig. 5.4 moves upward, meaning that the range of technology gaps at which catch-up occurs becomes smaller. Eventually, when the $\rho_i - \rho_j$ line shifts to a position above the net spillover curve, there will be no opportunity at all for catch-up. If, on the other hand, the exogenous rate of growth of the knowledge stock in the backward region is increased (e.g. by expanding research efforts) up to a level comparable with the advanced region, i.e., the $\rho_i - \rho_j$ line ultimately coincides with the horizontal axis, and the (stable) equilibrium gap is zero, implying complete converge in the long run.

Next, we consider the impact of the geographical distance between the two regions. A decrease in the geographical distance has the effect that the spillover curves S_i and S_j increase proportionally to the decrease in geographical distance (explained by Fig. 5.2) and the maximum of the $S_j - S_i$ curve in Fig. 5.5 moves upwards[7]. Fig. 5.6 displays the bifurcation diagram for this case.[8] The horizontal axis of the bifurcation diagram shows the values of the geographical distance parameter γ_{ij}. The vertical axis shows the equilibrium values of the technology gap. The line E_{sj} shows the stable equilibrium, while the line E_{uj} points to the unstable

---

[5] This assumption is not essential. Obviously, the case where region *j* is the leader is the mirror-image of the case we discuss.

[6] Verspagen (1991) estimates a simpler version of this model for a large sample of countries over the post-war period, and finds that falling behind is a frequent phenomenon.

[7] The maximum also moves a little bit away from the y-axis, but this is a very small effect.

[8] Note that the figure should show a discontinuous graph (in the model a geographical distance is either 1 or 2, not 1.5), however, for visual reasons the individual points are connected.

equilibrium. The line S_{maxj} represents the top of the net spillover curve in Fig. 5.5. This figure shows that for high values of γ_{ij} no equilibrium value of the technology gap exists. In terms of Fig. 5.5, this occurs when there are no intersection points between the curves. For a threshold value of γ_{ij}, one equilibrium appears. This occurs when the two curves in Fig. 5.5 are tangent. For values of γ_{ij} smaller than the threshold level, two equilibria exist, as described by the curves in the bifurcation diagram.

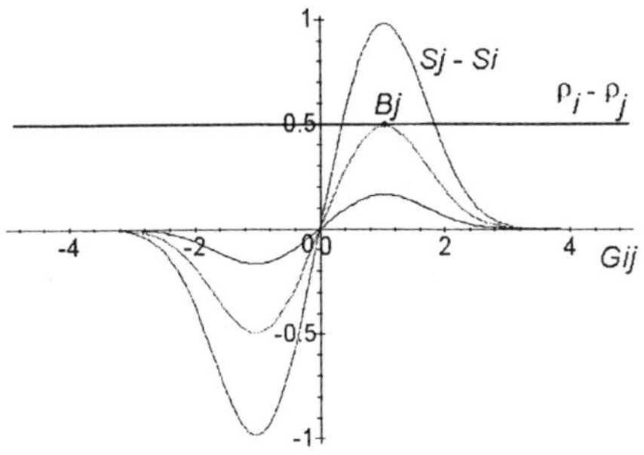

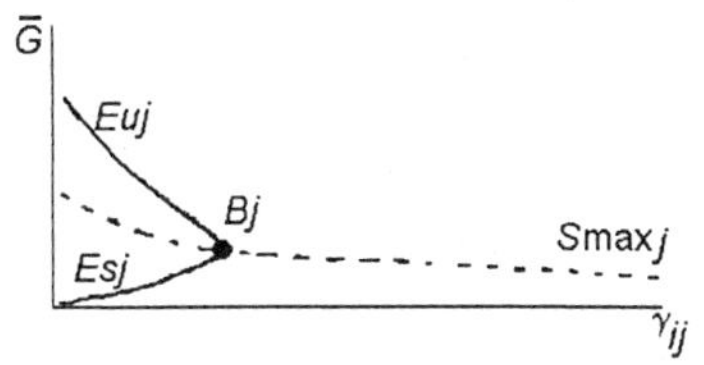

Fig. 5.5 The impact of geographical distance on the net spillover curve **Fig. 5.6** Bifurcation with respect to geographical distance

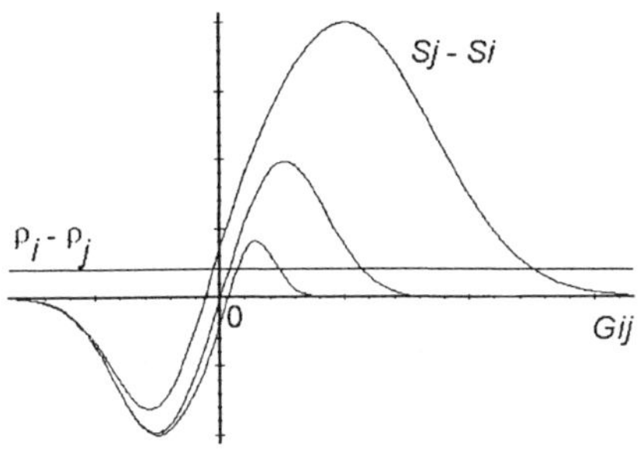

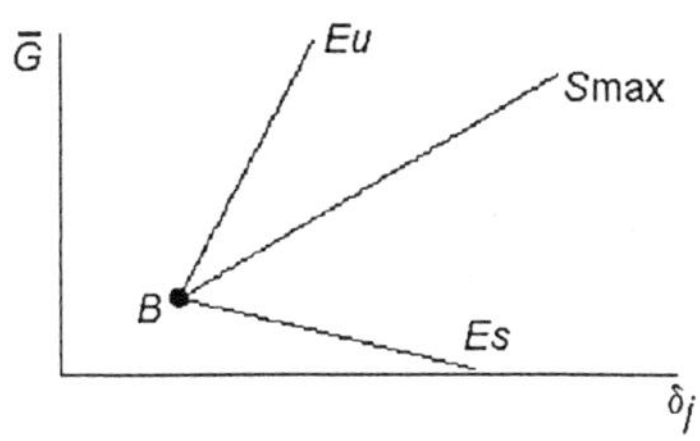

Fig. 5.7 The impact of the learning capability on the net spillover curve **Fig. 5.8** Bifurcation with respect to the learning capability

A similar bifurcation analysis can be performed for the parameter δ. The effect of an increase in the learning capability of the backward region j (δ_j) on the $S_j - S_i$ curve is displayed in Fig. 5.7. Note that δ_j is the only parameter that has changed, δ_i is kept constant. It can clearly be seen that on the right hand side of the figure the top of the curve has moved to the upper right of the figure and the curve does not intersect with the origin anymore. What has happened on the left-hand side is a bit more difficult to see. The minimum point has moved upwards so that it is closer to the horizontal axis. Also, there is a small movement of the minimum point away from the y-axis. The bifurcation diagram now looks as displayed in Fig. 5.8.

Note that the E_s line for the stable equilibrium can even go below the x-axis if the difference in exogenous growth rates of the knowledge stock is small enough, which illustrates an interesting special case of the model. This situation indicates a take-over in leadership by the (initially) lagging region. In terms of Fig. 5.7, this occurs when the horizontal line ($\rho_i - \rho_j$) intersects with the $S_j - S_i$ curve left from the y-axis, where the gap is smaller than zero, indicating that region j is the leader region. The combination of a large learning capability in the lagging region together with a small difference in the exogenous rate of growth between laggard and leader gives rise to a take over of the lead position by the backward region. Note that it is primarily learning capability that drives this process of take-over.

We omit the bifurcation analysis for the μ parameter, which is relatively straightforward, and jump to extend the model to a multi-regional case. Suppose we have a world with k regions, so that each region can be characterized by $k - 1$ technology gaps (we omit the trivial case of G_{ii}). Spillovers are received from each of the other regions, so that the S terms in Eq. (5.2) now become sums of spillovers over $k - 1$ regions. This gives rise to the following modified form of Eq. (5.5):

$$\dot{G}_{ij} = \frac{d}{dt} \ln \frac{K_i}{K_j} = \frac{\dot{K}_i}{K_i} - \frac{\dot{K}_j}{K_j} \qquad (5.5')$$

$$= \frac{\alpha}{1 - \alpha\beta\lambda}((\rho_i - \rho_j) + (\Sigma_n S_{in} - \Sigma_n S_{jn}) - (S_j - S_i))$$

$$\text{with } 0 < \alpha\beta\lambda < 1 \text{ and } n \neq i, j$$

in which $\Sigma_n S_{in}$ and $\Sigma_n S_{jn}$ denote the spillovers received by region i and j respectively from all regions n for which $n \neq i, j$ (this term is thus invariant to G_{ij}). Note that Eq. (5.5') specifies the (growth of the) gap between the two regions i and j only. There are k regions in total, thus every region i has $k - 1$ of these equations.

Eq. (5.5'), under the ceteris paribus assumption with respect to the knowledge stocks in regions other than i and j, gives rise to identical figures as Figs. 5.4–5.8. The only difference is that in the case of Eq. (5.5'), ($\rho_i - \rho_j$) and ($\Sigma_n S_{in} - \Sigma_n S_{jn}$) are lumped together into the horizontal lines that used to be determined by ($\rho_i - \rho_j$) only. A movement of this horizontal line (and therefore in the horizontal position of $E2$) can now be caused by two factors. First, a variation in the difference between the exogenous rates of growth of the knowledge stocks of two regions (as before), and, second, a difference across regions in the spillovers received from all other regions.

The latter term is largely determined by geographic location. The subset of regions to which this term refers does not differ between i and j, but when, for example, region i is closer to the advanced regions than region j is, this gives region i an advantage over region j. Also, the learning capability (δ and μ) has an impact on how ($\Sigma_n S_{in} - \Sigma_n S_{jn}$) differs between i and j.

5.3 The Influence of Geographic Distance on Regional Growth Differentials

When we specify a (symmetric) matrix of distances between regions, the model is fully specified, and time paths for the G variables result from any set of initial values. However, for more than two regions, these time paths are extremely tedious to work out analytically, which is why we resort to simulations to describe the outcomes of the model. By carrying out many simulations (with randomized initial conditions) it is possible to examine the general behavior of the model, and we find that certain patterns in the gaps of the knowledge stocks appear repeatedly. All simulations use a Pascal computer program that implements a Runge-Kutta algorithm to numerically solve the differential equations for G.

We use three different geographical spheres (distance matrices). These are a lattice of honeycombs, a column, and a globe. Appendix A gives an exact description and a map of these spheres. The regions on these spheres are assumed to be homogeneous areas. In other words, no differences of the relative importance (e.g., political) of the regions are assumed, nor do we assume differences in the degree of connectedness (e.g., the presence of harbors, mountains, roads and railways). Since this is a one-sector model, we also assume that the regions have homogenous economic structures. The first of these spheres is two-dimensional. A honeycomb pattern is chosen in order to provide an equal number of contiguous neighbors for each region, with each neighbor having an equally long border. This would not be the case when using a lattice of squares, which would have the additional difficulty of judging the importance of the different kinds of neighbors – queens, bishops or rooks[9] – by assigning weights to them. Because the lattice is flat and has a hexagonal shape in itself, there is always exactly one central region. This region has a favorable location, as will become clear from the experiments.

The second sphere used has the shape of a column. In the column, the single central region from the lattice is replaced by a belt of regions around the middle of the column. Finally, a globe is modeled. In the globe, no inherently central location is present. Both the column and the globe consist of a honeycomb pattern, in the case of the globe pentagons had to be added (the regions are constructed as the pattern on a soccer ball, i.e., 12 pentagons and 20 hexagons)[10]. The lattice of hon-

---

[9] These terms are borrowed from chess. A queen is allowed to move in all directions indicating that all 8 neighbors of a square are equally important. A lattice with these characteristics is called a Moore neighborhood. A bishop is only allowed to move in a diagonal way, while a rook is only allowed to move horizontally or vertically, meaning that one might want to assign a different (lower) weight to a neighbors, which do not share a border but only one point (the bishops-case) than to neighbors, which do share a border (the rooks-case). When only neighbors of the rook type are considered, the plain is called a von-Neumann neighborhood.

[10] It is impossible to construct a three-dimensional figure by the single use of hexagons. Hexagons will always produce a flat sphere, since the sum of the angles of three contingent hexagons is equal to 360 degrees. By adding pentagons, the total angle will be less than 360 and thus producing a three-dimensional figure. It would have been possible to

eycombs can be considered similar to a country, whereas the globe could be a model for a world. The column can be seen as an intermediate case, since it replaces the one central region of the lattice by a belt of centrally located regions, whereas on the globe every region is centrally located.

Geographic distance in the three geographical spheres is measured by assigning a weight of 1 to neighboring regions (in the sense that two regions share one border). Regions which do not share a border with a specific region are given a weight by using the concept of nearest neighbors, which means that a different (lower) weight is attributed to a second order neighbor. A second order neighbor does not share a border with a specific region, but does share a border with a neighbor of the specific region. Thus, no evaluation of relative importance of the connection between regions, based on ex ante known information (for example the presence of roads and railways) is taken into account. Only geographical distances are reflected. In this way, the distance d_{ij} is determined for every region in relation to every other region. Now, it is possible to construct a region-by-region matrix of shortest paths. Then, the corresponding weights (γ) are determined using the inverse of the orders (inverse shortest path, Haggett, Cliff and Frey 1977). Note that this way of measuring geographical distance is a special case of $1/(d_{ij}^{x})$ with x equal to 1.

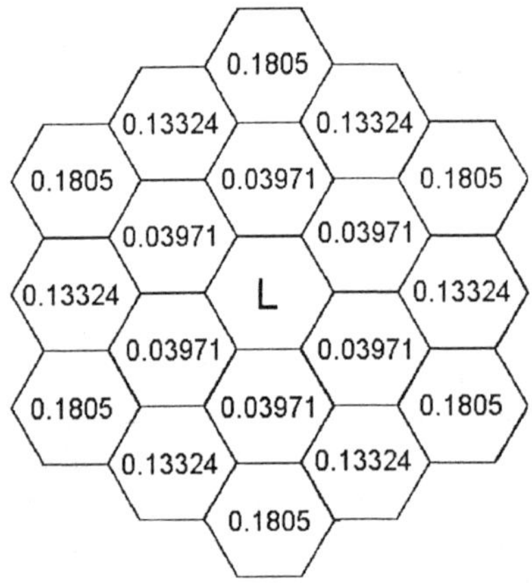

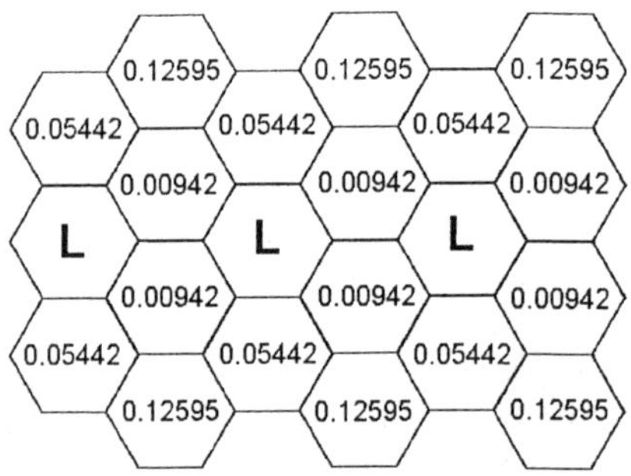

Fig. 5.9 Regional gaps for a lattice of honeycombs (L denotes a leader region) **Fig. 5.10** Regional gaps for a column (L denotes a leader region)

construct a three-dimensional sphere by using pentagons only, however, in that case the total number of pentagons (regions) used would be twelve. The globe that is used in the simulations consists of thirty-two planes (regions), which was considered to give more interesting interactions than a sphere containing only twelve planes.

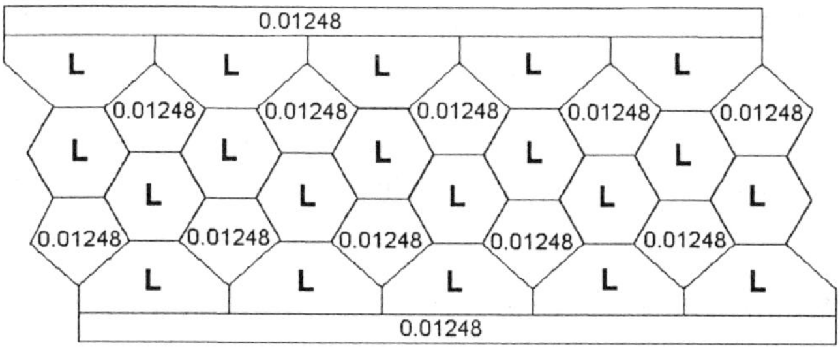

Fig. 5.11 Regional gaps for a globe (L denotes a leader region)

Our first simulation experiment assumes that all regions are initially equal, except for their geographic location[11]. Thus, $\delta_i = \delta$, $\mu_i = \mu$, and initially all values for G are equal to zero. The values for the parameters δ and μ remain fixed over the run, while the values for G, naturally, evolve according to Eq. (5.5'). Figs. 5.9, 5.10 and 5.11 show the resulting distribution of the gaps (after 10000 simulation periods) toward the leader of each region in the different geographic spheres.[12]

Fig. 5.9 displays the resulting pattern for the lattice of honeycombs. The numbers within each cell refer to the size of the technology gap a region has towards the leader region. Knowledge spills into to a region more easily when the distance towards a knowledge intensive region is small. Since the central region has the most favorable location for receiving knowledge spillovers, it will end up being the leader.

Fig. 5.10 displays the resulting gaps for the column. The regions located in the middle band of the column have the most favorable position with respect to receiving spillovers from neighboring regions, because they have more neighbors with a low order than the regions on the edge of the column. These three regions therefore come out as the joint leaders.

Finally, Fig. 5.11 displays the results for the globe. In this sphere, all hexagon shaped regions have an equal potential for receiving knowledge spillovers. The same holds for all pentagon-shaped regions. However, a hexagon-shaped region has a small advantage over a pentagon-shaped region with respect to receiving knowledge spillovers, since the honeycomb-shape provides a region with six instead of five direct neighbors and thus probably more spillovers. Fig. 5.11 indicates that this is indeed the case, all hexagon-shaped regions have a leading position, while all pentagon-shaped ones display a small gap.

Summarizing, the lattice shows a typical resulting pattern with the centrally located region being the leader (i.e., a strong polarization of economic growth). The column is characterized by a belt of leaders. The globe has many leaders, since all hexagon-shaped regions have the same favorable geographic location for receiving spillovers from other regions.

[11] The initial values of the data are described in Appendix B.

[12] 10000 periods is always enough to let the gaps reach their equilibrium values, i.e., no change has occurred for a long time when the system reaches period 10000.

These three figures have shown the pattern of gaps that is solely determined by geographical distances. Next, two sets of experiments have been set up in which systematic differences between regions are introduced. The first set considers a variation across regions of the intrinsic learning capability parameter δ. In the second set, the exogenous rate of knowledge generation is varied across regions.

5.3.1 Systematic Differences in the Learning Capability Parameter δ

In this set of experiments, a randomly generated value for δ is assigned to each region. The value for δ is drawn from a uniform distribution with upper boundary 2, and lower boundary varying from 1.8 until 2, taking steps of 0.01. The choice of these ranges is based on the calibrations of the model. In none of these ranges falling behind occurs, i.e., an equilibrium value of the gap always exists. Fifty runs were done for each range. Recall that the higher δ for a region, the more it is capable to implement the knowledge that spills over from other regions.

Figs. 5.12, 5.13, and 5.14 show the results of this experiment for each geographical sphere. The horizontal axis displays the lower boundary of the range out of which the learning capabilities are assigned. As we move from left to right on this axis, the range from which values for δ are drawn becomes smaller, and hence (on average) the regions become more equal with respect to their value of δ. On the extreme right hand side, the uniform range reduces to a single value (i.e., 2), and we are back in the situation where all parameters are equal (Figs. 5.9–5.11). On the vertical axis, the coefficient of variation of the resulting technology gaps over all regions (after 10000 simulated time periods) is shown[13]. The shades in the figure correspond to frequencies over the 50 runs, with complete black corresponding to a frequency of 50 (i.e., all runs). White shades indicate very low (sometimes zero) frequencies.

[13] We also did some simulations measuring polarization instead of disparity, using the measures proposed by Esteban and Ray (1994). These results show that polarization is rather constant over the horizontal dimension of Fig. 5.14. This implies that our model does not have anything specific to say about polarization as opposed to disparity.

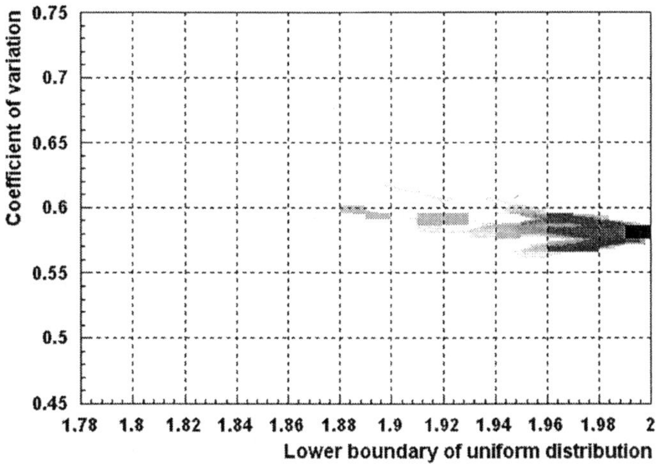

Fig. 5.12 Frequency diagram of the coefficient of variation at the end of the run. Initial variation in the learning capability simulated on the lattice

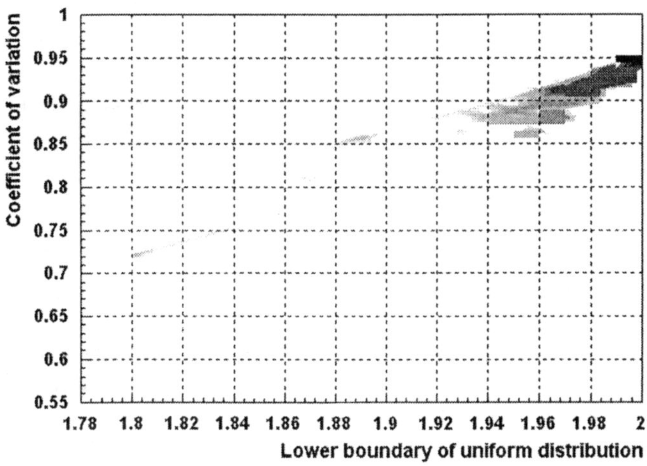

Fig. 5.13 Frequency diagram of the coefficient of variation at the end of the run. Initial variation in the learning capability simulated on the column

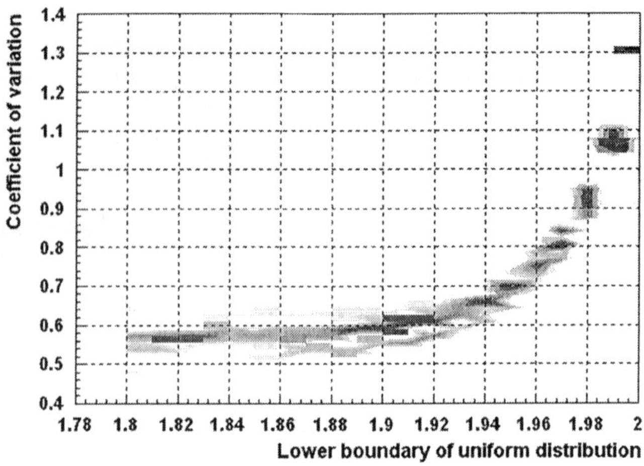

Fig. 5.14 Frequency diagram of the coefficient of variation at the end of the run. Initial variation in the learning capability simulated on the globe

The sensitivity to the lower boundary for the lattice is shown in Fig. 5.12. For large ranges, i.e., heterogeneous values for δ, the outcome of the process as characterized by the coefficient of variation varies greatly. No unique value, or range of values, emerges as an attractor of the process. When the range from which we draw δ becomes smaller, i.e., when we move to the right in the figure, the process 'zooms into' a narrowing range for the coefficient of variation. Thus, the result for the lattice appears as comet-shaped. The trail of the 'comet' disperses equally to all directions as the lower boundary is decreased.

For the column (Fig. 5.13) and the globe (Fig. 5.14) the same analysis is done. One striking difference as compared to Fig. 5.12 emerges from these two figures. The coefficient of variation for the column is also comet-shaped, but the trail extends in a downward direction for larger ranges (i.e., moving to the left). The globe shows the same downwardly sloping trend (from right to left) as the column does. In this case, however, the band does not widen very much when moving right to left, i.e., the dispersion of results remains relatively constant.

Note that the result of a downward slope (moving right to left) is somewhat counterintuitive. It says that higher diversity across regions in terms of the learning capability parameter δ causes lower levels of disparity in the resulting gaps at the end of the simulation, whereas less diversity in δ leads to more disparity in knowledge stocks. Thus, compared to a situation of identical regions, introducing heterogeneity in terms of learning capability leads to convergence. Compared to the intuition from the convergence theory briefly sketched in the introduction, this is a surprising result.

What causes this surprising result? When the range is limited to only one value (in this case this value is 2), the only factor that has an influence on the difference between regions is the geographical distance. As was seen before (Figs. 5.9–5.11).

When differences with respect to δ are increased, a second polarizing factor executes an influence on the regions. These two polarizing factors counteract. On average, a number of regions with unfavorable geographical locations will randomly obtain a high learning capability, while some centrally located regions will randomly obtain a low learning capability. This leads to lower levels of disparity *within* one run, but also leads to more disparity *across* the runs. Since there is more opportunity for the learning capability as a second polarizing factor with larger ranges (at the left end of the figure), the outcomes – in terms of coefficient of variation – of different runs become more distinct, causing the fan-shape in the lattice and the column.

This does not yet explain why the figures for the column (13) and globe (14) show this counterintuitive result, and the lattice does not. The difference between the spheres lies mainly in the fact that the column and globe have multiple geographically favored regions. For the lattice there is only one central region, while the six first ring regions have a relatively large gap.

This means that in the case of a globe or column, there is more than one 'neighborhood', with a 'local leader'. When such a 'local' leader is affected by a low learning capability (which is more likely for larger ranges of this parameter), this not only affects the local leader, but indeed the whole neighborhood. This causes an overall lower coefficient of variation. On the other hand, a lower coefficient of variation could be the result of higher learning capabilities for the peripheral regions. However, the chance that all (or at least a large part of the) peripheral regions (randomly) obtain(s) a higher learning capability than the potential local leaders is limited. The impact of differences in learning capability grows disproportionally with the range, causing the upward sloping pattern in Figs. 5.13 and 5.14. Since the globe has more potential local leaders than the column (20 versus 3), the chance is higher in case of the globe that a local leader (randomly) obtains a low learning capability. Therefore, Fig. 5.14 displays a steeper upward slope than Fig. 5.13.

What is also peculiar in Fig. 5.14 is that from 1.9 downwards there seems to be a horizontal trend. This can be explained as follows. At the left-hand side of the figure, the randomness in the parameter is apparently so high that all central regions are affected, and additional randomness does not increase the chance that an additional potential local leader obtains a relatively low learning capability. Therefore, the number of potential local leaders does not decrease any further at larger parameter ranges, resulting in a horizontal trend in the figure.

The conclusion that can be drawn from these figures (13 and 14) is an extraordinary one. Contrary to what is generally assumed, it is found that when differences in learning capabilities increase (decrease) across regions, there may be a tendency for the disparity between them to decrease (increase). This is a direct result from the inclusion of geographical distance in the model. Thus, the results show that leaving out distance from a model of technology gaps may indeed affect the results.

One may ask the question as to whether or not this result is robust to considering variations in other parameters, i.e., μ and ρ. Limitations of space do not permit us to present results for both μ and ρ. We will shortly discuss the results for variations in ρ. Readers interested in variations in μ, which yield similar patterns, may request these results from the authors.

5.3.2 Differences in the Exogenous Rate of Knowledge Generation ρ

In the following set of experiments the exogenous rate of growth of the knowledge stock is randomly chosen from different ranges, while the initial knowledge stocks (as well as other parameters) of the regions are identical across regions. In this experiment, it generally does matter for final leadership whether a region has an initial advantage. The region that becomes a leader early on will, in most of the cases, remain in this position. There are two determinants of leadership in the first stages of the simulation. First, the exogenous rate of growth of the knowledge stock is a very important influence in determining the leader in the first periods of simulation. Second, the geographical distance to centrally located regions is also important. In general, the region that has the highest ρ among the centrally located regions will most likely become a leader.

In order to investigate the sensitivity of the model to changes in the range out of which the exogenous rate of growth was chosen, several sets of 50 runs were carried out each with a different range. Figs. 5.15, 5.16, and 5.17 show the results when the lower boundary of the range is increased from 1.8 to 2 with steps of 0.01. The upper boundary is held constant at 2. The vertical axis shows the coefficient of variation over the last period in every run.

For the lattice (Fig. 5.15), the coefficient of variation displays a fan-shape (identical to the experiment with the learning capability). All coefficients of variation tend to one value as the gap between lower and upper boundary decreases. The explanation is the same as in the experiment with the learning capability. Increasing the difference in exogenous rates of growth of the knowledge stock introduces a second polarizing factor, next to the geographic distance. The two polarizing factors counteract, thereby causing a larger equality across gaps *within* one run and a larger disparity across coefficients of variation *across* several (50) runs.

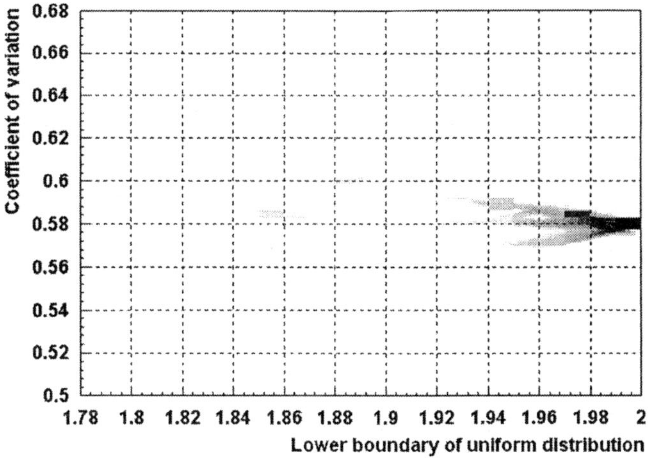

Fig. 5.15 Frequency diagram of the coefficient of variation at the end of the run. Initial variation in the exogenous rate of knowledge generation simulated on the lattice

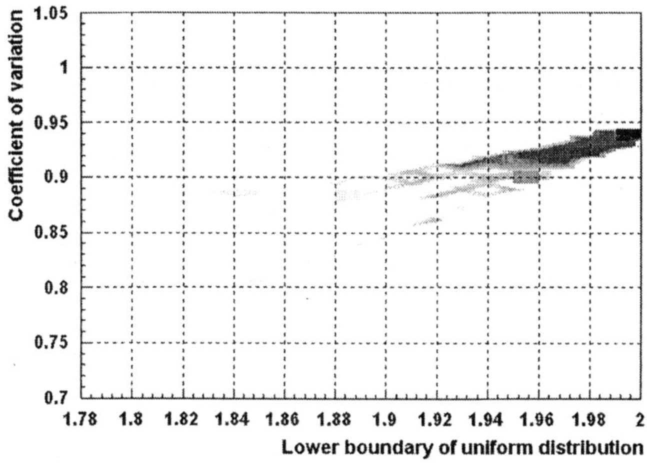

Fig. 5.16 Frequency diagram of the coefficient of variation at the end of the run. Initial variation in the exogenous rate of knowledge generation simulated on the column

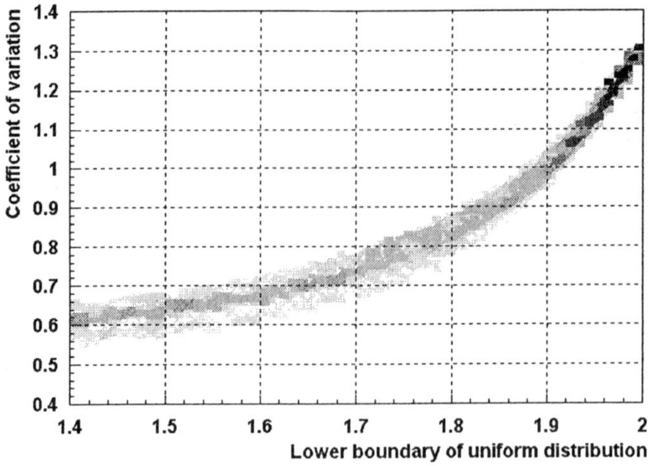

Fig. 5.17 Frequency diagram of the coefficient of variation at the end of the run. Initial variation in the exogenous rate of knowledge generation simulated on the globe

For the column, the sensitivity to the lower boundary of the interval is given in Fig. 5.16. Contrary to the lattice, this sphere seems to generate mostly coefficients of variation that are smaller than the one in which three leaders appear (the dark spot at the right of the figure)[14]. Again a fan-shape appears. The explanation for these features is the same as for the case in which the learning capability was varied across ranges. The fan-shape is generated by the growing inequality in the coefficient of variation across runs, as the difference in learning capabilities across regions counteracts the difference in geographic location (from right to left). The line is upward sloping because within each run the disparity across gaps increases, as the range is decreased, due to the characteristics of the column.

The figure for the globe (Fig. 5.17) shows a curve that is upward sloping and shows less of a fan-shape. There is only little dispersion at the left-hand side of the curve. Contrary to the former experiment with the learning capability, this curve does not yet show a horizontal part when the gap between lower and upper boundary is large. However, we see that the lowest observation has a coefficient of variation of about 0.6 at a lower boundary of 1.4. If the lower boundary would further be decreased, the figure would start to show a horizontal trend.

The general conclusion from this section follows the same directions as the former section in which the learning capability of each region was varied. The patterns in this section show similar trends, although there are a few differences. The graph for the column under a variation of the exogenous rate of growth shows the same fan shape, but the upward trend is much less steep than under a variation in the learning capability in the former section. This indicates that a variation in the

[14] Fig. 5.7 shows the distribution of the gaps across regions as occurs at the utter right of the figure.

learning capability has a stronger counteracting effect (to geographical distance) than a variation in the exogenous rate of growth of the knowledge stock. If the range in which a parameter can vary is small, there exists more dispersion of the gaps across runs in the case of the learning capability compared to the exogenous rate of growth of the knowledge stock.

Virtually the same holds for the globe. The effect of the learning capability on overall disparity is larger than in the experiment in which the exogenous rate of growth of the knowledge stock was varied in the same range.

5.4 Conclusions

This Chapter presented a model for knowledge spillovers based on geographical distance as well as technological distance. The regions in our model receive knowledge spillovers from other regions, and this enables them to grow rapidly. Our model is similar to some of the models found in the 'technology gap' tradition of analyzing convergence of GDP per capita. Compared to these models, we add the spatial distance effect on spillovers. The further away other regions are, the less strong spillovers from these regions are. Our research question is, what is the effect of including spatial distance on convergence? From the simulations, we can draw several conclusions. When regions are located on a two-dimensional lattice, the more differences there are across regions with regard to learning capability or the exogenous rate of growth of the knowledge stock, the larger is the variation in resulting gaps. The runs can show both higher and lower disparity than the 'perfect' distribution of gaps as regions initially have more differences.

For a column-shaped geographical structure, the case in which all regions are equal to each other generates the highest disparity. As soon as regions show differences with respect to the learning capability or the exogenous rate of growth, the disparity becomes less. Contrary to the situation for the lattice, there are no runs that have a larger coefficient of variation than the case with identical regions.

When regions are grouped on a globe, which has many central regions as compared to the column, a slightly different effect of the change in parameters is found. As with the column, the disparity decreases as regions show larger differences with respect to the value of the parameter. However, the slope is non-linear, and for a certain minimum level of disparity, the disparity will not decrease any further. Larger initial differences in parameters will induce results, which have a similar coefficient of variation.

This result stems from the process in which varying a parameter such as the learning capability or the exogenous rate of knowledge generation, provides a counteracting force relative to the effect of spatial distance. Increasing differences in learning capability or the (exogenous) rate of growth of knowledge (R&D) are in themselves polarizing factors, leading to the tendency that regions with high values for these parameters grow rapidly. The spatial structure on which regions are grouped is also a polarizing factor, leading, ceteris paribus other parameter differences, to high growth rates in centrally located regions. Combining differences in learning capability or exogenous knowledge growth with spatial distance leads to the result that the two forces counteract each other.

Appendix A

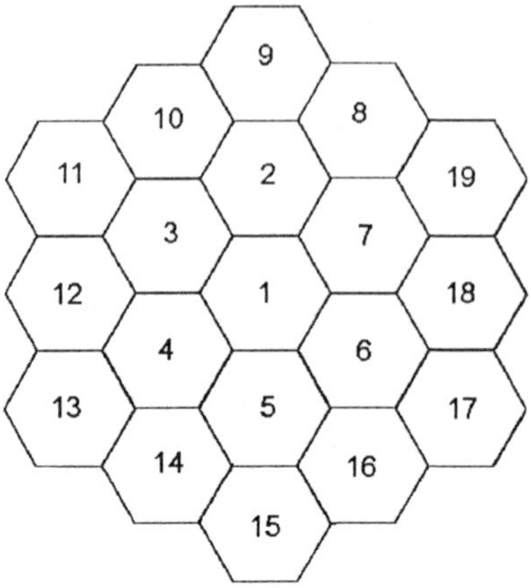

Fig. 5.18 Lattice of honeycombs

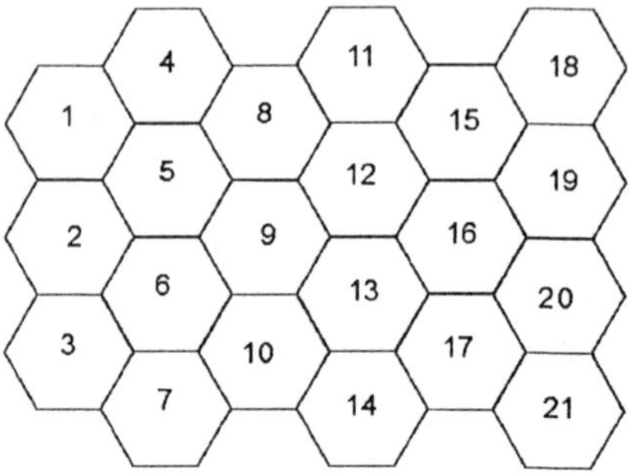

Fig. 5.19 A column

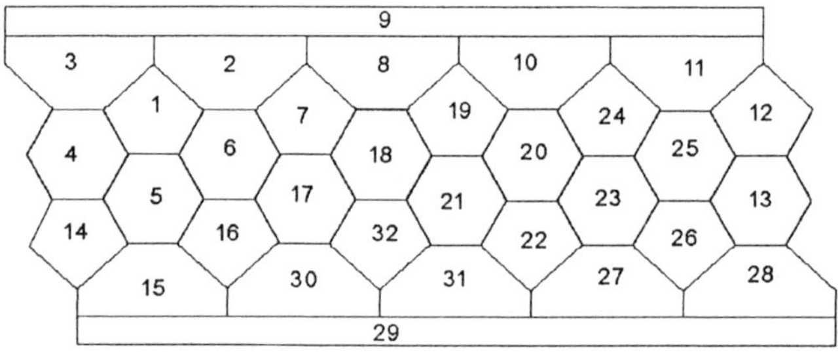

Fig. 5.20 A globe

Fig. 5.18 displays the topography of the regions on a lattice of honeycombs. The number within each hexagon was used to establish the geographical distances between all hexagons. Fig. 5.19 displays the column. Region 1 borders to regions 4, 5 and 2 as well as 18 and 19, region 2 to regions 1, 5, 6 and 3 as well as 19 and 20, and region 3 has next to region 2, 6 and 7 also regions 20 and 21 as its direct neighbors. Thus, if one would walk from region 20 to the right, one would reappear at the left of the figure in region 2 or 3.

The figure below (Fig. 5.20) represents a globe with 12 pentagons and 20 hexagons. For the graphical representation, we used the same principle that was applied in making a map of the world. Hence, the regions close to the poles look larger as they actually are, while the regions around the equator show their true proportions. At the bottom and at the top are regions 29 and 9. These are pentagons, for example region 9 borders to five regions, namely 3, 2, 8, 10 and 11. Region 29 and 9 are in reality as large as region 1. The graphic representation of a globe has also as a consequence that for example region 3 seems to differ in size from region 6. Again, this is not the case in reality, region 3 is an ordinary hexagon. The same goes for all the other regions bordering 9 or 29. It should also be noted that region 11 borders not only to regions 9, 10, 24, 25 and 12, but also to region 3. In this way, region 12 also borders to regions 3 and 4, region 13 has regions 4 and 14 as direct neighbors as well, whereas region 28 also shares a border with regions 14 and 15.

Appendix B

Default levels of the variables and values of the parameters:

10000	(Number of periods, t)
10	(Knowledge stock, K)
1	(Exogenous rate of growth of the knowledge stock, ρ)
1	(Learning Capability δ)
1	(Catch-up parameter, μ)
0.005	(β)
0.005	(α)
1	(Verdoorn parameter, λ)

γ (geographical distance) is constructed with the help of three different types of distance tables, one for each sphere.

References

Abramovitz MA (1994) The Origins of the Postwar Catch-Up and Convergence Boom.In: Fagerberg J, Verspagen B and N Von Tunzelmann (eds) *The Dynamics of Trade*, Technology and Growth. Edward Elgar, Aldershot, pp 21–52

Acs ZJ, Audretsch DB and Feldman MP (1992) Real effects of academic research: comment. *American Economic Review* 82(1): 363–367

Amable B (1993) Catch-up and convergence: a model of cumulative growth. *International Review of Applied Economics* 7(1): 1–25

Audretsch DB and Feldman MP(1996) Knowledge spillovers and the geography of innovation and production. *American Economic Review* 86(3): 630–640

Baptista R (1998) Clusters, innovation and growth: a survey of the literature. In: Swann PGM (ed) *The dynamics of industrial clustering – international comparisons in computing and biotechnology*. Oxford University Press, Oxford, pp 13–51

Barro RJ (1984) *Macroeconomics*. Wiley, New York

Barro RJ and Sala-i-Martin X (1991) Convergence across states and regions. *Brooking Papers Economic Activity* 1: 107–182
Barro RJ and Sala-i-Martin X (1992a) Convergence. *Journal of Political Economy* 100: 223–251
Barro RJ and Sala-i-Martin X (1992b) Regional growth and migration: A Japan-United States comparison. *Journal of the Japanese and International Economies* 6: 312–346
Barro RJ and Sala-i-Martin X (1995) *Economic growth*. McGraw-Hill, New York
Baumol W (1986) Productivity growth, convergence and welfare: What the long-run data show. *American Economic Review* 76: 1072–1085
Breschi S (1995) *Spatial patterns of innovation: evidence from patent data*. Paper presented at the workshop on New Research Findings: The Economics of Scientific and Technological Research in Europe, Urbino, Italy 24–25 February 1995
Caniëls MCJ and Verspagen B (2001) Barriers to knowledge spillovers and regional convergence in an evolutionary model. *Journal of Evolutionary Economics* 11: 307–330
De Long JB (1988) Productivity growth, convergence and welfare: comment. *American Economic Review* 78: 1138–1159
Esteban J and Ray D (1994) On the measurement of polarization. *Econometrica* 62: 819–851
Fagerberg J (1988) Why growth rates differ. In: Dosi G et al. (eds) *Technical change and economic theory*. Pinter Publishers, London, pp 432–457
Fagerberg J (1994) Technology and international differences in growth rates. *Journal of Economic Literature* 32: 1147–1175
Haggett P, Cliff AD and Frey A (1977) *Locational models*. Edward Arnold (publishers) Ltd, London
Jaffe A (1989) Real effects of academic research. *American Economic Review* 79: 957–970
Jaffe A, Trajtenberg M and Henderson R (1993) Geographic localization of knowledge spillovers as evidenced by patent citations. *Quarterly Journal of Economics* 108: 577–598
Freeman C (1991) Networks of innovators: a synthesis of research issues. *Research Policy* 20: 499–514
Malerba F and Orsenigo L (1995) Schumpeterian patterns of innovation. *Cambridge Journal of Economics* 19(1): 47–66
Pavitt K (1987) *On the nature of technology*. University of Sussex – Science Policy Research Unit, Brighton
Verspagen B (1991) A new empirical approach to catching up and falling behind. *Structural Change and Economic Dynamics* 2: 359–380
Von Hippel E (1988) *The sources of innovation*. Oxford University Press, New York
Von Hippel E (1994) "Sticky information" and the locus of problem solving: implications for innovation. *Management Science* 40(3): 429–439

6 The Spatial Distribution of Labour Productivity in the European Regions: A Study in Spatial Econometrics

Esther Vayá and Jordi Suriñach

University of Barcelona, Spain

6.1 Introduction

In recent years, as a consequence, among other factors, of the emergence of the New Economic Geography (Krugman 1991a, b), interest in the location of economic activity and, more specifically, its concentration has grown. The models developed in this branch of economics seek to account for the eventual location of industry in terms of competing centripetal and centrifugal forces (see Ottaviano and Puga 1997; Krugman 1998; and Martin 1999 for a survey of models developed in the New Economic Geography). Interestingly, many of these models support the existence of an inverted-U shape relationship between the degree of integration and the level of concentration of activity.

The identification of this relationship has led several empirical studies to analyse the effects of the process of European integration on both the location of the industrial sector (Greenaway and Hine 1991; Fluvia and Gual 1993; Molle 1996; Brülhart and Torstensson 1996; Amiti 1997) and the evolution of regional disparities and convergence (Suarez-Villa and Cuadrado-Roura 1993; Esteban 1994; Neven and Gouyette 1994; Armstrong 1995; Molle and Boeckhout 1995; Fingleton, 1999; López-Bazo et al. 1999). This first group of studies has typically computed sectoral concentration and regional specialisation indices, while the second group presents measures of inequality such as the Gini or the sigma-convergence index. Several of these convergence studies have complemented these indices with the application of a density function (see Quah, 1993, 1996a, 1996b), which is a useful tool for obtaining a rapid impression of the dynamics of the regional distribution of a specific variable.

However, all of these indices and tools have one major shortcoming: they fail to take into account the space in which each region is located, considering it as an isolated unit and ignoring any possible links with its *neighbouring*[1] regions. The absence of a spatial dimension has major implications for analyses of location or inequality. First, both types of index are unable to reveal whether the variable being analysed is randomly distributed in space or whether, on the contrary, there is a significant trend towards a clustering of similar (positive spatial dependence) or dissimilar (negative spatial dependence) values of the variable under study. And second, pockets of regions with significantly high or low values of the variable, as well as potential spatial non-stationarities, cannot be detected. This can easily give rise to situations in which different spatial distributions of one variable present similar index values, or even one in which the spatial relocation of the most dynamic region goes undetected because the global dispersion of the variable remains constant.

Given these inherent weaknesses in the traditional indices applied in location and inequality studies, such analyses have turned to the techniques of Exploratory Spatial Data Analysis (ESDA), including global and local spatial dependence tests and Moran Scatterplots (Anselin 1998; Haslett et al. 1990; Haslett et al. 1991). The mapping of variables, after all, depicts quite clearly the existence of a correlation between neighbouring regions or the location of nearby regions with high or low levels of this variable in question. Yet, while this is generally true, the information derived from a map is somewhat subjective and highly dependent on, for example, the range of intervals selected for mapping the variable. On the contrary, spatial dependence tests provide a statistical framework for considering such questions.

The aim of this Chapter, therefore, is to show how traditional location or inequality analyses can be usefully complemented by techniques of Exploratory Spatial Data Analysis, both from a theoretical and an empirical point of view. In seeking to illustrate this, we organise the rest of the Chapter as follows. Section two briefly describes some of the better known ESDA techniques. In Section three, the sensitivity of these techniques to five patterns of spatial distribution is evaluated. To illustrate this, and after reviewing the consequences of an integration process on the location of activity predicted by models from New Economic Geography, a simulation exercise is performed whereby different samples of a variable (here, "level of activity") are generated for various hypotheses of spatial distribution. Each simulated spatial distribution can then be linked to a range of scenarios predicted by models developed in the New Economic Geography during a process of integration. The next step involves the computation of spatial dependence tests for each case. The values obtained for the various distribution patterns allows the most likely evolution of the spatial tests during the integration process to be established. In section four, this framework is applied to the case of the distribution of labour productivity in the European Union (EU) regions from 1975-1996. Finally, section five concludes.

---

[1] Neighbouring here is used as a generic concept, and is not limited to its purely geographical sense of contiguity.

6.2 ESDA Techniques: the Spatial Dependence Tests

ESDA incorporates a set of statistical and visualisation techniques that enable the researcher to determine the presence of spatial dependence in the spatial distribution of a particular variable, the existence of spatial clusters and other forms of spatial instability (see among others Anselin 1998; Haslett et al. 1990, 1991). There are many such tools, but in this section we shall focus on those that are perhaps best known.

6.2.1 Global Spatial Dependence Tests

Spatial dependence, or spatial autocorrelation, is said to exist when the values observed at one location (for instance, in one region) depend on the values observed in its *neighbouring* regions. Hence the location of the values of one variable cannot be changed without affecting the information in the sample (Anselin 1988; Anselin and Florax 1995). Various statistics have been proposed for verifying the existence of spatial autocorrelation in a specific variable and one of the most widely used is the Moran I test (Moran 1948). It is computed as:

$$I = \frac{N}{S}\frac{\sum_i \sum_h w_{ih} z_i z_h}{\sum_i z_i^2} \qquad (6.1)$$

where N is the number of observations, w_{ih} is the element of the spatial weights matrix W that expresses the potential interaction between two regions *i* and *h*, S is the sum of all the weights (all the elements in the weights matrix) and z_i represents the normalised value of a variable x being analysed in region *i*. Although there is no consensus on the specification of W, the contiguity criterion is usually applied. Then, w_{ih} will be 1 if regions *i* and *h* are neighbours and 0 if otherwise. In our analysis, we used a rows standardised (so that each row sums 1) contiguity matrix.

Once standardised, a significant and positive value for this statistic indicates that similar values of the variable tend to cluster in space (positive spatial dependence). So, for example, if we considered level of activity, positive spatial dependence would indicate that neighbouring regions possess similar levels of activity, revealing that the location of activity follows a marked spatial pattern. On the other hand, when the test is significant and negative, the tendency is for dissimilar values to cluster in neighbouring locations (negative spatial dependence). In this latter case, this might illustrate a situation where the strength of centripetal forces within the region is such that it prevents the diffusion of manufacturing activities to its neighbours.

Non-significance of the Moran I test implies the 'acceptance' of the null hypothesis: the non-existence of spatial autocorrelation, that is, the prevalence of a random distribution of the variable throughout space.

Finally, although here we have defined a first order contiguity matrix, it would be interesting to determine whether these patterns of spatial autocorrelation would be the same among regions that are not strictly adjacent one to another. Spatial correlograms (Cliff and Ord 1973, 1981) easily summarise this information by means of a graph in which the values of the standardised Moran I test for different orders of contiguity[2] are displayed.

6.2.2 Local Spatial Dependence Tests

Global spatial dependence tests consider spatial autocorrelation in the whole sample. As a result, they cannot supply information about situations that are characterised by the existence of just a few clusters located in specific parts of the study area. This means such tests are unable to identify specific significant hotspots or macro-regions where activity might be greater or lower than the average, i.e. spatial concentration of regions with values significantly above or below those expected in an homogeneous distribution. This weakness in the global tests, however, can be overcome by applying local spatial dependence tests. For example, the local Moran I_i test (Anselin 1995) indicates the extent to which each region is surrounded by regions with similar or dissimilar values of the analysed quantity. The local Moran statistic, I_i can be defined as:

$$I_i = \frac{z_i}{\sum_i z_i^2 / N} \sum_{h \in H_i} w_{ih} z_h \tag{6.2}$$

where z_i is the observation for region *i* in deviations from the mean and H_i is the set of neighbouring regions to region *i.* In this case, a significant positive (negative) result indicates the existence of a cluster of similar (dissimilar) values of the variable composed of region i and its neighbours. However, this statistic does not enable us to know whether we are dealing with a cluster of high or low variable values. One way to complement the results of the Local Moran test is by computing the Moran Scatterplot map for the variable (Anselin 1996). After standardising the values of variable x for each region and its neighbours, the Moran Scatterplot map informs us whether a region and its neighbours show similar high or low values or dissimilar values of the variable. Then, the combined information derived from the local Moran test and the Moran Scatterplot map allows us to know whether one region and its neighbours constitute a cluster of high or low values of the variable and, also, whether one region possesses significantly high levels of activity while for its neighbours the reverse is the case (or *vice versa*).

---

[2] One region *h* would be second order contiguous to another region *i* if it is first order contiguous to another unit that itself is first order contiguous to the region *h*.

6.3 Behaviour of the Spatial Dependence Tests Under Different Distributions of Economic Activity

This section describes the spatial distribution patterns for a number of potential situations regarding the location of economic activity. First, we briefly outline the endogenous explanation of the agglomeration process provided by models proposed in New Economic Geography. Second, the spatial distributions of economic activity generated in the simulation exercise are presented, and an attempt is made to link each of them with the potential stages of an integration process. Finally, the results of the simulation exercise, in terms of the behaviour of the spatial statistics, are discussed.[3]

6.3.1 Location of Activity and New Economic Geography

The earliest models developed in New Economic Geography offered an endogenous explanation for the agglomeration of activity in a territory. Thus, following the seminal papers of Krugman (1991a,b), models of this kind have explained the regional distribution of industry as being the result of the tug-of-war between forces that tend to promote geographical concentration (centripetal forces) and those that tend to oppose it (centrifugal forces).

Among the forces that contribute to the agglomeration, we include the three traditional Marshallian sources of external economies: backward and forward linkages between firms[4] (consequence of input-output linkages), the existence of a thick local labour market (under the assumption of inter-regional mobility[5]) and the presence of external economies via information spillovers[6]. By contrast, the presence of immobile factors (certain land and natural resources and even, in an international context, people as well), the high land rents in places characterised by a notable concentration of economic activity, and external diseconomies (such as congestion) are the main centrifugal forces working against industrial concentration. Whatever the case, it should be pointed out that economic agglomerations can

---

[3] The authors are grateful for the comments and suggestions made by Enrique López-Bazo during the elaboration of the simulation exercise.

[4] See Krugman and Venables (1995) and Venables (1996).

[5] Puga (1996) introduces the hypothesis of intersectoral mobility as well as inter-regional mobility to Krugman's model (1991a,b). The benefit obtained by the R&D labs from input differentiation plays a role in Martin and Ottaviano (1999) which is equivalent to that of inter-regional labour mobility in Krugman (1991b) and that of input-output linkages in Venables (1996) and Krugman and Venables (1995).

[6] In the words of Ottaviano and Puga (1997), technological externalities arising from personal interactions matter more for small scale agglomerations, while to explain large scale agglomerations, we must either look to other technological externalities whose effectiveness does not decrease sharply with distance, or turn to pecuniary externalities.

occur at various levels of aggregation: from small scale agglomerations of clearly defined sectors (as, for instance, highly specialised industrial districts) to large scale agglomerations that cut across state and country boundaries.[7]

Turning to this last case, economic integration might have an impact on the eventual spatial location of economic activities given that it affects the equilibrium between the forces of dispersion and agglomeration. Thus, in the first stages of integration (characterised by high trade costs), the need to supply markets locally encourages firms to locate in different regions, leading to a stable, symmetric equilibrium. However, as the integration processes advance (leading to intermediate values of trade costs) and under the hypothesis of increasing returns, the incentives for self-sufficiency weaken, leading to the concentration of the industry in the region with the largest market-size. Then, the pecuniary externalities take over and firms and workers cluster together, and the cumulative causation process begins,[8] leading regions to differentiate endogenously into an industrialised core and a deindustrialised periphery.[9]

However, the relationship between the integration process and agglomeration is far from monotonous or linear, but rather presents an inverted U-shape. Thus, the prices of local factors (land rents and wages) and goods tend to rise wherever agglomeration takes place. This is particularly relevant if they involve certain immobile factors that are important for production (for instance, labour) or non-tradable goods that are important for consumption (for instance, housing). In this case, as further integration reduces the importance of pecuniary externalities, differences in the prices of immobile goods and factors take over. As a consequence, industry spreads to less developed regions, reducing the previous spatial divergence in the distribution of activity throughout the territory. In this sense, Puga and Venables (1996) presented a gradual process of industrialisation where, having reached a critical mass, firms move from the core to other regions at some distance from this core to avoid the high wages present in the centre, but near enough to it in order to benefit from its advantages in terms of agglomeration economies (positive pecuniary externalities created by inter-firm linkages). The process may then repeat itself, so industrialisation takes the form of a sequence of waves, with industry spreading from country to country.

6.3.2 Five Theoretical Spatial Distributions

For the simulation exercise, a number of different spatial distributions of a generic variable, namely "level of activity", were generated. While this variable can be as-

---

[7] See Krugman (1991a,b; 1998).

[8] As to where agglomeration takes place, in the presence of agglomeration economies, historical accidents can trigger self-reinforcing agglomeration processes.

[9] Puga and Venables (1996) point out that the timing of the emergence a core-periphery scheme depends on the intensity of the scale economies and input-output links.

sociated both with employment in the industrial sector (based on a traditional analysis of location) and output (per inhabitant or per worker, based on a study of the level of regional disparities), the results obtained here are more closely linked to the latter.

In order that each simulated sample might be identified with a real territory, the regions of the European Union were used as a reference. Specifically, a set of 108 EU regions incorporating the entire EU12 territory was selected. The regions comprised the NUTS II regional classification level for Portugal (5 regions), Spain (17), France (21), Italy (29) and Greece (13), and the NUTS I level for the UK (11), Belgium (3), Holland (4) and Germany (11). Denmark, Ireland and Luxembourg were each considered as comprising a single region.

Assuming the existence of an inverted U-shape relationship between integration and agglomeration, as proposed by New Economic Geography models and explained in the previous section, in the simulation exercise we considered five situations, each corresponding to a different hypothesis concerning the spatial distribution of activity during an integration process.

In the first of these distribution patterns, called the *homogenous distribution,* activity is assumed to be homogeneously distributed throughout the territory as a consequence of, among other possibilities, the existence of similar regional endowments in a range of factors, the presence of very high transport costs (as is the case in, for instance, the first stages of an integration process) or a situation characterised by autarchy. In addition, this *homogenous distribution* could be observed during the last stage of the integration process, when transports costs were very low and centrifugal forces were dominant, leading to a more balanced distribution of activity throughout the territory.[10]

The next three scenarios seek to reflect the spatial distributions of activity that might appear during the intermediate stages of an integration process, when centripetal forces are stronger than centrifugal ones, resulting in the agglomeration of activity in certain regions.[11] The first scenario considers the *traditional core-periphery distribution* characterised by a marked geographical polarisation. Here, the core was identified as lying in the northern regions (those of France, the UK, Germany, Holland, Luxembourg, Belgium and North Italy) where most of the activity is concentrated;[12] the opposite situation prevails in the South.[13] The second scenario is that of the *monocentric distribution.* Here, instead of a clear spatial division between northern and southern regions as in the previous case, the core is

---

[10] Besides, the homogenous distribution could be considered as a steady state distribution resulting from a neo-classical model with absolute convergence. The authors would like to thank an anonymous referee for this comment.

[11] We will assume increasing returns, and backward and forward linkages to be the main source of the external economies.

[12] We assume that these regions are characterised by larger markets and better endowments of infrastructure and skilled labour.

[13] However, the decision to locate the core in the North was taken for illustrative purposes only and any qualitative results do not depend on the specific location of the core in the simulation exercise.

located in a small group of neighbouring regions (the *monocentre,* located in the south of Germany, the north of Italy and the east of France), with activity being randomly distributed throughout the rest of the territory. The third scenario is that of *polycentric distribution*. Here, we also considered high concentrations of activity but in regions (*polycentres*) scattered through the territory, the location of which was randomly determined.[14] Furthermore, an additional hypothesis was incorporated whereby as a consequence of the strong attraction of these polycentres, the activity levels in their neighbours were considered to be significantly lower than those in the other regions.

Finally, the last scenario, called a *hierarchical distribution*, seeks to identify the possible diffusion of activity during advanced stages of the integration process. Thus, and in line with the approach taken in Puga (1996), and Puga and Venables (1996), with the monocentric distribution, we consider the existence of diffusion of activity into the regions lying closest to the monocentre. Thus, we assumed that, as a consequence of the presence of diseconomies of scale in the core regions, activity would initially spread towards the contiguous neighbouring regions of the monocentre and subsequently to the neighbours in a second order of contiguity with the monocentre.[15]

A hundred samples were generated for each of the scenarios in the series, and the same average value of the simulated variable was assigned to each sample. In addition, the simulation exercise analysed the sensitivity of the spatial dependence tests to the degree of deviation of the series (see Table 6.1). In order to do this, two types of deviation were identified: *between group* and *within group*, as well as two degrees of deviation: *small* and *medium*. Thus if we take the *core-periphery distribution* as an example, the *between group* deviation is the deviation in the level of activity between the northern regions (the *core*) and the southern regions (the *periphery*), whereas the *within group* deviation is the deviation in the level of activity inside the group of regions forming the core or the periphery. Similarly, this concept of *within* and *between* deviation was applied to the *monocentric distribution* and *hierarchical distribution*. In case of *homogeneous distribution* and *polycentric distribution*, differences were defined only for *within* deviation.

6.3.3 Results of the Simulation Exercise

The simulation exercise sought to respond to two main questions. The first was concerned with the usefulness of spatial analysis as a complement to traditional non-spatial analyses of location or inequality based on the computation of concen-

---

[14] Historical accidents could have triggered off a self-reinforcing agglomeration process that might explain the location of these polycentres.

[15] Finally, during the last stage of the integration process, a *homogeneous distribution* might reappear.

tration/dispersion indices, while the second was concerned with the sensitivity of the spatial dependence test to the characteristics of the distribution of the analysed series (specifically, their sensitivity to *within* and *between* deviations). In response to the first of these questions, in section 6.3.3.1, we present the results of both the traditional non-spatial analysis and the spatial analysis of the generated series. Here, we include one example for each of the five simulated distribution patterns, but do not show the detailed characteristics of the series in terms of their deviation. In response to the second question, in section 6.3.3.2, we focus on the sensitivity of the spatial dependence tests (specifically, the global dependence test) to changes in the *within* and *between group* deviation of the series.

6.3.3.1 Non-Spatial and Spatial Analysis: the Optimum Combination

In seeking to demonstrate how a spatial analysis might complement a traditional non-spatial analysis of location (or inequality), in this section we present one example for each of the five spatial distribution patterns of economic activity generated. Figures 6.1.a, 6.2.a, 6.3.a, 6.4.a, and 6.5.a depict each of the selected series in map form.

In order to determine the degree of concentration of the variables generated across space, standard deviation (σ) was computed for each case. In addition, the density function corresponding to each hypothesis was also calculated. The results are shown in Figures 6.1.b, 6.2.b, 6.3.b, 6.4.b, and 6.5.b.

Spatial analysis for homogeneous distribution

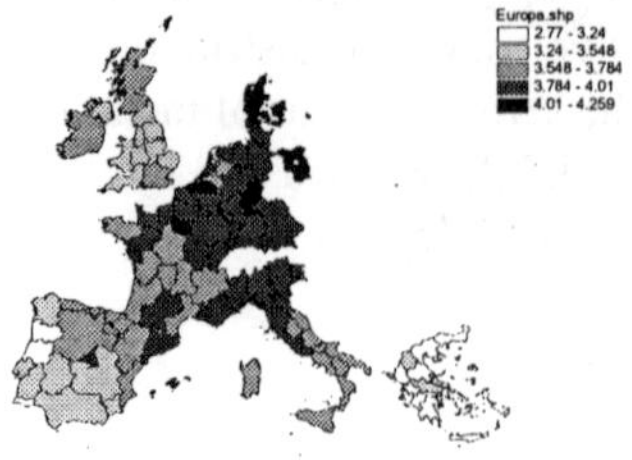

Fig. 6.1.a Spatial distribution

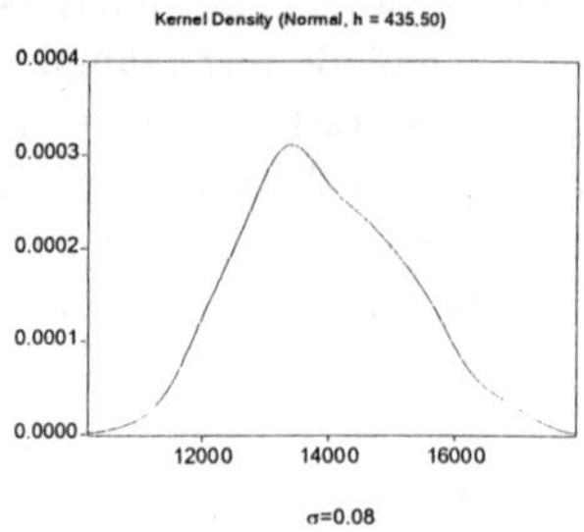

Fig. 6.1.b Density function

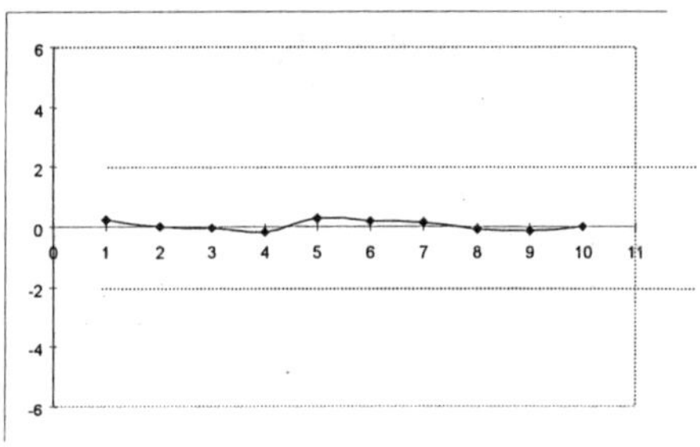

Fig. 6.1.c Spatial correlogram

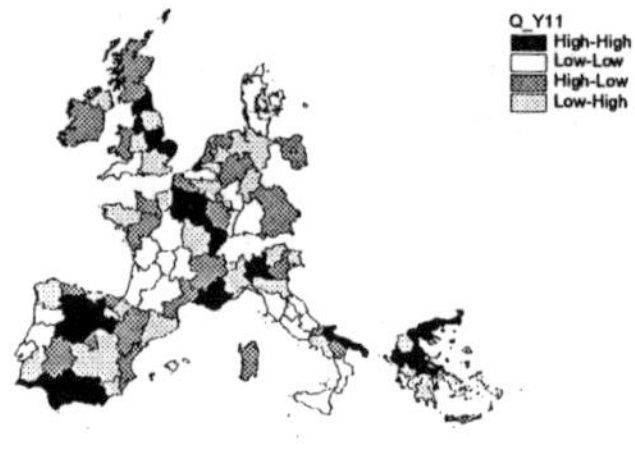

Fig. 6.1.d Moran scatterplot map

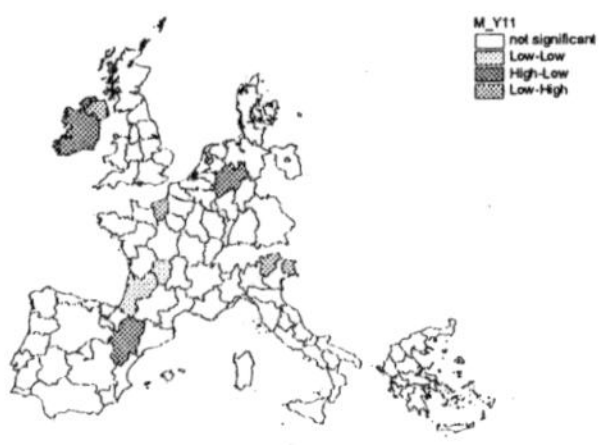

Fig. 6.1.e Local analysis

Spatial analysis for core-periphery distribution

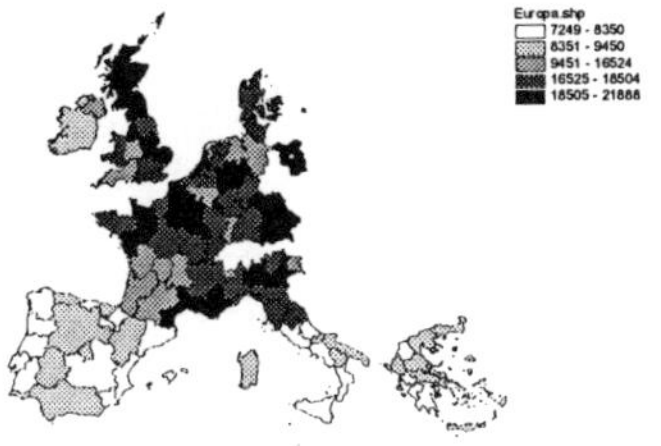

Fig. 6.2.a Spatial distribution

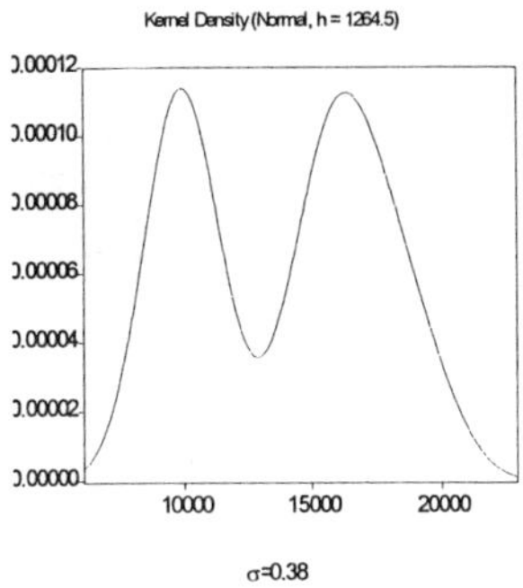

Fig. 6.2.b Density function

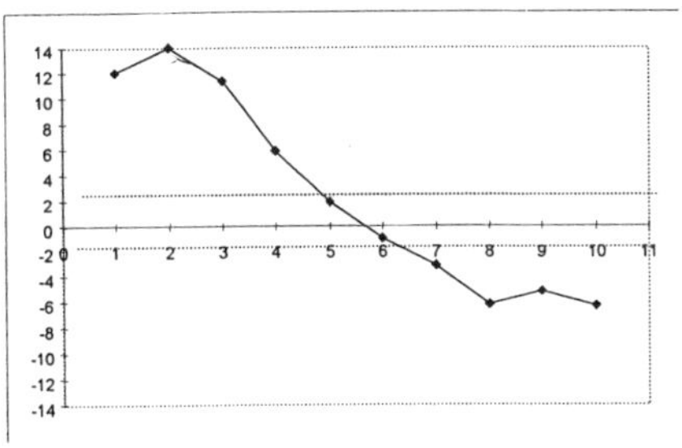

Fig. 6.2.c Spatial correlogram

Fig. 6.2.d Moran scatterplot map

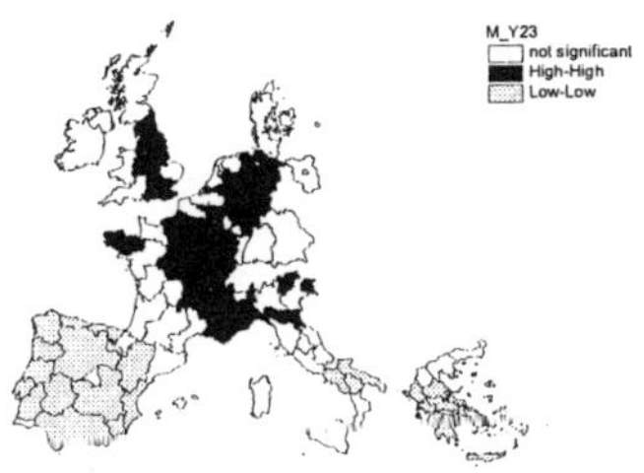

Fig. 6.2.e Local analysis

Spatial analysis for monocentric distribution

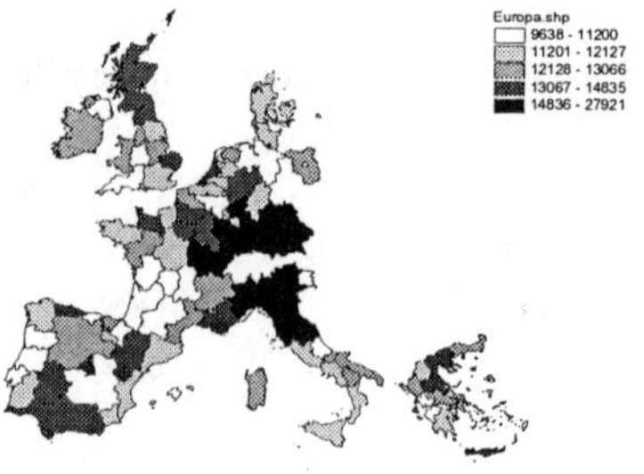

Fig. 6.3.a Spatial distribution

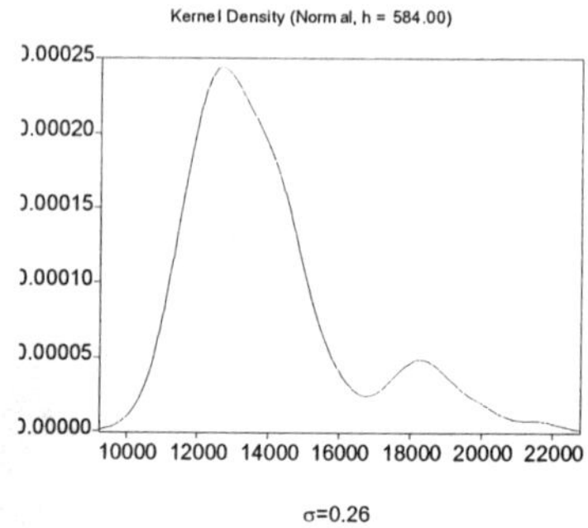

Fig. 6.3.b Density function

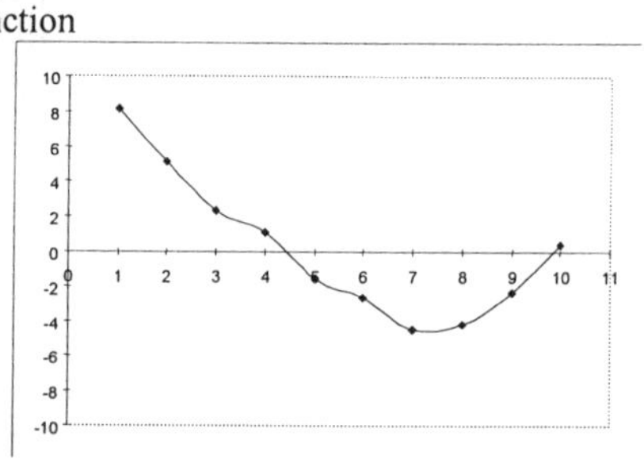

Fig. 6.3.c Spatial correlogram

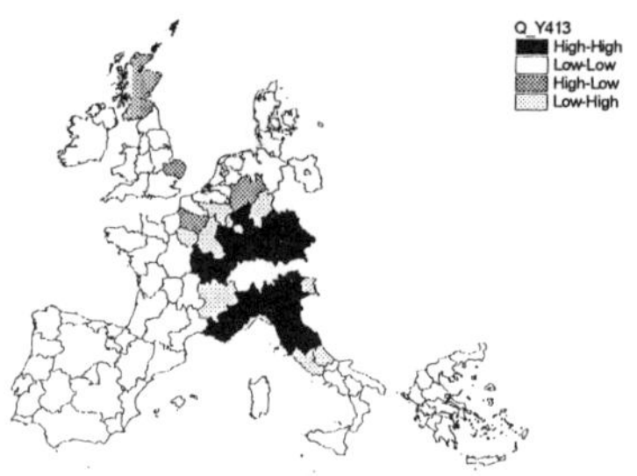

Fig. 6.3.d Moran Scatterplot map

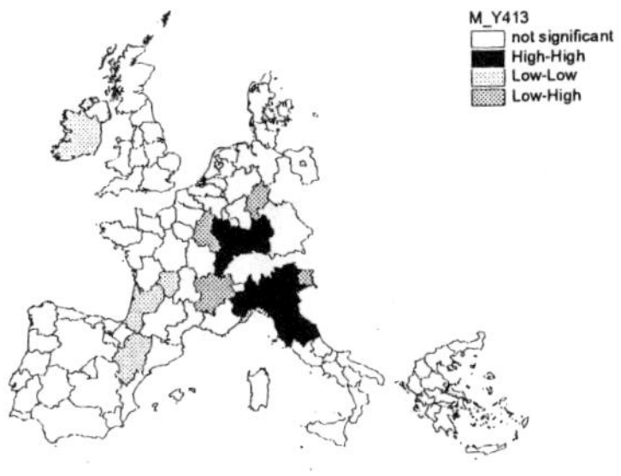

Fig. 6.3.e Local Moran

Spatial analysis for polycentric distribution

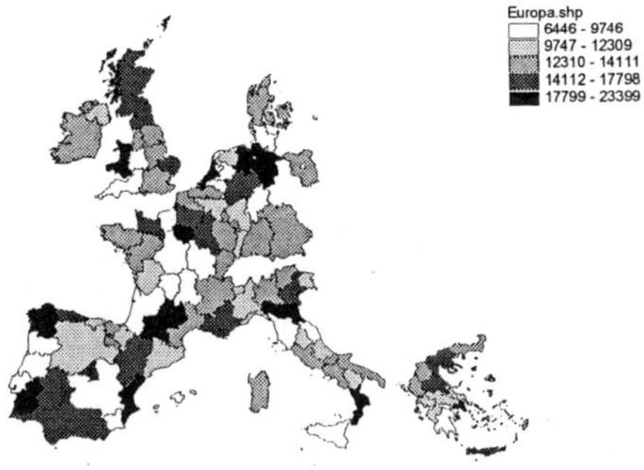

Fig. 6.4.a Spatial distribution

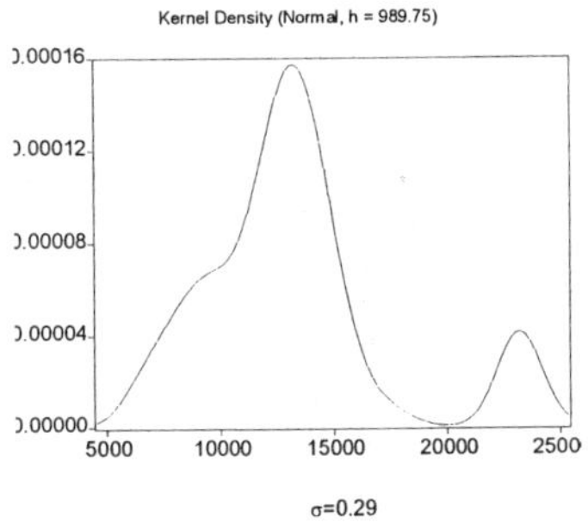

Fig. 6.4.b Density function

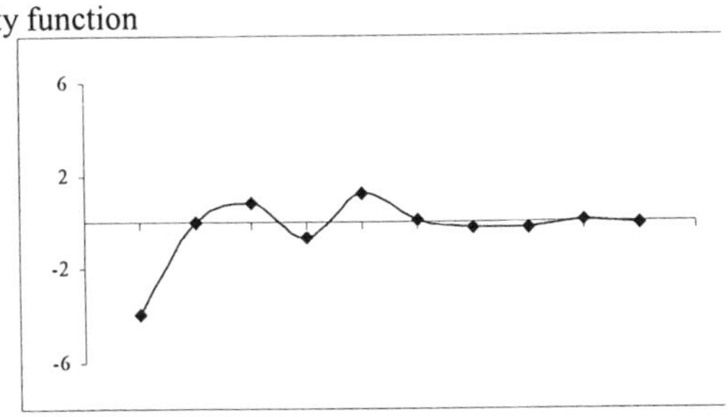

Fig. 6.4.c Spatial correlogram

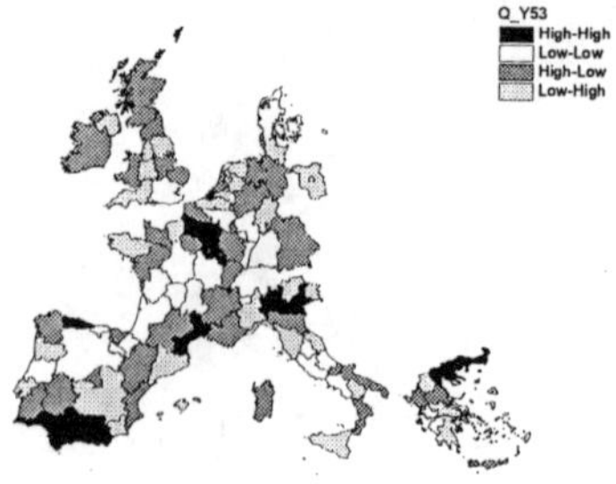

Fig. 6.4.d Moran Scatterplot map Local analysis

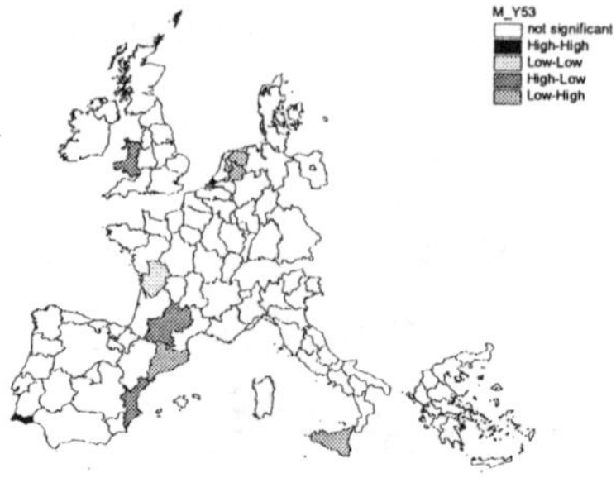

Fig. 6.4.e Local analysis

Spatial analysis for hierarchical distribution

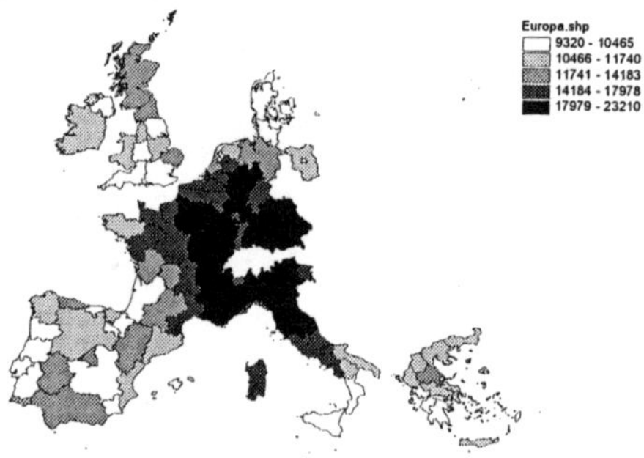

Fig. 6.5.a Spatial distribution

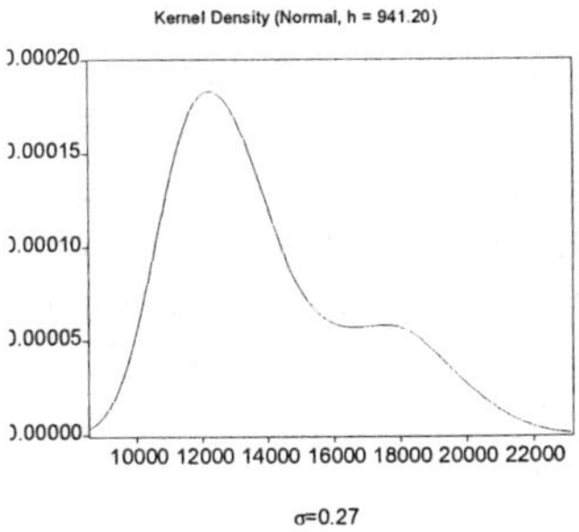

Fig. 6.5.b Density function

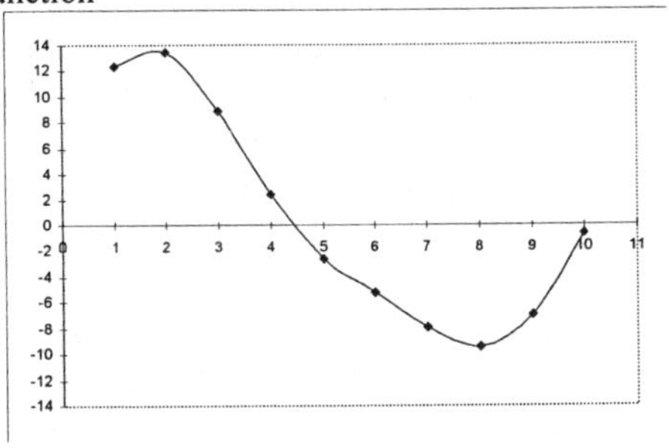

Fig. 6.5.c Spatial correlogram

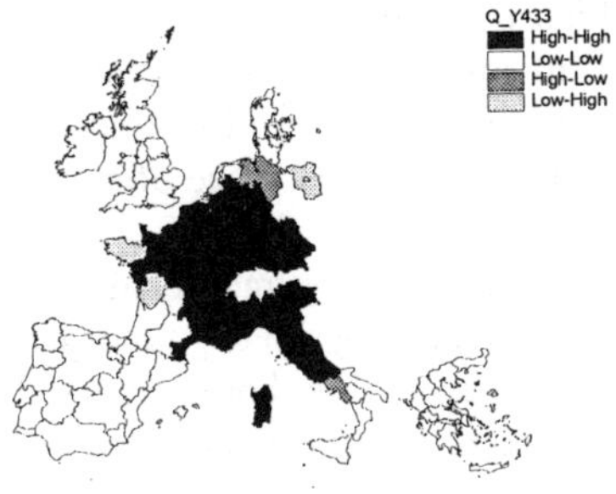

Fig. 6.5.d Moran scatterplot map

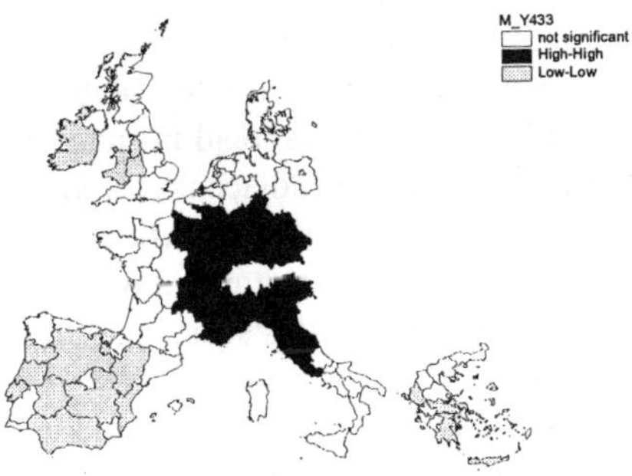

Fig. 6.5.e Local analysis

As it can be seen, the *homogenous distribution* presented the lowest standard deviation, indicating a fairly even distribution of the activity compared to that of the other spatial distributions. In contrast, the distribution for a *core-periphery* pattern presented the highest values on this statistic, reflecting a highly significant degree of dispersion. Finally, the results obtained for a *monocentric, hierarchical* and *polycentric distribution* were very similar (though the degree of dispersion was slightly higher in the case of the *polycentric distribution*).

As for the density functions, Figure 6.1.b (corresponding to *homogenous distribution*) shows the existence of only one mode with a high concentration of probability, indicating a very low degree of regional dispersion. At the opposite extreme is the external shape presented by the distribution for the *core-periphery* pattern (Figure 6.2.b). Here, two distinct modes can be distinguished, pointing to the existence of a significant degree of heterogeneity and, hence, a high degree of inequality in the distribution (a finding in line with the value recorded on the σ statistic). Furthermore, it was found that when there was no hierarchical distribution around the monocentre (*monocentric distribution*), there was only one significant mode with a large concentration of probability in most of the regions in the sample (Figure 6.3.b). A notable tail to the right of this mode was also observed, reflecting the existence of a small group of regions in which high levels of activity were concentrated. Once the diffusion process had been triggered (*hierarchical distribution*), the concentration of probability in the mode seemed to decrease and the tail appeared to widen (Figure 6.5.b). Finally, the density function for the *polycentric distribution* revealed an important mass of probability concentrated around the average value of the variable (Figure 6.4.b). However, a sizeable tail to the left of the mode appeared, indicating the existence of a small group of regions with low levels of activity.

As these results show, although the traditional non-spatial analysis of inequality can provide information concerning the degree of dispersion, it sheds no light on the spatial distribution of the variable. For example, density function for the *core-periphery distribution* (Figure 6.2.b) cannot tell us whether the regions included in each mode are close in space (and hence there is spatial heterogeneity) or not. Similarly, while the standard deviation for the last three distributions are very similar (albeit there are some differences in their density functions), the location of the activity responds to quite distinct criteria in each case: a group of neighbouring regions account for most of the activity in *monocentric and hierarchical distributions*, while a number of regions, widespread in space, account for most of the activity and attract the activity of their adjacent regions in *polycentric distribution*. This result is particularly significant if we consider what the consequences might be of one region located in the monocentre and of one of the polycentres improving in relation to their neighbours. In this sense, the maps of the series tell us more about the spatial distribution of activity in each case. Yet at the same time, the information derived from these maps is subjective and highly dependent on, for example, the range of intervals used in depicting the variable.

So, it would seem that a spatial analysis of each of the generated distributions is justified in order to obtain a more complete understanding of the characteristics of these patterns.[16] For this reason, the Moran I index was computed using a standardised binary contiguity matrix for ten different orders of contiguity. The results are depicted in the spatial correlograms in Figures 6.1.c, 6.2.c, 6.3.c, 6.4.c, and 6.5.c. In addition, the results of the Moran Scatterplot maps (Figures 6.1.d, 6.2.d, 6.3.d, 6.4.d, and 6.5.d) and the local spatial association tests (Figures 6.1.e, 6.2.e, 6.3.e, 6.4.e, and 6.5.e) are presented. Only the regions with a significant (10% significance level) value on the I_i test are marked on these figures, indicating whether there was a cluster of similarly high levels of activity, a cluster of similarly low levels of activity or a cluster of dissimilar levels of activity between this region and its neighbours.[17]

Firstly, in the case of the *homogeneous distribution*, it can be seen that the Moran I test, using a first order contiguity matrix, detected the randomness of the distribution of the activity in space perfectly. The same result was obtained for the other weight matrices, showing the non-existence of any pattern of spatial dependence for any order of contiguity (in all cases the two discontinuous lines were not exceeded). However, the pattern detected at a global level was not reproduced at a local level (Figure 6.1.e) since some regions had a significant value on the I_i test. This highlights the need to compute both global and local spatial association tests in order to obtain a more accurate picture of the real spatial distribution of the variable.

In contrast, the spatial correlogram for the *core-periphery distribution* (Figure 6.2.c) showed a completely different situation: positive and significant spatial dependence for the first orders of contiguity (from 1 up to 5) and a clear pattern of negative spatial dependence for the higher orders (from order 7 upwards). The latter pattern maintained its intensity despite the distance. This result shows, therefore, how the clear bimodality and heterogeneity detected previously using the density function (Figure 6.2.b) follows a marked spatial pattern. At the local level (Figure 6.2.e), some clusters were apparent in some regions of the South, with lower levels of activity than expected. The opposite scenario was detected in certain regions in the North, reinforcing the idea of a clear spatial division in the distribution of activity in this case.

In the case of the *monocentric distribution*, its spatial correlogram was similar to the previous case but revealed, unlike this previous case, a convex shape for the higher orders of contiguity, indicating that the intensity of the negative spatial dependence decreased with distance. So, while the negative spatial dependence was maintained for the higher orders of contiguity when there were two clear spatial regimes (that is, a spatial heterogeneity based on a North-South division), a positive spatial autocorrelation reappeared among distant regions in the case of a

---

[16] The spatial analysis was conducted using the Space Stat (version 1.9) software linked to ArcView (version 3.1).

[17] In these figures, "High-Low" means that region i shows a significant concentration of high levels of activity, while the contrary is observed in its neighbours. "Low-High" shows the opposite distribution.

monocentric distribution. Regarding the local analysis, it seems that the regions that differed most from the general trend presented significant values of the local test (remember that only 16 of the 108 regions were in the monocentre). Thus, as shown by Figure 6.3.e, only the regions within the monocentre presented positive and significant values of the I_i test, showing the existence of clusters of similarly high levels of concentration of activity.

In the case of a *polycentric distribution,* as can be seen from its spatial correlogram (Figure 6.4.c) and unlike the earlier cases, there was a significant negative spatial dependence between adjacent regions, reflecting the effect of the core-periphery patterns contained therein (a marked trend towards the clustering of dissimilar levels of activity in space). In addition, non significant spatial autocorrelation was detected for the remaining orders of contiguity, demonstrating how only those regions that can strictly be defined as neighbours are affected by the attraction of the activity to the core regions The local analysis (Figure 6.4.e) revealed the presence of a number of clusters with a negative and significant value on the I_i test, reflecting the existence of clear small-scale core-periphery patterns. However, it is interesting to note that, despite the global negative spatial dependence detected, there were two clusters with a positive and significant value for the I_i test.

Finally, and comparing the results for the *monocentric* and *hierarchical distributions*, the effects of a diffusion process on the spatial association tests were observed. Thus, as regards the spatial correlograms, it seems that as the distribution of activity became more hierarchical, the number of orders of contiguity with a positive spatial dependence increased (Figures 6.3.c and 6.5.c). This result reflects the greater similarities between the regions in the monocentre with their neighbours of first and second order of contiguity as the diffusion process advanced. In addition, as the distribution was more hierarchical, less distance was needed to find a negative and significant spatial dependence pattern between regions. Furthermore, it is interesting to note that the diffusion process caused an increase in the number of clusters with a high level of activity, spreading towards the neighbouring regions of the monocentre. However it should be noted that, although a diffusion process was in operation, the regions in the monocentre maintained their key role, accounting for most of the activity. For this reason, in the local analysis, the diffusion process meant that the clusters detected at the beginning remained to the end. In contrast, the results obtained would have been different had the diffusion process led to the loss of the privileged position of the core-regions (resulting in a more balanced distribution of the activity). Had this occurred, it is likely that, at a local level, the clusters detected before disappeared after the diffusion process.

At this point in the study, and based on the results from the spatial analysis, three observations might usefully be made. The first point concerns the expected behaviour of the spatial dependence tests during an integration process (which indeed followed the inverted-U shape as proposed by a number of models in New Economic Geography). Thus, in the first stages of integration, the non-existence of a significant global spatial dependence pattern for any order of contiguity is to be expected, given the low level spatial concentration of the activity.

Subsequently, as the integration process develops, the centripetal forces would be expected to prevail over the centrifugal forces, making it more advantageous for the activity to concentrate in space. In such a case, we would expect the appearance of a positive and significant spatial dependence pattern for the first orders of contiguity and maybe a negative spatial autocorrelation between the more distant regions, thereby highlighting the differences in the spatial distribution of the activity among regions. In addition, a local analysis should show the existence of a number of clusters with a higher (or lower) concentration of the activity than that which is expected in the case of an homogeneous distribution (or, clusters of dissimilar values as in case of the *polycentric distribution*).

When a high degree of integration is reached, positive links between distant regions arise as a consequence of the spatial diffusion of activity. Finally, during the last stage of integration, when the eventual outcome of the integration process is a return to a fairly even or a random distribution of the activity among the regions, the pattern of spatial autocorrelation should disappear (returning to a *homogenous distribution*).

The second observation concerns the level of aggregation used in this analysis. In the simulation described here it was assumed that the concentration process was supraregional so that certain neighbouring regions would display advantages and, as a result, attract activity. Similarly, the diffusion process was also assumed to be supraregional, affecting those neighbouring regions of first and second order contiguity with the monocentre. Although the level of aggregation used in our sample (NUTS II) was low enough to make these assumptions, in those cases where the concentration process appears only at a local scale (such as within industrial districts), a regional spatial analysis would not detect, for instance, a diffusion process or the relocation of the core.

Finally, it is worth noting that, in order to obtain a good description of the spatial distribution of one variable, both global and local spatial analyses are needed. This is particularly the case when there is a spatial relocation of the most dynamic activity across regions since global spatial pattern indicators might be unaffected, and so significant changes in the spatial distribution of the activity would go undetected. Moreover, local spatial analysis would provide a more accurate explanation of the changes that had actually occurred and better inform us about the presence of local patterns that might differ markedly from the global spatial dependence scheme that happened to be detected.

6.3.3.2 Sensitivity of Spatial Dependence Tests to the Within and Between Group Deviation

To conclude this study of the simulation exercise, the sensitivity of the spatial dependence tests to the distribution of the generated series was analysed. To do this, we examined the response of the Moran I test to changes in the *within* and *between* group deviation. Table 6.2 presents a summary of these results, including both the percentage number of cases in which the Moran I test was significant (5% level of significance) and the average value for each distribution.

Three main conclusions can be drawn from this table. Firstly, in relation to the *homogeneous distribution*, it can be seen that changes in the definition of the deviation of the series did not affect the Moran I test, indicating the non- existence of a spatial dependence pattern in all cases. Here, only the computation of the traditional concentration/dispersion indices allowed the differences between the generated series to be detected.

Secondly, after analysing the behaviour of the Moran I test in the other distributions, we can conclude that the higher the *between group* deviation, the greater the likelihood of finding a significant spatial dependence pattern and, hence, obtaining a higher average value for the Moran I test. In contrast, the higher the *within group* deviation, the lower the likelihood of finding a significant spatial dependence pattern. Thus, it would seem that spatial dependence increases when the degree of heterogeneity between the groups of neighbouring regions increases and the degree of homogeneity shown by the regions inside each group also increases.

Thirdly, after comparing the results for *monocentric* and *hierarchical distributions*, it can be concluded that the more hierarchical the distribution of the activity, the higher the average value of the Moran I test and the lower the influence of the *within group* deviation on the likelihood of finding a significant spatial dependence pattern.

6.4 Empirical Exercise: the Spatial Distribution of Labour Productivity

Having established the response of the global and local spatial dependence tests to various spatial distributions of a generated variable, namely "level of activity", here we apply the analysis outlined to the GDP per worker for the same European regions as those described above in the period 1975-1996. We used the REGIO database (EUROSTAT), together with the database produced by the ESOC-Lab at the London School of Economics.

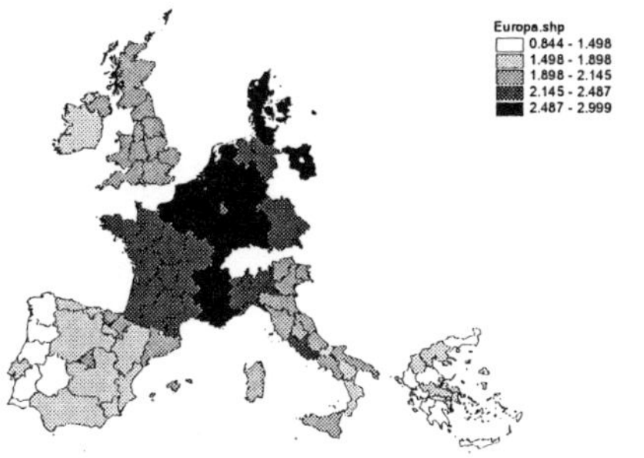

Fig. 6.6.a Spatial distribution Log GDP pw, 1975

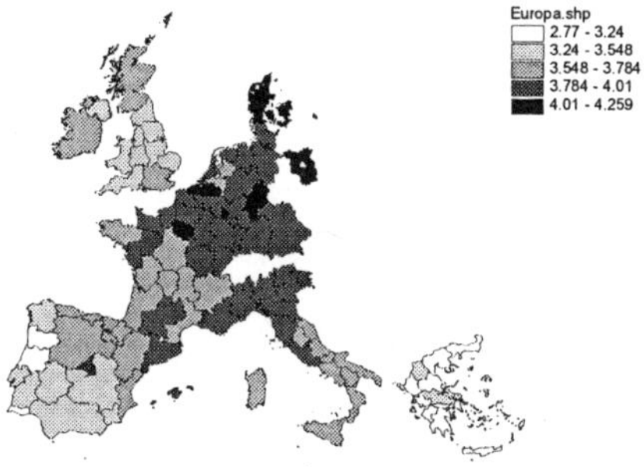

Fig. 6.6.b Spatial distribution Log GDP pw, 1996

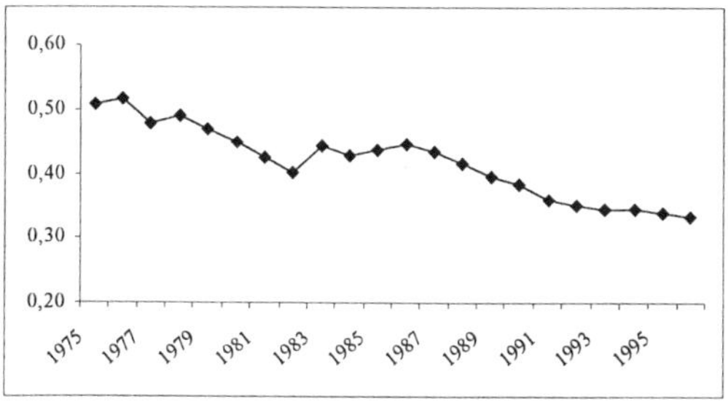

Fig. 6.7 Sigma convergence, 1975-1996

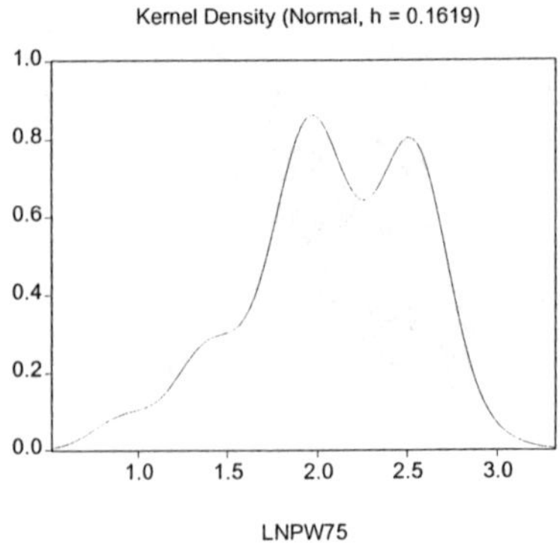

Fig. 6.8.a Density function, 1975

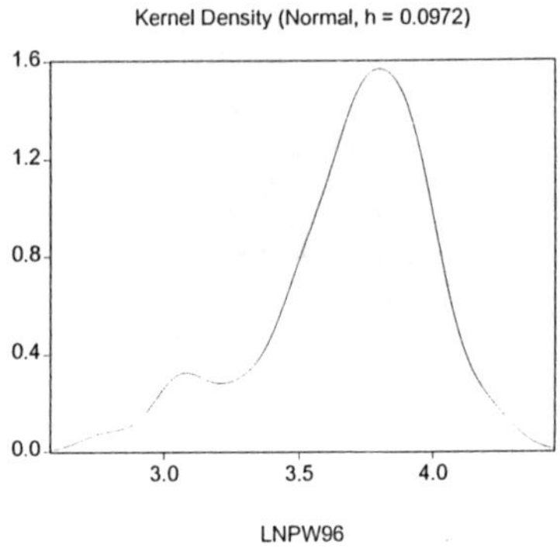

Fig. 6.8.b Density function, 1996

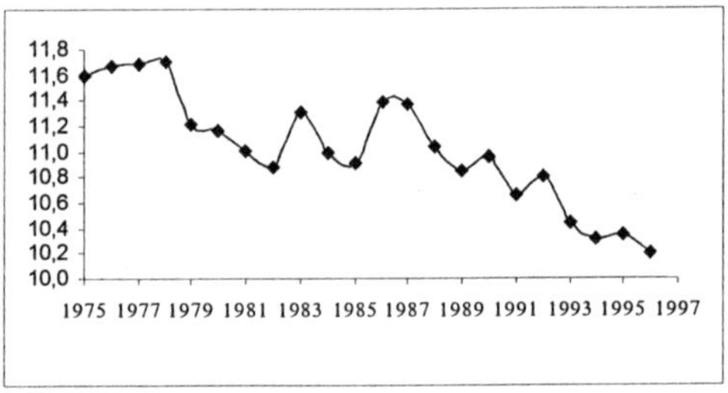

Fig. 6.9 Moran's I standardised, 1975-1996

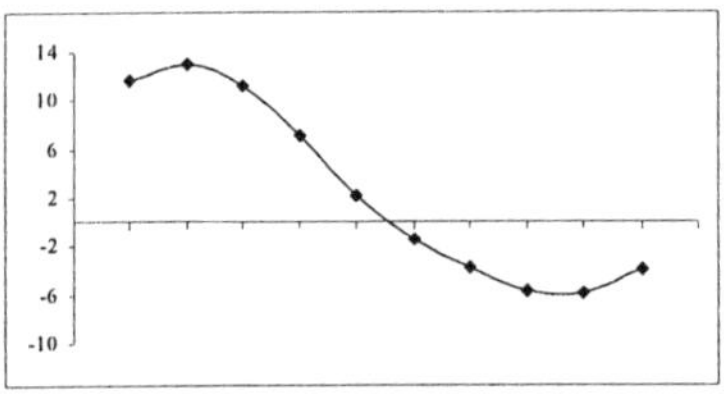

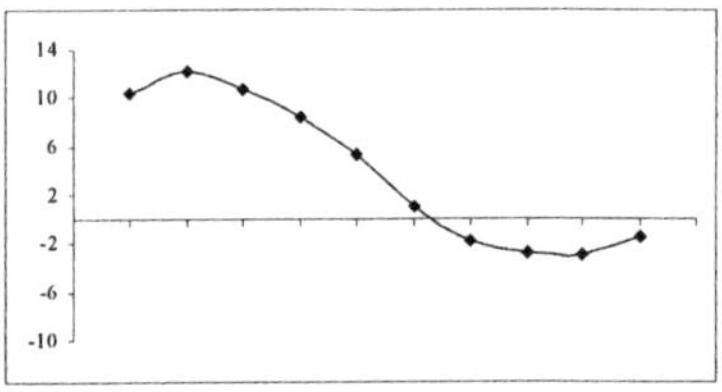

Fig. 6.10.a,b Spatial correlogram, Log GDP pw 1975, 1996

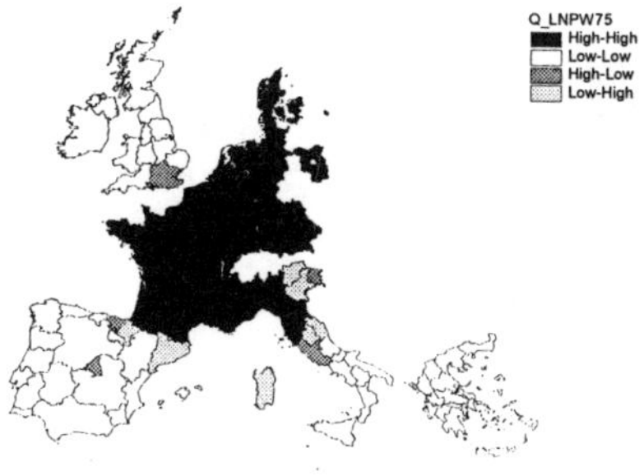

Fig. 6.11.a Moran scatterplot map, 1975

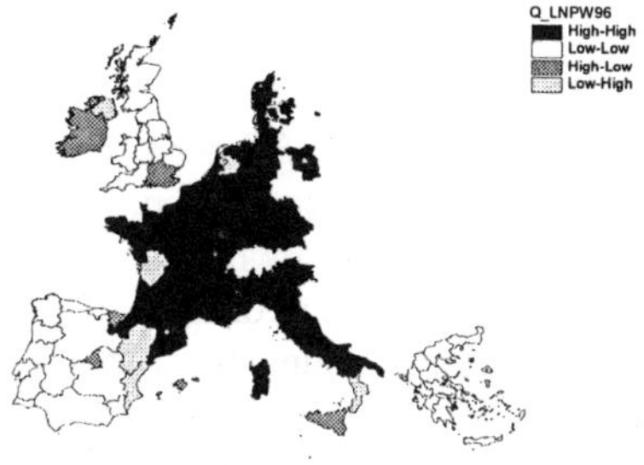

Fig. 6.11.b Moran scatterplot map, 1996

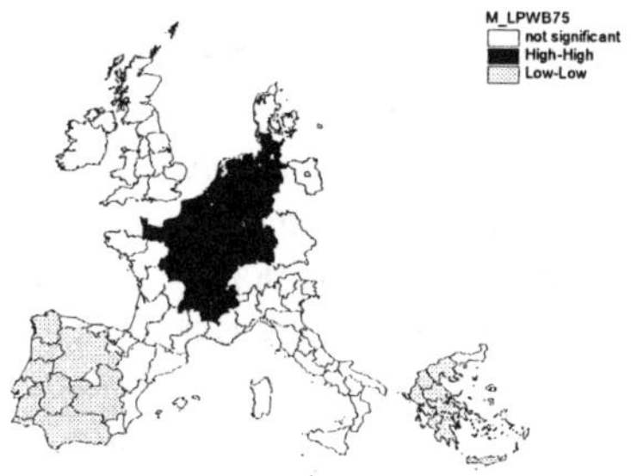

Fig. 6.12.a Local Moran Ii, 1975

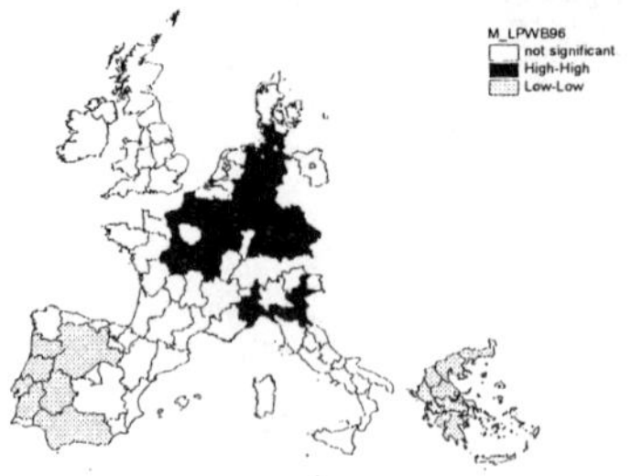

Fig. 6.12.b Local Moran Ii, 1996

Figures 6a and 6b show the spatial distribution of log GDP per worker for 1975 and 1996. As can be seen, all of the regions of Portugal and Greece, most regions in the UK (together with Ireland), some Spanish (except its north-eastern regions and Madrid), and southern Italian regions (at the beginning of the period) showed the lowest labour productivity values (in fact, at the beginning of the period, the Portuguese region of Alentejo was a clear outlier because of the low value achieved by this variable). In contrast, in 1975 the highest values were recorded in all the German, French, Dutch and Belgian regions and Luxembourg, together with some northern Italian regions. However, by the end of the period, the French regions had lost their initial advantage, while those of northern Italy had seen their fortunes reversed.

In order to analyse the degree of inequality in this variable and its evolution during the period 1975-1996, the standard deviation statistic (σ, sigma-convergence) was computed (Figure 6.7). This computation revealed a decreasing trend in the evolution of dispersion throughout the period (despite the slight increase observed from 1982 to 1987). From the density functions (Figures 6.8.a and 6.8.b), it can be seen that at the beginning of the period there is a slight bimodality in the distribution of labour productivity among the European regions. A sizeable tail is also detected to the left of the first mode, showing the existence of a small group of regions with very low levels of labour productivity (basically, the Portuguese and Greek regions). At the end of the period, the previous bimodality seems to have disappeared, revealing a fall in the degree of inequality (in line with the results of the evolution of the σ statistic). However, a sizeable tail to the left of the mode persists.

As a result of the non-spatial analysis two conclusions can be drawn about the distribution of labour productivity among European regions and its evolution during the period 1975-1996. Thus, firstly, the slight bimodality detected in the density function for 1975 alerts us to the presence of a certain heterogeneity in the distribution of labour productivity. In addition, from Figure 6.6.a, it can be seen that this heterogeneity follows a spatial scheme, similar to the *core-periphery distribu-*

tion analysed in the simulation exercise. Second, from the evolution of the sigma-convergence and after comparing the density function for 1975 and 1996, it seems that there has been a reduction in the disparities between European regions, with the previous bimodality disappearing. In addition, by the end of the period, the spatial North-South division detected in 1975 becomes less clear, and the traditional European core seems to move towards the south of Germany and the north of Italy (more similar to the *monocentric distribution* analysed in the simulation exercise).

Having reached this point, the next step would be to try to show how the previous analysis might be complemented by means of ESDA tools. In this sense, it is interesting to note that, although an inclination for similar values of productivity to cluster among neighbouring regions is detected in Figures 6.6.a and 6.6.b (rather than a random distribution), this result is subjective and highly dependent on the range of intervals selected for mapping the variable. So, based on this analysis it is not possible to determine whether there is significant and positive spatial dependence in the distribution of labour productivity. Neither is it possible to determine whether there are significant links between distant regions and whether the intensity of spatial dependence decreases or increases during the period. In addition, it is not possible to detect significant clusters of similar or dissimilar values of labour productivity.

In order to clarify these doubts, a spatial analysis was conducted. First, the Moran I test was computed using a first order contiguity matrix (Figure 6.9). Figures 6.10.a and 6.10.b show the spatial correlograms for ten different orders of contiguity matrices for 1975 and 1996 and Figures 6.11.a and 6.11.b present the Moran Scatterplot maps for these years. From these figures, three conclusions can be drawn. Firstly, a significant and positive spatial dependence was detected among neighbouring regions throughout the period, highlighting a trend towards the clustering of similar levels of GDP per worker in space. It would also seem that there were certain factors or advantages that extended beyond the regional frontiers, resulting in the concentration of high levels of labour productivity in some neighbouring regions. This trend towards the clustering of similar levels of labour productivity in space was corroborated by the results of the Moran Scatterplot maps for 1975 and 1996. Thus, most of the regions displayed GDP per worker values that were very similar to those of their neighbours. The exceptions to this general trend included Madrid, País Vasco, Sicilia, Lazio and Friuli-Venezia Giulia, Ireland and South East, which presented higher levels of productivity than those of their neighbours. The opposite was true for Navarra, Catalunya, Sardegna, Trentino Alto Adige, Umbria and Marche in 1975 and Aragón, Valencia, Calabria, Northern Ireland and Poitou-Charentes in 1996.

Secondly, the spatial correlograms showed a significant positive spatial dependence in 1975 for the first five orders of contiguity, while a significant negative spatial dependence was predominant in the last five. Thus, if we take into consideration the results obtained in the simulation exercise, it would seem that at the beginning of the period, there was a highly marked degree of heterogeneity in the distribution of labour productivity following a clear spatial pattern (similar to the *core-periphery distribution* in the simulation exercise). When this analysis was repeated for 1996, a significant negative spatial dependence was only found for the last three orders. This result shows the loss of intensity of the previous North-South spatial division by the end of the period (with the final distribution being more similar to the *monocentric distribution* proposed in the simulation exercise).

Finally, in spite of the high values recorded on the Moran I test throughout the period, it would appear that the intensity of this spatial dependence decreased from the mid-eighties onwards. This result was obtained for all ten orders of contiguity. On this point it is worth noting that from the previous simulation exercise we concluded that the spatial dependence increased with increasing heterogeneity among the groups of neighbouring regions and with increasing homogeneity among the regions inside each group. Thus, the fall detected in the Moran I values during the period for all the orders of contiguity might indicate that the marked spatial division observed in 1975 between the low levels of labour productivity in the *south* and the high levels in the *north* was not so clear by the end of the period, with the inter-group deviation decreasing. This finding would be consistent with the decrease in global dispersion detected by means of sigma convergence and with the disappearance of the initial bimodality detected in the density function. Furthermore, it might also be the case that northern and southern European regions became less homogenous between themselves. In any case, the decrease in the spatial dependence intensity might well be in line with the evolution predicted by the simulation exercise as the integration process advanced (from intermediate to low transport costs).

Finally, a local spatial analysis was conducted. Figures 6.12.a and 6.12.b show those regions with a significant value on the local Moran I_i test in 1975 and 1996, indicating whether the clusters detected presented similar (high or low) or dissimilar values of labour productivity between regions and their neighbours. Two conclusions can be drawn from this analysis. First, despite the result of the Moran Scatterplot maps, in none of the regions was the level of labour productivity sufficiently different from that of its neighbours to show a significant and negative value of the local Moran test. Secondly, two large groups of clusters were observed in 1975. One of these displayed very low levels of GDP per worker, and included all the Portuguese regions, their Spanish neighbours (Galicia, Extremadura, Castilla-León, Castilla-La Mancha and Andalucía) and the Greek regions (except to Anatoloki Makedonia). The second cluster, with high values of labour productivity, included all the Dutch and Belgian regions, Luxembourg and most of the German and Northern French regions. When we repeated the analysis for 1996, two results were obtained. First, at the end of the period the clusters of low levels of productivity were once again located in the Greek and Portuguese regions, along with the adjoining Spanish regions, revealing the persistence of their poor situation in terms of labour productivity. Second, a certain relocation of the core was detected, as Dutch and Northern French regions lost their positions as significant clusters of high levels of productivity, while the reverse was true with some of the Northern Italian regions such as Piemonte, Veneto or Emilia-Romagna. Clearly, this shows the efficacy of the local spatial analysis since this trend was not detected using global spatial association tests.

6.5 Final Remarks and Conclusions

The main goal of the Chapter has been to demonstrate the extent to which the findings of traditional, non-spatial, location or inequality analyses might be improved with the use of Exploratory Spatial Data Analysis techniques.

In seeking to demonstrate this, a simulation exercise was first conducted. This exercise showed that spatial analysis served as a useful complement to the results obtained from the more usual dispersion indices. In addition, the sensitivity of the spatial dependence tests to different spatial distributions of one generic variable, namely "level of activity", was studied, allowing us to draw some general conclusions about the behaviour of these tests. Thus, among other results, it would seem that the intensity of spatial dependence increases when the degree of heterogeneity between the groups of neighbouring regions or/and the degree of homogeneity shown by the regions inside each group increase. In addition, the more hierarchical the distribution of the activity, the higher the average value of the Moran I test.

Having completed the simulation, an empirical exercise was performed whereby the spatial distribution of GDP per worker in the European regions between 1975 and 1996 was analysed. The results allowed four conclusions to be drawn. First, there was a clear positive spatial dependence between neighbouring regions throughout the period, indicating that labour productivity does not have a random spatial distribution. Such a finding has major economic as well as econometric repercussions (given that problems of spatial dependence might appear in those econometric models that use labour productivity as an endogenous or exogenous variable).

Second, a significant and negative spatial association between distant regions was detected, indicating the existence of, at least at the beginning of the period, two main spatial regimens in the distribution of GDP per worker in the regions of the EU.

Third, the initial intensity of this positive (and negative) spatial dependence pattern fell during the period. This trend, together with the fall detected in the global dispersion of this variable, might suggest that the spatial division between low levels of labour productivity in the *south* and high levels in the *north* observed at the outset was less marked by the end of the period (although the differences between the most peripheral and the most advanced regions persisted). In addition, it could be possible that *northern* and *southern* European regions might be becoming less homogenous between themselves, causing spatial dependence test values to fall.

Finally, the local spatial analysis revealed the presence of two large groups throughout the period: one of these displayed very low levels of GDP per worker (and included Portuguese regions, their Spanish neighbours and all of the Greek regions) while the other displayed high values of labour productivity (and included

Dutch and Belgian regions, Luxembourg and most of the German and Northern French regions). Thus, it would seem that the advantages and disadvantages experienced by these regions in terms of labour productivity have persisted over the years. However, by the end of the period, a degree of relocation of the high labour productivity clusters from the Northern French regions to those of Northern Italy was detected, leading to some extent to a repositioning of the European core.

References

Amiti M (1997) Specialization patterns in Europe. Discussion Paper N°363, Centre for Economic Performance, LSE.

Anselin L (1988) Spatial econometrics: methods and models. Kluwer Academic Publishers, The Netherlands.

Anselin L (1995) Local indicators of spatial association-LISA. Geographical Analysis 27: 93-115.

Anselin L (1996) The Moran Scatterplot as an ESDA tool to assess local instability in spatial association. In Fisher M, Scholten H and Unwin D (eds) Spatial analytical perspectives on GIS. London, Taylor and Francis.

Anselin L (1998) Exploratory spatial data analysis in a geocomputational environment, Regional Research Institute, West Virginia University, Research Paper 9801.

Anselin L, Florax R (1995) New directions in Spatial Econometrics. Springer-Verlag, Berlin.

Armstrong HW (1995) Cross-sectional analysis of the regional growth process. In Armstrong and Vickerman (eds) Convergence and Divergence among European Regions, London, pp 40-65.

Brülhart M, Torstesson J (1996) Regional integration, scale economies and industry location. Discussion Paper nº. 1435, Centre for Economic Policy Research.

Cliff A, Ord J (1973) Spatial Autocorrelation. London, Pion.

Cliff A, Ord J (1981) Spatial process. Models and applications. Pion Ltd., London.

Esteban J (1994) La desigualdad interregional en Europa y en España: descripción y análisis. In: *Crecimiento y Convergencia Regional en España y Europa*, vol 2 IAE, Barcelona, pp 13-84.

Fingleton B (1999) Estimates of Time to Economic Convergence: An Analysis of Regions of the European Union. International Regional Science Review 22(1):5-30.

Fluvià M, Gual J (1993) Comercio Internacional y Desarrollo Regional en el Marco de la Integración Económica Europea. In: *Crecimiento y Convergencia Regional en España y Europa*. vol 2, pp 85-123. Instituto de Análisis Económico, CSIC.

Greenaway D, Hine RC (1991) Intra-industry specialization and adjustment in the European economic space. Journal of Common Market Studies, 29(6): 603-22.

Haslett J, Wills G, Unwin A (1990) SPIDER: An interactive statistical tool for the analysis of spatially distributed data, International Journal of Geographical Information Systems, 4:285-296.

Haslett J, Bradley R, Craig P, Wills G, Unwin A (1991) Dynamic graphics for exploring spatial data with applications to locating global and local anomalies, The American Statistician, 45: 234-242.

Krugman P (1991a) Increasing returns and economic geography. Journal of Political Economy 99: 484-499.

Krugman P (1991b) Geography and Trade. MIT Press, Cambridge MA.

Krugman P (1998) What's new about the new economic geography? Oxford Review of Economic Policy 14(2):7-17.

Krugman P, Venables A (1995) Globalization and the inequality of nations. Quarterly Journal of Economics 110: 857-880.

Krugman P, Venables A (1996) Integration, specialization and adjustment. European Economic Review 40: 959-967.

López-Bazo E, Vayá E, Mora AJ, Suriñach J (1999) Regional economic dynamics and convergence in the European Union, The Annals of Regional Science, 33(3): 343-370.

Martin R (1999) The new "geographical turn" in economics:some critical reflections. Cambridge Journal of Economics 23:65-91.

Martin P, Ottaviano GIP (1999) Growing locations:Industry location in a model of endogeneous growth. European Economics Review 43(2):281-302.

Molle W (1996) The regional economic structure of the European Union: an analysis of long term developments. In: Karin Peschel (eds) Regional growth and regional policy within the framework of European integration., Physica-Verlag.

Molle W, Boeckhout S (1995) Economic disparity under conditions of integration-A long term view of the European case, Papers in Regional Science, 74:105-123.

Moran P (1948) The interpretation of statistical maps. Journal of the Royal Statistical Society B 10:243-251.

Neven DJ, Gouyette C (1994) Regional convergence in the European Community, CEPR Discussion Paper. Series 914, London.

Puga D (1996) The rise and fall of regional inequalities. Discussion Paper Nº 314, Centre for Economic Performance, LSE.

Puga D, Venables AJ (1996) The spread of industry: spatial agglomeration in economic development. Journal of the Japanese and International Economics 10:440-464.

Ottaviano G, Puga D (1997) Agglomeration in the global economy: A survey of the 'new economic geography'. Discussion Paper Nº 356, Centre for Economic Performance, LSE.

Quah DT (1993) Empirical cross-section dynamics in economic growth. European Economic Review 37:426-434

Quah DT (1996a) Regional convergence clusters across Europe. European Economic Review 40:951-958

Quah DT (1996b) Twin Peaks: growth and convergence in models of distribution dynamics. Economic Journal 106:1045-1055.

Suarez-Villa L, Cuadrado JR (1993) Regional Economic Integration and the evolution of Disparities. Papers in Regional Science 72 (4): 369-387.

Venables A (1996) Equilibrium locations of vertically linked industries. International Economic Review 37: 341-359.

Table 6.1 Description of the generated distributions (1/2)

	Characteristics of the generated series
Homogenous distribution	
Small deviation	$X \sim N(\overline{X}, 0.20\overline{X})$
Medium deviation	$X \sim N(\overline{X}, 0.40\overline{X})$
Core-Periphery distribution[1]	
Small within group deviation, Small between group deviation	$X_C \sim N(\overline{X}_C, 0.10\overline{X}_C)$ $X_P \sim N(\overline{X}_P, 0.10\overline{X}_P)$ $\overline{X}_P = 0.6\overline{X}_C$
Medium within group deviation, Small between group deviation	$X_C \sim N(\overline{X}_C, 0.30\overline{X}_C)$ $X_P \sim N(\overline{X}_P, 0.30\overline{X}_P)$ $\overline{X}_P = 0.6\overline{X}_C$
Small within group deviation, Medium between group deviation	$X_C \sim N(\overline{X}_C, 0.10\overline{X}_C)$ $X_P \sim N(\overline{X}_P, 0.10\overline{X}_P)$ $\overline{X}_P = 0.45\overline{X}_C$
Medium within group deviation, Medium between group deviation	$X_C \sim N(\overline{X}_C, 0.30\overline{X}_C)$ $X_P \sim N(\overline{X}_P, 0.30\overline{X}_P)$ $\overline{X}_P = 0.45\overline{X}_C$
Monocentric Distribution [2]	
Small within group deviation, Small between group deviation	$X_C \sim N(\overline{X}_C, 0.10\overline{X}_C)$ $X_P \sim N(\overline{X}_P, 0.10\overline{X}_P)$ $\overline{X}_P = 0.70\overline{X}_C$
Medium within group deviation, Small between group deviation	$X_C \sim N(\overline{X}_C, 0.30\overline{X}_C)$ $X_P \sim N(\overline{X}_P, 0.30\overline{X}_P)$ $\overline{X}_P = 0.70\overline{X}_C$
Small within group deviation, Medium between group deviation	$X_C \sim N(\overline{X}_C, 0.10\overline{X}_C)$ $X_P \sim N(\overline{X}_P, 0.10\overline{X}_P)$ $\overline{X}_P = 0.50\overline{X}_C$
Medium within group deviation, Medium between group deviation	$X_C \sim N(\overline{X}_C, 0.30\overline{X}_C)$ $X_P \sim N(\overline{X}_P, 0.30\overline{X}_P)$ $\overline{X}_P = 0.50\overline{X}_C$

Note: $\overline{X}$..average of the generated series
[1]Number of obs.in the core: $N_C=63$; Number of obs. in the periphery: $N_P=45$
[2]Number of obs. in the monocentre: $N_C=16$ Number of obs. outside the monocentre: $N_P=92$

Table 6.1 Description of the generated distributions (2/2)

Polycentric distribution[(3)]	
Small within group deviation	$X_C \sim N(\overline{X}_C, 0.10\overline{X}_C)$ $X_{C1} \sim N(\overline{X}_{C1}, 0.10\overline{X}_{C1}); \overline{X}_{C1} = 0.50\overline{X}_C$ $X_P \sim N(\overline{X}_P, 0.10\overline{X}_P); \overline{X}_P = 0.70\overline{X}_C$
Medium within group deviation	$X_C \sim N(\overline{X}_C, 0.30\overline{X}_C)$ $X_{C1} \sim N(\overline{X}_{C1}, 0.30\overline{X}_{C1}); \overline{X}_{C1} = 0.50\overline{X}_C$ $X_P \sim N(\overline{X}_P, 0.30\overline{X}_P); \overline{X}_P = 0.70\overline{X}_C$
Hierarchical Distribution[(4)]	
Small within group deviation, Small between group deviation	$X_C \sim N(\overline{X}_C, 0.10\overline{X}_C)$ $X_{C1} \sim N(\overline{X}_{C1}, 0.10\overline{X}_{C1}); \overline{X}_{C1} = 0.90\overline{X}_C$ $X_{C2} \sim N(\overline{X}_{C2}\ 0.10\overline{X}_{C2}); \overline{X}_{C2} = 0.85\overline{X}_{C1}$ $X_P \sim N(\overline{X}_P, 0.10\overline{X}_P); \overline{X}_P = 0.85\overline{X}_{C2}$
Medium within group deviation, Small between group deviation	$X_C \sim N(\overline{X}_C, 0.30\overline{X}_C)$ $X_{C1} \sim N(\overline{X}_{C1}, 0.30\overline{X}_{C1}); \overline{X}_{C1} = 0.90\overline{X}_C$ $X_{C2} \sim N(\overline{X}_{C2}\ 0.30\overline{X}_{C2}); \overline{X}_{C2} = 0.85\overline{X}_{C1}$ $X_P \sim N(\overline{X}_P, 0.30\overline{X}_P); \overline{X}_P = 0.85\overline{X}_{C2}$
Small within group deviation, Medium between group deviation	$X_C \sim N(\overline{X}_C, 0.10\overline{X}_C)$ $X_{C1} \sim N(\overline{X}_{C1}, 0.10\overline{X}_{C1}); \overline{X}_{C1} = 0.90\overline{X}_C$ $X_{C2} \sim N(\overline{X}_{C2}\ 0.10\overline{X}_{C2}); \overline{X}_{C2} = 0.85\overline{X}_{C1}$ $X_P \sim N(\overline{X}_P, 0.10\overline{X}_P); \overline{X}_P = 0.70\overline{X}_{C2}$
Medium within group deviation, Medium between group deviation	$X_C \sim N(\overline{X}_C, 0.10\overline{X}_C)$ $X_{C1} \sim N(\overline{X}_{C1}, 0.30\overline{X}_{C1}); \overline{X}_{C1} = 0.90\overline{X}_C$ $X_{C2} \sim N(\overline{X}_{C2}\ 0.30\overline{X}_{C2}); \overline{X}_{C2} = 0.85\overline{X}_{C1}$ $X_P \sim N(\overline{X}_P, 0.30\overline{X}_P); \overline{X}_P = 0.70\overline{X}_{C2}$

[(3)] Number of Polycentres: N_C=16 ; Number of first order contiguity neighbours of the polycentres: N_{C1}=32; Number of the rest of the observations: N_P=65

[(4)] Number of observations in the monocentre: N_C=16; Number of first order contiguity neighbours of the monocentre: N_{C1}=15; Number of second order contiguity neighbours of the monocentre: N_{C2}=17; Number of the rest of the observations in the sample: N_P=60.

Table 6.2 Results of the Moran I for the simulation exercise

Distribution	% with significant Moran's I standardised	Average of Moran's I (standardised)
Homogenous Distribution		
Small deviation	5%	0.25*
Medium deviation	5%	0.25*
Core-Periphery Distribution		
Small within group deviation, Small between group deviation	100%	11.05
Medium within group deviation, Small between group deviation	100%	4.9
Small within group deviation, Medium between group deviation	100%	11.94
Medium within group deviation, Medium between group deviation	100%	6.97
Monocentric distribution.		
Small within group deviation, Small between group deviation	100%	6.08
Medium within group deviation, Small between group deviation	40%	1.82*
Small within group deviation, Medium between group deviation	100%	8.16
Medium within group deviation, Medium between group deviation	97%	4.51
Polycentric Distribution		
Small within group deviation	100%	-4.78
Medium within group deviation	53%	-2.18
Hierarchical Distribution		
Small within group deviation, Small between group deviation	100%	10.07
Medium within group deviation, Small between group deviation	87%	3.61
Small within group deviation, Medium between group deviation	100%	12.32
Medium within group deviation, Medium between group deviation	99%	6.12

*Not significant at 5% level.

7 Trade, Human Capital and Innovation: The Engines of European Regional Growth in the 1990s

Harald Badinger and Gabriele Tondl

Vienna University of Economics and Business Administration, Austria

7.1 Introduction

Regional disparities have been a major policy concern in the European Union, in particular since its Southern enlargement. Widely known as the Union's cohesion problem, it gave rise to the implementation of EU regional policy supported by substantial financial assistance from the Union's budget. Although regional disparities have shown a tendency to decrease gradually, the regional convergence process has been interrupted several times. After the post 1975 period of divergence, a period of convergence set in with the accession of Southern countries to the EU and the warm-up phase of the EU's Internal Market 1986-1992. Thereafter, growth developments of the 1990s again suggest a modest convergence.

Given this background, this Chapter wishes to discover what has determined growth of EU regions in the 1990s and to offer an empirical assessment of this process. Hereby we shall particularly focus on endogenous growth factors, potential channels of technology transfer, and trade. These specific growth aspects have been only partially analyzed in regional growth studies for the EU (Fagerberg and Verspagen 1996; Fagerberg et al. 1997; Vanhoudt et al. 2000; Paci and Pigliaru 2001; Tondl 2001). A systematic assessment of regional growth based on an explicit endogenous growth model and trade theory is missing. Therefore, we shall combine theoretical aspects of the human capital and innovation growth literature (Lucas 1988, Romer 1990), catching-up theory (Abramovitz 1986, 1989) and trade and integration theory (Grossman and Helpman 1991; Baldwin 1993) in order to examine growth factors and mechanisms in EU regions. For our empirical analysis we have compiled a large dataset, with many variables not used thus far, for 159 regions in the 1990s, and will test the importance for regional growth of educational attainment, patenting activity, regional trade, and technological catching-up by estimating a spatial cross-section growth accounting model. In this model, growth is explained by factor accumulation, endogenous factors, technology transfer conditioned by endowment factors and transmitted by trade, and finally the growth of other regions within a certain distance. As to the estimation, we use a spatial lag model (Anselin 1988) which optimally specifies the growth process of EU regions and use the general spatial two stages least squares estimator (GS2SLS) as suggested by Kelejian and Prucha (1998).

Our results indicate that the income growth of EU regions is above all determined by the fundamental growth factors physical capital and labour participation, but it is also to a great extent related to the growth performance of surrounding regions. High growth regions are close to other dynamic regions. Further, higher education attainment levels and changes are clearly associated with higher growth. Higher education is also an important prerequisite for technological catching-up by lagging regions. Finally, we find that a high foreign trade share contributes to regional growth, also by promoting technological catching-up.

The rest of the Chapter is structured as follows: Section 7.2 discusses the main issues raised in the Chapter and reviews the literature. Section 7.3 describes the model we estimate. Section 7.4 gives data definitions, discusses econometric procedures and presents the results of the estimation. Section 7.5 concludes.

7.2 Theoretical Arguments: How may Common Growth Factors Together with Trade and Economic Geography Factors Determine the Growth of EU Regions?

The growth of European regions exhibits no distinct convergence mechanism, i.e. a growth surplus of poor regions vis-à-vis rich ones. Rather, it is the case that one can find high growth associated with both rich as well as poor regions.

This suggests that multiple growth paradigms govern regional growth in Europe. Evidently, a neoclassical convergence mechanism has helped poor regions to converge (European Commission 2000; Tondl 1999). Endogenous growth based concepts permitted rich regions to maintain a leading income position (Tondl 2001). In addition, European regions have become increasingly integrated with other European as well as global markets, particularly in the 1990s after the creation of the Single Market and the WTO Uruguay Round. Trade flows and other kinds of international interactions seem to have become substantial for EU regions and are likely to favour technology transfer on an international scale. Finally, since regions are not isolated economic entities, there is reason to assume that their growth may be linked to that of other regions.

We will examine these possible sources of European regional growth more closely and review the findings of the existing literature in order to extract the working hypotheses for our investigation.

First, given European regional income disparities and considering the neoclassical convergence hypothesis, one may suppose that physical capital accumulation should play an important role for the growth of lagging regions, especially as investment in a new generation of capital is also linked with technological advance (Kaldor 1961). Moreover, as a part of capital consists of public capital which is associated with external effects (Aschauer 1989), the accumulation of total capital stock would have an even higher growth impact. Empirical evidence on the role of investment for EU regions is mixed. While Bacchetta (1994) finds no clear impact on growth, Vanhoudt et al. (2000) detect an important growth effect of private but not of public capital.

A second conjecture is that human capital plays an important role for EU regional growth. This idea has heavily influenced EU regional policy in the past decade, although there is only minor empirical support for this hypothesis for EU regions so far. The traditional arguments on the role of human capital go back to Lucas (1988) who views human capital – in the sense of knowledge – as a central factor of production, which enables sustained growth due to its non-decreasing returns. Mankiw et al. (1992), in an influential paper, extend the neoclassical growth model by human capital as an additional accumulable factor and provide an empirical test for the OECD countries. They conclude that changes in human capital translate into significant changes of growth rates.

Yet there are other channels by which human capital can influence the growth rate. First, human capital is a central prerequisite for innovation activity as set out in Romer (1990). Second, human capital influences the capacity to adapt technological advances from abroad (technological catching-up, see below).

Empirical country studies on the growth effect of human capital, however, provide rather mixed results, largely due to the variety of different, often problematic indicators of human capital and to measurement problems. Researchers used the average number of years in education (e.g., Barro 1991), the share of primary/secondary/third level educational attainment, or sometimes school enrolment as a proxy. Prominent studies such as Benhabib and Spiegel (1994) and Pritchett (1996) suggest that it is the level of educational attainment that influences growth but not its change. More recent studies also find a positive impact of human capital growth. As Krueger and Lindahl (1999) and de la Fuente and Domenech (2000) demonstrate, human capital data sets may show considerable measurement errors, which can lead to particularly erratic observations when looking at human capital growth. The recent estimates of de la Fuente and Domenech (2000) are based on a carefully revised dataset and suggest that the elasticity of output with respect to human capital is about 0.27 for OECD countries. Studies of Bassannini et al. (2001) and Cohen and Soto (2001), both based as well on revised data sets, confirm these results. Furthermore, Temple (2001) replicates the Benhabib and Spiegel and the Pritchett study assuming a different functional relationship between income and human capital and concludes that changes in human capital do have an effect on growth rates.

Due to the even more pronounced problems of data availability for regions, there is only little evidence on the effects of human capital at the EU regional level. Tondl (1999, 2001) shows that incomes and productivity of Southern EU regions are positively linked to school enrolment. Further, Vanhoudt et al. (2000) find an output elasticity for human capital, measured as educational attainment, for EU regions of 0.18 when estimating a production function for 1996.

Third, given that European regions belong to fairly rich, knowledge and R&D-driven economies we may expect that innovation activities are an important source for regional growth. On the theoretical scale, we can base our arguments on the familiar models of Romer (1990) and Aghion and Howitt (1992) of R&D based endogenous growth. These models look at R&D efforts as the starting point for innovations, and suggest that growth is a function of resources employed in the R&D sector.

Empirical studies in this area have shown that research activity can indeed explain a high growth performance by rich countries or regions. Fagerberg (1987) verifies a positive relation between the number of patent applications by a country and growth. De la Fuente (1998) finds that R&D expenditures are an important cause of high growth in rich EU countries. Bassanini et al. (2001) in contrast, in a recent OECD growth study, identify a positive growth impact only for business R&D spending but not for public R&D spending. For EU regions, Fagerberg et al. (1997) find that the superior growth performance of rich regions in the 1980s can be explained by the share of business sector workforce employed in R&D. Similarly, Paci and Pigliaru (2001) detect a significant correlation between patent applications of European regions and their productivity growth. These studies show that it is not simply the total quantity of resources devoted to R&D, either researchers or expenditures, but the share of labour force employed in innovation activity, or an indicator for innovation output as patents which preferably explain its growth impact, – a fact that has first been pointed out by Jones (1995) when watching ailing US growth rates despite its growing number of researchers.

Fourth, with fairly open and integrated economies like the EU countries, we may expect that trade and integration is an important factor for the growth of their regions. First, openness manifests itself as substantial trade flows. Recent studies suggest that international trade relations of EU regions are quite large, particularly for regions in small EU countries (Tondl 2001). In addition, openness includes more than physical trade flows, namely all kind of interactions that ease technology transfer on an international scale.

As argued in the modern trade literature and integration theory, trade triggers important supply side effects (Grossman and Helpman 1991; Balassa 1961; Baldwin 1993; Breuss 1996; Keuschnigg and Kohler 1996), which induce efficiency improvements in the enterprise sector and finally lead to additional growth. On the one hand, entry into foreign markets and opening of the home markets means increased competition which forces firms to improve their efficiency in order to become more competitive. Firms will have to close their productivity gap vis-à-vis the leaders to stay in the market (pro-competitive effect). On the other hand, as argued by Adam Smith, firms can face a larger sales market if operating internationally, which enables them to benefit from economies of scale (market size effect). As a consequence, free trade and integration lead to improvements in productivity. Economies with a high trade share will therefore show a higher growth performance than non-traders.

A number of empirical studies for different sets of countries have shown that foreign trade is promoting growth (e.g. Balassa 1978; Kormendi and Meguire 1985; Dollar 1992). For EU countries this has been verified e.g. by Baldwin and Seghezza (1996), Ben-David (1996), Breuss (1998), and Badinger (2001). Since there is very limited data available on the external trade of regions, there are no studies which address the important issue of trade-induced growth effects for EU regions.

In addition to the described supply side effects, openness of EU regions should facilitate productivity increases through international technology transfers. Openness should permit regions to close their technology gap with the leader more easily.

As to theory, on the one hand the early catching-up literature based on the neoclassical growth model suggests that technological transfer is an important source of technological advance of poor economies (Gerschenkron 1962; Nelson and Phelps 1966; Abramovitz 1986, 1989; Bernard and Jones 1996).

On the other hand, endogenous growth models with trade integration, as in Romer and Rivera-Batiz (1991), attribute a central role to international technology spillovers. In their theoretical set-up knowledge spillovers from abroad add to national knowledge and thus increase the knowledge stock to affect R&D. Furthermore, as innovations become more profitable for a larger sales market a higher innovation rate is to be expected. Thus integration can yield a growth bonus. In addition, also in the Rivera-Batiz and Romer model, economies importing goods from other countries/regions with a higher technological level can import technological progress and may give up their own innovation activity (Rivera-Batiz and Xie 1993).

The possibility of technological transfer is influenced by several factors, such as the social capacity of an economy and particular transmitters to which new technologies are attached. Social capacity is largely determined by the human capital available in an economy and its own engagement in R&D, since knowledge and expertise make it more likely to adopt technologies from abroad. Transmitters of technology may be imported goods or foreign direct investment (FDI).

Technological catching-up depends on the ability of an economy to make use of internationally available technologies. Human capital in the lagging economy is important, as pointed out initially by Gerschenkron (1962) and Nelson and Phelps (1966), and later by Abramovitz (1986), Parente and Prescott (1994) and Benhabib and Spiegel (1994). Griffith et al. (2000) find that domestic R&D improves the absorptive capacity for technology transfer by lagging OECD countries.

An important contribution which attempted to measure the scope of international trade related technology spillovers is Coe and Helpman (1995) who stress the importance of imported intermediaries as a means of technology spillovers. This is measured by the import volume of a country weighted by R&D of the export country. Keller (1998) offers a similar approach. Another group of studies attempts to measure international technology spillovers by patent citations, e.g. Jaffe and Trajtenberg (1996) for countries and Maurseth (2001) for EU regions. Yet a different and growing branch of literature investigated the potential for technology spillovers via FDI (Lichtenberg and van Pottelsberghe de la

Potterie 1996; Baldwin et al. 1999).[1] Unfortunately, there is practically no evidence on these effects for EU regions.

Finally, since regions are not isolated economies but may be linked by a number of not easy-to-measure relationships with each other, we also wish to account for the possibility of spatial dependencies between the growth of EU regions. From an economic theory perspective, new economic geography models (Krugman 1991, Krugman and Venables 1996 Fujita, Krugman and Venables 1999) propose a core-periphery pattern which may persist but finally diminish in the course of integration. In that case, the core hosts the high income agglomeration consisting of a number of regions with dynamic growth. The appearance of a new location advantage may lead to the creation of a new fast growing agglomeration. Consequently, new economic geography predicts that high growth regions are close to each other and so are low growth regions. High and low growth regions would be clustered.

The spatial econometric literature (Anselin 1988, Anselin and Florax 1995, Kelejian and Prucha 1998) offers appropriate models and estimation techniques to account for spatial autocorrelation (e.g. between dependent variables). Recently, a few very innovative empirical studies have appeared which have employed this econometric approach. Fingleton (2001), investigating productivity growth of EU regions in the period 1975-1995, found strong spatial autocorrelation which has increased over time. Vayá et al. (2002), estimating convergence equations for a smaller sample of EU regions but a similar time period (1975-1992) likewise find an important impact by the growth of neighbouring regions on a region´s performance. (See also Baumont, Ertur and Le Gallo in this volume, Chapter3, who estimate convergence in EU 12 regions for 1980-1995 and find a strong impact of neighbouring regions growth). Given the findings of this literature, we therefore propose that growth rates are influenced by other regions growth and that this dependency decreases with distance. The exact transmission mechanism

---

[1] Apart from international technology transfer, regional economists have particularly emphasized the possibility of regional technology spillovers. That technology spillovers are geographically rather limited because of the importance of face-to face contacts was suggested by Audretsch and Feldman (1996) and Krugman (1991). Empirically, Paci and Pigliaru (2001) found that productivity growth of an EU region is highly correlated with those of its neighbouring regions when estimating spatial lag models. Paci and Usai (2000) detect R&D spillovers between Italian adjacent regions. Funke and Niebuhr (2000) investigate R&D spillovers with spatial interaction models for West German regions and find a significant contribution of R&D spillovers to productivity growth which decay fairly fast with distance. Bottazzi and Peri (1999) regard EU regions and similarly find that local clustering, i.e. spillovers, is important for R&D results, while R&D spillovers quickly fade with distance. Branstetter (2001) tests the likeliness of intranational versus international technology spillovers with U.S. firm level data. These studies concluded that technology spillovers are geographically rather limited and not effected on an international scale. What we are interested in this study is not the existence of regional spillovers but rather of international technology transfer, i.e. international spillovers.

behind the growth dependency may be manifold. Factors such as labour market linkages, investment flows, trade relations with inputs and final products, and technology spillovers underlie the observed growth spillovers. In terms of economic geography, these relationships are typical of agglomerations. Agglomerations are made up of a number of equally growing, dynamic regions which benefit from this kind of spillover.

Following the above discussion there are six hypotheses on regional growth determinants in the EU which will be examined in this Chapter:

1. The accumulation of capital enhances the growth of per capita income. We shall test for the impact of total capital stock accumulation including private and public capital.
2. A (change in) the high educational level in the population leads to higher growth. This will be tested by looking at medium and higher level educational attainment rates (upper secondary schooling, university) in the population.
3. A high innovation rate, as given by patent applications should also permit higher growth.
4. International technology transfer is important for growth and should be facilitated by a region's educational standards, domestic R&D volume and transmission through imports.
5. International technological catching-up is stimulated by a strong export engagement.
6. Regional growth is related to other regions' growth provided they are located within a certain distance.

Of course, we also expect that an increase in labour participation implies a higher income growth.

7.3 The Benchmark Model

Taking our hypotheses formulated in section 7.2, we now derive our empirical model which will then be tested in section 7.4. We have chosen a model in a growth accounting framework, an approach which we prefer to a convergence model since we are not interested in movements towards the steady state but in the issue of what determines growth. Empirical growth studies now increasingly return to this growth accounting approach (see e.g., Temple 1999). The model we propose will also be compatible with different growth theories so as to nest both a neoclassical model and an endogenous growth model as well as technological catching-up.

Our point of departure is the following augmented Cobb-Douglas production function

$$Y = AK^{\alpha} H^{\beta} L^{\gamma} \tag{7.1}$$

where Y = gross value added (GVA), K = stock of physical capital, H = stock of human capital and L = labour. We assume constant returns to scale, which implies that $\gamma = 1-\alpha-\beta$. As we are ultimately interested in explaining gross value added per capita (and not per employee), we rewrite Eq. (7.1) as

$$Y = AK^{\alpha} H^{\beta} POP^{\gamma} PART^{\gamma} \tag{7.2}$$

where POP = population and $PART$ = participation rate (L/POP). In intensive form and log-differences we have

$$\Delta \ln y_t = \Delta \ln A_t + \alpha \, \Delta \ln k_t + \beta \Delta \ln h_t + \gamma \Delta \ln PART_t \tag{7.3}$$

where $y = Y/POP$ = GVA per capita, A = total factor productivity (technological progress), $k = K/POP$ = physical capital per capita and $h = H/POP$ = human capital per capita.

As outlined above, technological progress (A) can be considered as a result of either innovation activity or technological catching-up. Additionally, some exogenous technical progress may also exist, which we will neglect for the moment. First, let us take the case that technological progress is driven by a region's own (domestic) R&D effort. A familiar formalization of this argument appears in the model by Romer (1990), where knowledge (A) is generated in the R&D sector according to the production function $\Delta \ln A_t = \phi H_A$, where ($H_A$) denotes the level of human capital employed in the R&D-sector. Due to the limited availability of regional research personnel data, we relate technological progress to the human capital share h_t (share of the population with higher level education).

$$\Delta \ln A_t = \varphi_1 \, h_t \tag{7.4a}$$

Benhabib and Spiegel (1994) similarly use an intensive measure of human capital (mean years of schooling), instead of the level of human capital (H), which appears in the Romer model. A further possibility to measure innovation activity would be to look at R&D expenditures. Again, limited data availability prevented us from using this indicator. Therefore, we shall test the relationship between technological progress and innovation by using educational attainment shares, i.e. the share of the population with a certain level of education.

Most empirical studies measure innovation by an input factor, which clearly gives a distorted picture. Innovations are not simply a function of inputs but subject to a number of further characteristics of the R&D process, such as the probabilistic nature of invention, the possibility that returns in the R&D sector are not constant (depletion of ideas), etc., as formalized for instance in the models of Grossman and Helpman (1991) or Aghion and Howitt (1992) and Segerstrom (1998).

As an alternative to the purely input oriented measure of domestic R&D efforts, one could focus more on the output side of the R&D sector. An eligible measure is the number of patent applications per employee (PAT), yielding the following specification for the growth rate of A

$$\Delta \ln A_t = \varphi_2 \, Pat_t \tag{7.4b}$$

Recently, a growing number of regional studies viewed innovation activity as a matter of patent applications (Paci and Usai 2000; Maurseth 2001) although there are also some problems with this kind of innovation measure. For instance, patents are typically registered at the site of the mother company and not at the place where research is actually done.

Besides domestic R&D effort, a second source of technological progress is technology transfer which permits catching-up against the technology leader. The economic rationale here is that a large technology gap is ceteris paribus associated with a large potential for catching-up and thus a higher growth rate of A. Ideally, the technology gap would be measured in terms of total factor productivity (A). However, as A is not directly observable, we use the labour productivity (y^*) as a proxy variable. The catching-up process then implies that the growth of A is described by the following relationship

$$\Delta \ln A_t = \eta_1 \left(\frac{y^*_{MAX,t} - y^*_t}{y^*_t} \right) = \eta_1 GAP_t \tag{7.5a}$$

where $y^*_{MAX,t}$ is the productivity level of the technology leader and the parameter η_1 indicates, how fast the technology gap (GAP) can be closed. A possible objection against Eq. (7.5a) is that catching-up does not come by itself. As Nelson and Phelps (1966) and others have shown, it requires human capital (H) for adapting and implementing new technologies. Originally, Nelson and Phelps argued that the catching-up parameter (here η_1) is a non decreasing function of H. Benhabib and Spiegel (1994) took up this argument and modelled the catching-up process as

$$\Delta \ln A_t = \eta_2 h_t \left(\frac{y^*_{MAX,t} - y^*_t}{y^*_t} \right) = \eta_2 h_t GAP_t \tag{7.5b}$$

Eq. (7.5b) states that holding human capital constant, economic units with a lower initial productivity level experience higher rates of productivity growth. Equivalently, for a given gap in technology, higher levels of human capital are associated with a faster catching-up process and thus a higher growth of A.

Trade has also played a prominent role in the discussion of the determinants of convergence (Ben-David and Kimhi 2000; Griffith et al. 2000). Thus, a straightforward modification of Eq. (7.5b) is to replace (H) by a measure of trade

$$\Delta \ln A_t = \eta_3 Tr_t \left(\frac{y^*_{MAX,t} - y^*_t}{y^*_t} \right) = \eta_3 Tr_t \; GAP_t \tag{7.5c}$$

Trade (Tr) could either stand for imports (m) or exports (x) in percent of GVA. The economic rationale, underlying Eq. (7.5c) is thus twofold. A higher import share would allow more spillovers from abroad (Coe and Helpman 1995), thereby fostering the catching-up process. Alternatively, a high export share may be associated with a pro-competitive effect of trade, which implies that the requirement to compete in the export market stimulates the catching-up against the technology (respectively market) leader.

Finally, a region's own R&D activities can also be thought as prerequisite for closing the technology gap. Drawing on Quah (1999) and Griffith et al. (2000) we can argue that an economy carrying out its own R&D possesses the specific technological know how that would make it more easy to adopt new technologies of the same class. This rationale can be expressed more formally as

$$\Delta \ln A_t = \eta_4 Pat_t \left(\frac{y^*_{MAX,t} - y^*_t}{y^*_t} \right) = \eta_4 Pat_t \; GAP_t \tag{7.5d}$$

Finally, we assume that regional growth is dependent on the growth performance of other regions and that this effect diminishes with increasing distance:[2]

$$\Delta \ln y_t = \mathbf{W} \Delta \ln y_t \tag{7.6}$$

where $\mathbf{W}$ is a N x N weighting matrix of the dimension equivalent to the number of regions N with elements w_{ij}

$$w_{ij} = d_{ij}^{-\delta} \quad \text{where i, j are regions} \tag{7.7}$$

$$\text{i} = 1, \ldots, \text{N}; \text{j} = 1, \ldots, \text{N}$$

$$\text{with } w_{ij} = 0 \text{ for i} = \text{j}$$

and d_{ij} indicates the distance between the centres of region i and region j. δ is the distance decay parameter. Thus, in this specification, the influence of other regions growth diminishes at a power function.[3] The spatial weighting matrix is row standardized, a transformation that has been argued as important for ensuring a well-defined statistical problem, but which complicates the economic interpretation (Bell and Bockstael 2000, p. 80). An important point is that the transformed weighting matrix is not symmetric any more, meaning that i may have a greater influence on j than vice visa. The implementation is that the growth spillover effect is conditional on the spatial context of each region (Fingleton 2001).

---

[2] Of course, the appropriateness of such a spatial lag model has to be tested econometrically, as will be done below. However, as this econometric specification is also motivated economically (see section 2), we set up this relation already here.

[3] An alternative specification for the weighting matrix is to have an exponential distance decay function $w_{ij} = \exp(-d_{ij})$. We use the power function for specifying the weighting matrix, following the argument of Fingleton (2001), who finds that both specifications yield similar results.

The spatial dependency of the growth rate represents the outcome of a number of interactions which occur between nearby regions: Trade relations, investment flows, labour market dependencies, technology spillovers, and other externalities which are not easily measured. Most of these interactions cannot be quantified due to a lack of data.

Inserting (7.4a) and the different specifications of (7.5) into Eq. (7.3) yields alternative testable specifications for the growth of GVA per capita (y), which are given by

$$\Delta \ln y = \lambda + \alpha \Delta \ln k_t + \beta \Delta \ln h_t + \gamma \Delta \ln PART_t + \mathbf{W} \Delta \ln y_t \quad (7.8)$$

$$\Delta \ln y_t = \lambda + \quad \ldots\ldots \quad + \varphi_1 h_t \quad (7.8a)$$

$$\Delta \ln y_t = \lambda + \quad \ldots\ldots \quad + \varphi_1 h_t + \eta_1 GAP_t \quad (7.8b)$$

$$\Delta \ln y_t = \lambda + \quad \ldots\ldots \quad + \varphi_1 h_t + \eta_2 h_t GAP_t \quad (7.8c)$$

$$\Delta \ln y_t = \lambda + \quad \ldots\ldots \quad + \varphi_1 h_t + \eta_3 Tr_t \; GAP_t \quad (7.8d)$$

$$\Delta \ln y_t = \lambda \quad \ldots\ldots \quad + \varphi_1 h_t + \eta_4 Pat_t GAP_t \quad (7.8e)$$

where growth is explained by factor accumulation ($\Delta \ln k$, $\Delta \ln h$) and growth of technology (A), which is in turn determined by an exogenous rate of technical progress λ, a region's own ability to innovate (here measured by Pat) and a catching-up component conditioned on either human capital, trade or patents. $\mathbf{W}\Delta \ln y_t$ represents the growth spillovers from other regions. The variable $\Delta \ln PART$ controls for changes in the participation rate, as (8) is specified in per capita terms. Alternatively we estimate Eqs. (7.8a) to (7.8e) with the variables Pat instead of h (see (7.4a), (7.4b)).

The models given by Eqs. (7.8) permits us to relate estimation results directly to economic theory and thus distinguishes what we are doing from the ad-hoc growth regressions which are often found in the literature. These have been criticized for good reason, notably by Durlauf who states that "it is no exaggeration to say that the theoretical and empirical growth literatures are evolving with little interaction." (Durlauf 2001, p. 65).

7.4 Estimation and Results

7.4.1 Data

We test Eqs. (7.8) for a cross-section of 159 European regions over the period 1993-1999. The sample covers practically all EU regions with the exception of the

UK which had to be excluded due to data problems. Further territories outside Europe and obvious outliers were excluded (French Overseas Departments, Ceuta y Melilla, Canaries, Azores, Ireland[4], Murcia, Uusimaa). Table 7.1 summarises the variables and their respective indicators as used in the regression.

Table 7.1 – Definition and sources of the variables

Variable	Definition	Units	Source
$\Delta \ln y_i$	average growth rate of GVA per capita	per cent p.a.	CamEcon
$\Delta \ln k_i$	average growth rate of physical capital (private and public) per capita	per cent p.a.	CamEcon, o.c.
$\boldsymbol{h_i}$	average attainment rate in the population – high education (tertiary)	per cent	Eurostat, o.c.
$\boldsymbol{h\_m_i}$	average attainment rate in the population – medium education (secondary)	per cent	Eurostat, o.c.
$\boldsymbol{Pat_i}$	average number of patent applications per employee	number per th. persons	Eurostat
$\boldsymbol{x_i\ (m_i)}$	avg.export (import) quota in percent of GVA	per cent	UN, o.c.
$\boldsymbol{GAP_i}$	$GAP = \left(\frac{y^*_{MAX,93} - y^*_{i,93}}{y^*_{i,93}} \right)$ = gap to technology leader[1)] in 1993 (initial level)	per cent	CamEcon, o.c.

i = 1, . . ., 159 = index of regions. – CamEcon ... Cambridge Econometrics, o.c. ... own calculations. – Averages refer to the period 1993-1999 or to a subperiod, as part of the data were not available for the full period (in particular for the variables *h*, *PAT*). – [1)] technology leader in 1993 was Ile de France. More details on the data used are given in the appendix.

The data set contains several additional variables which so far have not been investigated in other regional growth studies. First, our data set uses computed capital stock data, the variable required by the growth model, and not investment rates which is the common proxy. Second, we use educational attainment rates computed from labour force surveys as an indicator of human capital, distinguishing also between medium and higher level attainment (see also Chapters 1 and 13). Since data on educational attainment is not easily available for EU regions for several periods, other studies have used either enrolment rates (Tondl 2001) or could not investigate the impact of changes in educational attainment (Vanhoudt et al. 2000). For the reasons outlined in section 7.3 and due to better data coverage, we chose to take patent applications as an indicator for innovation and not research expenditures or research personnel.[5] Finally, we also

[4] While Ireland offers one of the most interesting growth stories to study separately, we have to exclude it from our sample as an exceptional outlier.

[5] Nevertheless, as shown by Bottazzi and Peri (1999) patents and R&D expenditures are highly correlated for EU regions.

could include a proxy for international trade of regions in our data set which allows us to analyse this novel aspect in regional growth research. A detailed description of our data set is given in the appendix.

7.4.2 Econometric Aspects

In econometric terms, Eq. (7.8) specifies a first order spatial lag model, which can be generalised to allow for spatial correlation also in the error term (Anselin 1988, Kelejian and Prucha 1998). In matrix notation we have

$$y = \rho_1 \mathbf{W} y + \mathbf{X}\beta + u \tag{7.9}$$

$$u = \rho_2 \mathbf{W} u + \varepsilon$$

$$\text{with } |\rho_1| < 1, \; |\rho_2| < 1$$

where y (N×1) is the dependent, endogenous variable and $\mathbf{X}$ (N×K) is the vector of K explanatory variables, $\mathbf{W}$ is a row-standardized the spatial weighting matrix with $w_{ij} = 0$ for $i=j$, and $w_{ij} = d^{-\delta}$ (see above). ρ_1 and ρ_2 are (scalar) spatial correlation parameters, and the vector β (Kx1) contains the parameters of the explanatory variables. u is a non-spherical disturbance term, while ε is assumed to be distributed NID $(0, \sigma^2)$.

In the estimating of Eq. (7.9), two econometric issues arise (Kelejian and Prucha 1998):

1. Since $\mathbf{W}y$ is correlated with the error term u, the OLS estimator of (7.9) is inconsistent. Therefore an instrumental variable estimator has to be used, where $\mathbf{X}$, $\mathbf{WX}$, $\mathbf{W}^2\mathbf{X}$ can be used as instruments for $\mathbf{W}y$.
2. However, this instrumental variable estimator does not consider the spatial correlation in the error term u $(\sim (0, \sigma^2\Omega)$, with $\Omega = (I-\rho_2\mathbf{W})^{-1}(I-\rho_2\mathbf{W}')^{-1})$ and thus yields inefficient estimates of the parameters and biased estimates of the standard errors. The appropriate answer is a feasible GLS estimation, which is equivalent to running unweighted least squares on the transformed series $y^* = (I-\rho_2\mathbf{W})y$, $\mathbf{W}y^* = (I-\rho_2\mathbf{W})\mathbf{W}y$, and $\mathbf{X}^* = (I-\rho_2\mathbf{W})\mathbf{X}$; for the estimation of ρ_2 we use the GM estimator suggested by Kelejian and Prucha (1999), which is based on the residuals u, obtained in step (1).[6] Again, (the same) instruments are used for $\mathbf{W}y^*$.

This two-stage procedure, developed by Kelejian and Prucha (1998), is referred to by the authors as feasible generalised spatial two stages least squares estimator (FGS2SLS).

---

6 As opposed to alternative estimators of the spatial autoregressive parameter (like the ML estimator recommended by Anselin (1988) for spatial models), whose asymptotic properties are not known, consistency was proven for this GM estimator by Kelejian and Prucha (1999).

As a general test for spatial dependence, we use the Moran´s I statistic. Further evidence on the exact form of the spatial dependence (substantive vs. nuisance dependence) is provided by two Lagrange-multiplier tests (LM_{ERR} and LM_{LAG}; see Anselin and Rey 1991, Anselin and Bera 1998, Anselin and Florax 1995).

7.4.3 Estimation Results

The models specified in Eqs. (7.8) are estimated with OLS respectively GS2SLS. The results are presented in Tables 7.2 and 7.3.

We start with a parsimonious model with factor accumulation, which is first estimated by OLS and then with GS2SLS with different distance decay parameters of the spatial weighting matrix (Table 7.2). The estimation indicates that per capita income growth is significantly positively related to the growth of both factors, physical capital and human capital. These results are in line with the recent findings of country level estimations (Bassannini et al. 2001; de la Fuente and Domenech 2000). Note however, that growth of EU regions is only sensitive to the attainment of a higher level of education. Differences in medium level educational attainment or changes of it do not explain growth (results not shown in Table 7.2). The size of the coefficients indicates an output elasticity of about 0.20 for physical capital and 0.045 for human capital. The estimates indicate that an increase in the labour participation rate is very important for growth. Interestingly, the size of the coefficients of the basic variables is practically unchanged under OLS and GS2SLS estimation.

Table 7.2 Estimation results of EU regional growth 1993-1999 without and with spatial effects

dependent variable: $\Delta \ln y$

	OLS		GS2SLS									
			distance decay parameter δ									
			1.25		1.5		**1.75**		2		2.5	
intercept	1.281***	(10.66)	0.234	(0.46)	0.493	(1.31)	**0.669****	(2.13)	0.781***	(2.81)	0.892***	(3.71)
$\Delta \ln k$	0.189***	(6.15)	0.187***	(5.58)	0.183***	(5.55)	**0.181*****	(5.56)	0.181***	(5.58)	0.179***	(5.59)
$\Delta \ln h$	0.045***	(2.78)	0.045***	(2.74)	0.045***	(2.80)	**0.046*****	(2.82)	0.046***	(2.83)	0.045***	(2.82)
$\Delta \ln part$	0.458***	(10.91)	0.335***	(5.68)	0.340***	(5.68)	**0.347*****	(5.73)	0.354***	(5.79)	0.363***	(5.90)
$W \Delta \ln y$			0.600**	(2.26)	0.441**	(2.21)	**0.342****	(2.09)	0.281*	(1.96)	0.221*	(1.81)
ρ_2			0.648		0.373		**0.231**		0.161		0.097	
z (I)			6.07		5.27		**4.65**		4.20		3.68	
LM_{ERR}			16.68***		15.56***		**13.66*****		12.03***		9.97***	
LM_{LAG}			15.24***		14.94***		**14.21*****		13.50***		12.43***	
R^2	0.65		0.68		0.68		**0.68**		0.67		0.67	
no.obs.	159		159		159		**159**		159		159	

Notes: t-statistics in parenthesis, GS2SLS is the general spatial two stage least squares procedure given in Kelejian/Prucha 1998, z (I) is the standardized z-value of Moran´s I, LM_{ERR} is the Lagrange multiplier test for residual spatial autocorrelation, LM_{LAG} is the Lagrange multiplier test for endogenous spatial lag, *** significant at the 1 % level, ** significant at the 5 % level, * significant at the 10 % level

In the spatial model, the coefficient of spatially lagged growth is positive, statistically significant and quite high. Obviously it has to vary with the distance decay parameter chosen for the spatial weighting matrix. If the spatial influence is permitted to decay almost linearly ($\delta = 1.25$), the coefficient of lagged growth is 0.60. If spatial dependence is considered to decay very fast ($\delta = 2.5$), the coefficient becomes much lower with 0.22. We select the model with a distance decay parameter of $\delta = 1.75$ as the basic model for the estimation with additional variables. This specification with $\delta = 1.75$ specifies a spatial lag model, as indicated by the LM_{LAG} statistic, and gives a large coefficient for the spatial lag with a quite high Moran's I statistic. Also, the value of the coefficient ρ_2 is of reasonable size (it increases towards 1 for smaller δ and decreases for larger δ).

Next we start from the spatial model with $\delta = 1.75$ and extend the estimated model to include innovation activity (Table 7.3, columns 2 and 3). If innovation is measured by patent applications, we get a positive but insignificant coefficient. Alternatively, if innovation is proxied by the share of higher level education, – note that both indicators of innovation show a high level of correlation –, the coefficient is positive and becomes statistically significant. We conclude that innovation is important for the growth of EU regions.

There is a possible explanation as to why the patents coefficient performs so poorly. Patents are assigned to the applicant's region. This is problematic with large companies since the head office will apply for the patent while the inventor may be located in another region. Thus the patent applications indicator as an innovation measure tends to underestimate the innovation activity of peripheral regions. Our result for patent applications is unsatisfactory considering the positive impact found in other studies. For example, Paci and Pigliaru (2001) investigate the period up to 1993 for EU 12 regions and verify a statistically significant and positive impact of patents on EU regional growth. However, they use patenting data corresponding to the residence of the innovator. This kind of data is not available for a larger sample and the more recent time period.

Table 7.3 Estimation results of EU regional growth 1993-1999 in a spatial lag model by GS2SLS
dependent variable Δ ln y , distance decay δ = 1.75

	(1)		(2)		(3)		(4)		(5)		(6)		(7)	
intercept	0.669**	(2 13)	0.859**	(2.36)	0.092	(0.34)	-0.147	(-0.56)	-0.122	(-0.46)	-0.132	(-0.49)	-0.031	(-0.13)
Δ ln k	0.181***	(5.56)	0.192***	(5.61)	0.152***	(5.37)	0.121***	(4.13)	0.127***	(4.40)	0.104***	(3.08)	0.169***	(5.69)
Δ ln h	0.046***	(2.82)	0.046***	(2.79)	0.064***	(4.06)	0.067***	(4.39)	0.063***	(4.12)	0.060***	(3.88)	0.065***	(4.13)
Δ ln part	0.347***	(5.73)	0.356***	(5.94)	0.314***	(5.55)	0.322***	(5.80)	0.333***	(5.89)	0.324***	(5.88)	0.291***	(5.22)
ln pat			0.025	(0.52)										
h					0.031***	(3.15)	0.035***	(3.70)	0.034***	(3.62)	0.024**	(2.49)	0.026***	(2.57)
x													0.009**	(2.40)
xGAP							0.797***	(2.38)						
mGAP									0.407**	(2.18)				
hGAP											1.259**	(1.97)		
W Δ ln y	0.342**	(2.09)	0.258	(1.51)	0.420***	(2.87)	0.472**	(3.36)	0.488***	(3.44)	0.578***	(3.57)	0.371**	(2.44)
ρ_2	0.231		0.306		-0.148		-0.262		-0.224		-0.359		0.016	
z(I)	4.65		4.48		2.11		2.00		2.29		2.25		2.95	
LM_{ERR}	13.66***		10.89***		1.65		1.17		1.68		1.65		3.57*	
LM_{LAG}	14.21***		11.78***		6.88***		7.42***		8.12***		8.99**		3.99**	
R^2	0.68		0.67		0.70		0.71		0.70		0.70		0.70	
no. obs.	159		159		159		159		159		159		159	

Notes: t-statistics in parenthesis, GS2SLS is the general spatial two stage least squares procedure given in Kelejian/Prucha 1998, z (I) is the standardized z-value of Moran´s I, LM_{ERR} is the Lagrange multiplier test for residual spatial autocorrelation, LM_{LAG} is the Lagrange multiplier test for endogenous spatial lag, *** significant at the 1 % level, ** significant at the 5 % level, * significant at the 10 % level

Next we start from our basic model including innovation, measured by the share of persons with higher level education and include the possibility of technological catching-up. We find that the technology gap indicating the possibility for technological catching-up does not enter significantly in the estimation (not shown in Table 7.3). However, if we condition the catching-up potential on human capital endowment, the coefficient becomes positive and its statistical significance improves (Table 7.3, column 6). This indicates that there is some technological catching-up taking place in EU regions which is based on the human capital endowment of a region. Regions can achieve productivity improvement through this channel, which becomes quite important for their income growth. The next estimates consider the possibility of closing the technology gap as a result of being engaged in foreign trade (Table 7.3, columns 4 and 5). The results show a positive and significant coefficient for the export weighted gap and thus indicate that exports can be considered as a stimulus to become more competitive and to close the technology gap. The coefficients of the import weighted technology GAP are also positive, but only half the size and statistically less significant. This may indicate that the possibility of technology transfer through imports is less important for regional growth. However, the lower significance of imports may also be due to measurement errors since our indicator for imports may be less reliable (see the appendix for a description of the methods used to proxy regional trade).

We also tested the possibility that technology transfer is facilitated by a region's own research efforts and can contribute to growth in this way. This hypothesis had to be rejected. A high domestic general R&D activity is no precondition to enhance technology transfer. If comparing that result with those of Griffith et al. (2000), R&D would have to take place in specific sectors to mediate a transfer of technologies in the same domain.

Note that the size of the technological catching-up coefficient cannot be compared with the other indicators since the gap variables constitute interacting variables.

The fact that we find evidence of international technology transfer via different channels complements the existing literature of regional technology spillovers, which considered it to be either geographically limited (Paci and Pigliaru 2001) or to take only place between similar regions (Maurseth 2001).

Finally, we also estimated the impact of foreign trade on growth treating exports and imports as a separate term not linked with technological catching-up. In both cases (results only shown for exports) trade enters positively in the growth estimation at the 5 per cent level. However, its impact is not of a significant size.

In general, we should emphasize that our basic model parameters are rather stable, except for that of capital accumulation which is significantly lower in the specifications with innovation and technological catching-up. We can interpret this result as an indication that the growth of EU regions is determined more by technological progress than by physical capital accumulation. Once we explicitly capture the different ways of making technological progress (domestic innovation, pro-competitive effect, technology transfer) their strong impact becomes evident and technological progress, previously probably related to capital goods and contained in its coefficient, is now explained by the appropriate variables. In addition, the relatively low coefficient of physical capital may be due to the fact that it is composed of private and public capital. It seems to be the case that only private investment, rather than both private and public investment, is linked to regional growth, as found by Vanhoudt et al. (2000). Therefore our coefficient estimate may be being depressed in value by the public capital component. The impact of human capital is quite important according to our results. Higher level educational attainment enters significantly into the growth regression in two ways. Changes in higher level education have a coefficient equal to 0.06, while the coefficient on the level of higher level educational attainment is 0.03. However, since the human capital variable does not include medium level education, at first sight our coefficients appear smaller than in other growth studies.

7.5 Conclusions

This Chapter analyzes the growth pattern of EU regions in the 1990s and tests the hypothesis that endogenous growth factors together with trade play a major role as causes of high growth performance. We tested this hypothesis performing cross-section growth regressions in a growth accounting framework and accounting for spatial dependencies. The estimated model includes the central production factors physical and human capital, and captures technological progress through a region's own innovation activity as well as international technology transfer conditioned on endowment factors and trade. Spatial spillover effects of regional growth are considered as well. Unfortunately, the lack of time series data does not permit us to estimate a panel data model. That analysis must be left to future research.

Our results indicate that capital accumulation as well as higher educational attainment levels and changes are important and robust factors for growth of EU regions. These results help to clarify the findings of previous regional growth literature that used simpler indicators and achieved mixed results.

The results also show that EU regions tend to cluster with respect to their growth performance. Regional growth is highly dependent on the growth performance of surrounding regions. Consequently, high growth regions are close to other high growth areas whereas low growth regions are close to other weakly growing areas.

Our analysis offers another important result: High growth EU regions generate technological progress through their own innovation activity and through international technology transfer. The latter can become a source of technological catching-up if a region possesses sufficient human capital in order to adopt available technologies. Finally, our results show that foreign trade plays an important role for technological catching-up of EU regions. Regions with high imports register a significant growth effect from technological catching-up which can be explained by import-related technology transfer. On the other hand, regions with high exports also show a significant growth effect of technological catching-up which can be interpreted as a pro-competitive effect of trade.

These findings suggest that trade openness as fostered by European Integration and international free trade agreements have become an important growth factor. Openness offers a growth potential both to rich and poor regions. Further, we can conclude that higher education standards and research activity are correct targets of EU policies as presently pursued with EU regional, technology and social policy since they are important for the growth of regions in advanced economies.

References

Abramovitz M (1986) Catching-up, forging ahead or falling behind. *Journal of Economic History*, 46: 385-406

Abramovitz M (1989) *Thinking about growth.* Cambridge University Press, Cambridge

Aghion P, Howitt P (1992) A model of growth through creative destruction. *Econometrica*, 60: 323-351

Anselin L (1988) *Spatial econometrics: Methods and models.* Dordrecht, Kluwer

Anselin L, Florax JGM (1995) *New directions in spatial econometrics.* Berlin et al., Springer

Anselin L, Bera AK (1998) Spatial dependence in linear regression models with an introduction to spatial econometrics. In: Giles D, Ullaah A (eds) *Handbook of applied economic statistics*, New York, Dekker, pp 237-289

Anselin L, Rey S (1991) Properties of tests for spatial dependence in linear regression models. *Geographic Analysis*, 23: 112-131

Aschauer D (1989) Is public expenditure productive? *Journal of Monetary Economics*, 23(2),pp. 177-200

Audretsch D, Feldman M (1996) R&D spillovers and the geography of innovation and production. *American Economic Review*, 86: 631-640

Bacchetta P (1994) Regional investment and growth in the European Community. *Department d'Economia i d'Historia Economica Barcelona discussion papers*, WP 257.94, Barcelona

Badinger H (2001) Growth effects of economic integration – The case of the EU member states (1950-2000). *IEF working paper,* no. 40, Research Institute for European Affairs, University of Economics and Business Administration, Vienna

Balassa B (1961) *The Theory of Economic Integration.* London: Allen & Unwin.

Balassa B (1978) Exports and economic growth: Further evidence, *Journal of Development Economics*, 5: 181-189

Baldwin R, Braconier H, Forslid, R (1999) Multinationals, endogenous growth and technology spillovers: Theory and evidence. *CEPR discussion paper*, no. 2155

Baldwin RE (1993) On the measurement of dynamic effects of integration. *Empirica*, 20(2): 129-144

Baldwin RE, Seghezza E (1996) Growth and European Integration: Towards an empirical assessment. *CEPR discussion paper*, no. 1393, London, Centre for Economic Policy Research

Barro R (1991) Economic growth in a cross-section of countries. *The Quarterly Journal of Economics*, 106(2): 407-443

Bassanini A, Scarpetta S. and Hemmings, P (2001) Economic Growth: The role of policies and institutions. Panel data evidence from OECD countries. *OECD working paper*, STI 2001/9

Bell KP, Bockstael N.E (2000) Applying the generalized-moments estimation approach to spatial problems involving microlevel data. *The Review of Economics and Statistics*, 82(1): 72-82

Ben-David D (1996) Trade and convergence among countries. *Journal of International Economics*, 40: 279-298

Ben-David D, Kimhi A (2000) Trade and the rate of income convergence. *NBER working paper,* no. 7642, Cambridge MA, National Bureau of Research.

Benhabib J, Spiegel M (1994) The role of human capital in economic development: Evidence from aggregate cross-country data. *Journal of Monetary Economics*, 34(2): 143-73

Bernard A, Jones C (1996) Technology and convergence. *The Economic Journal*, 106: 1037-1044

Bottazzi L, Peri G (1999), Innovation, demand and knowledge spillovers: Theory and evidence from European regions. *CEPR discussion paper*, no. 2279, London, Centre for Economic Policy Research

Branstetter LG (2001) Are knowledge spillovers international or intranational in scope? Microeconometric evidence from the U.S. and Japan. *Journal of International Economics*, 53: 53-79

Breuss F (1996) Austria's approach towards the EU. *IEF working paper*, no. 18, Research Institute for European Affairs, University of Economics and Business Administration, Vienna

Breuss F (1998) Die Nachhaltigkeit der volkswirtschaftlichen Entwicklung der MOEL. In: Fritz Breuss (ed) *Reifegrad der mittel-und osteuropäischen EU Beitrittswerber*, Vienna, WIFO-Austrian Institute for Economic Research

Coe DT, Helpman E (1995) International R&D Spillovers. *European Economic Review*, 39(5): 859-87

Cohen D, Soto M (2001) Growth and human capital: Good data, good results. *CEPR discussion paper*, no. 3025, London, Centre for Economic Policy Research

de la Fuente A (1998) Convergence Equations and Income Dynamics: The Sources of OECD Convergence, 1970-95. *CEPR discussion paper*, no. 1794, London, Centre for Economic Policy Research

de la Fuente A, Domenech R (2000) Human capital in growth regressions: How much difference does data quality make? *CEPR discussion paper*, no. 2466, London, Centre for Economic Policy Research

Dollar D (1992) Outward oriented developing countries really do grow more rapidly: Evidence from 95 LDCs, 1976-1985. *Economic Development and Cultural Change*, 40(3): 523-544

Durlauf SN (2001) Manifesto for a Growth Econometrics. *Journal of Econometric*s, 100: 65-69

Durlauf SN, Quah, D (1998) The new empirics of economic growth. *NBER working* paper, no. 6422, Cambridge, Ma: National Bureau of Economic Research

European Commission (2000) The EU Economy. 2000 Review. *European Economy*, no. 71, Office for Official Publications of the EC, Luxembourg

Fagerberg J (1987) A technology gap approach to why growth rates differ. *Research Policy*, 16(2-4): 87-99

Fagerberg J, Verspagen B (1996) Heading for divergence? Regional growth in Europe reconsidered. *Journal of Common Market Studies*, 34(3): 431-48

Fagerberg J, Verspagen B, Caniels M (1997) Technology, growth and unemployment across European regions. *Regional Studies*, 31: 457-466

Fingleton B (2001) Theoretical economic geography and spatial econometrics: dynamic perspectives. *Journal of Economic Geography*, 1: 201-225

Fujita M, Krugman P, Venables A (1999) *The spatial economy: Cities, regions, and international trade.* Cambridge MA, MIT Press

Funke M, Niebuhr A (2000) Spatial R&D spillovers and economic growth – Evidence from West Germany. *HWWA discussion paper*, no. 98, Hamburg, Hamburgisches Welt-Wirtschafts-Archiv

Gerschenkron A (1962) *Economic backwardness in historical perspective.* Cambridge, Harvard University Press

Griffith R, Redding St, Van Reenen J (2000) Mapping the two faces of R&D: Productivity growth in a panel of OECD industries. *CEPR discussion paper,* no. 2457, London, Centre for economic policy research

Griliches Z (1980) R&D and the Productivity Slowdown. *American Economic Review*, 70: 343-348

Grossman GM, Helpman E (1991) *Innovation and growth in the global economy.* Cambridge MA, MIT Press

Hallet M (1999) The regional impact of the single currency. In: Fischer M, Nijkamp P (eds) *Spatial dynamics of European Integration.* Heidelberg et al., Springer: pp 94-109

Islam N (1995) Growth empirics: A panel data approach. *Quarterly Journal of Economics*, 4: 1127-1170

Jaffe A, Trajtenberg M (1998) International Knowledge Flows: Evidence from patent citations. *NBER working paper*, no. 6507, Cambridge MA, National Bureau of Economic Research

Jones CI (1995b) R&D-Based Models of Economic Growth. *Journal of Political Economy*, 103(4): 759-784

Kaldor N (1961) Capital accumulation and economic growth. In: Lutz F (ed), *The theory of capital*, London, Macmillan

Kelejian H, Prucha I (1998) A generalized two-stage least squares procedure for estimating a spatial autoregressive model with autoregressive disturbances. *Journal of Real State Finance and Economics*, 17: 99-121

Kelejian H, Prucha I (1999) A generalized moments estimator for the autoregressive parameter in a spatial model. *International Economic Review*, 40(2): 509-533

Keller W (1998) Are international R&D spillovers trade-related? Analyzing spillovers among randomly matched trade partners. *European Economic Review*, 42: 1469-1481

Keuschnigg C, Kohler W (1996) Austria in the European Union. Dynamic Gains from integration and distributional implications. *Economic Policy*: 155-211

Kormendi R, Meguire Ph (1985) Macro-economic determinants of growth: Cross country evidence. *Journal of Monetary Economics*, 16: 141-163

Krueger A, Lindahl M (2000) Education for growth: Why and for Whom. *NBER working paper*, no. 7591

Krugman P (1991) *Economic geography and trade.* Cambridge MA, MIT Press.

Krugman P (1991) Increasing returns and economic geography. *Journal of Political Economy*, 99: 484-99

Krugman P, Venables A (1996) Integration, specialization and adjustment. *European Economic Review*, 40: 956-67

Lichtenberg F, van Pottelsberghe de la Potterie B (1996) Does foreign direct investment transfer technology across borders. *NBER working paper*, no. 5668, Cambridge MA, National Bureau of Economic Research

Lucas RE (1988) On the mechanics of economic development. *Journal of Monetary Economics*, 22(1): 3-42

Mankiw G, Romer D, Weil D (1992) A contribution to the empirics of economic growth. *Quarterly Journal of Economics*, 107: 407-437

Maurseth PB (2001) Convergence, geography and technology, Norwegian Institute of international affairs. *Structural Change and Economic* Dynamics, 12: 247-276

Nelson R, Phelps E (1966) Investment in humans, technological diffusion, and economic growth. *American Economic Review*, Papers and Proceedings, 61: 69-75

Niebuhr A (2001) Convergence and the effects of spatial interaction. *Jahrbuch für Regionalwissenschaft*, 21: 113-133

Paci R, Pigliaru F (2001) Technological diffusion, spatial spillovers and regional convergence in Europe. University of Cagliari, *CRENoS Working Paper,* no. 01/01

Paci R, Usai St (2000) Externalities, knowledge spillovers and the spatial distribution of innovation. *GeoJournal*, 49(4): 381-390

Parente St, Prescott E (1994) Barriers to technology adoption and development. *Journal of Political Economy*, 102(2): 298-321

Podrecca E, Carmeci G (2001) Education and growth. The causality links. University of Trieste, mimeo

Pritchett L (1996) Where has all the education gone? *World Bank working papers,* no. 1581

Quah D (1999) Cross-country growth comparison: Theory to empirics. *CEPR discussion paper*, no. 2294, London, Centre for economic policy research

Rey S, Montouri B (1999) US regional income convergence: A spatial econometric perspective. *Regional Studies*, vol. 33(2): 145-156

Rivera-Batiz LA, Romer PM (1991) Economic integration and endogenous growth. *Quarterly Journal of Economics*, 106: 531-555

Rivera-Batiz, L.A. and Xie, D (1993) Integration among unequals. *Regional Science and Urban Economics*, 23: 337-354

Romer P (1990) Endogenous technological change. *Journal of Political Economy*, 98(5), part II: 71-S102

Segerstrom PS (1998) Endogenous growth without scale effects. *American Economic Review*, 88: 1290-1310

Sinha D, Sinha T (1998) Cart before the horse? The saving-growth nexus in Mexico. *Economics Letters*, 61: 43-47

Temple J (1999) The New Growth Evidence. *Journal of Economic Literature*, 37(1): 112-156

Temple J (2001) Generalizations that aren't? Evidence on education and growth. *European Economic Review*, 45: 905-918

Tondl G (1999) The changing pattern of regional convergence in Europe. *Jahrbuch für Regionalwissenschaft*, 19(1): 1-33

Tondl G (2001) Convergence after divergence? Regional growth in Europe. Wien/New York, Springer

Vanhoudt P, Mathä T, and Smid B (2000) How productive are capital investments in Europe? *EIB papers*, 5(2)

Vayá E, López-Bazo E, Moreno R, Suriñach M (2002) Growth and externalities across economies. An empirical analysis using spatial econometrics. *Series on Spatial Econometrics and Statistics*, Berlin et al., Springer (forthcoming)

White H (1980) A Heteroscedasticity-Consistent Covariance Matrix Estimator and a Direct Test for Heteroscedasticity. *Econometrica*, 48: 817-838

Appendix

Data Description

Y_i = real gross value added in million Euro (1990 prices, 1990 exchange rate (ECU)), taken from the Cambridge Econometrics 2001 data set.

K_i = real capital stock (private and public capital) in million Euro (1990 prices, 1990 exchange rate (ECU)), calculated using a perpetual inventory method: $K_t = K_{t-1}\ (1-\delta) + I_t$, with I_t = investment (investment expenditure, taken from the Cambridge Econometrics 2001 data set), δ ... depreciation rate (assumption: 5%). Initial level K_0 calculated as $K_{1975} = I_{75}/(g_{I,75\text{-}01})$ where I_{75} = investment in 1975, $g_{I,75\text{-}01}$ = average growth rate p.a. of investment from 1975 to 2001 (see Griliches, 1980).

POP = population in 1000 persons, taken from the Cambridge Econometrics 2001 data set.

L_i = employment in 1000 persons, taken from the Cambridge Econometrics 2001 data set.

h_i = educational attainment in percent: share of population with attainment "high", calculated from Eurostat Labor Force Surveys.

h_m_i = educational attainment in percent: share of population with attainment "medium", calculated from Eurostat Labor Force Surveys.

patents = number of patent applications, taken from the New Cronos/Regio from Eurostat.

x_i (m_i) = exports (imports) of goods in percent of gross value added (actual prices); data is calculated from UN world trade statistics. Following the approach of Hallet (1999), exports and imports are approximated by assuming that the trade intensity in branch j (j=1, . . , 17; NACE-Clio, RR17) is the same for all regions of a country and allocating the national trade flow of a branch to a region according to the regional share of that branch in national GVA. Total trade of a region i, which belongs to country I, is thus given as sum of $\sum_j T_{Ij}\ w_{ij}$, where T_{Ij} is country I's trade in sector j and w_{ij} is the share of region i in country I's gross value added in sector j.

i = 1, . . ., 1 (region index): Our analysis covers 159 NUTS II regions for the 1990s. The sample does not include regions from the UK and Ireland. The UK was excluded due to data problems. Ireland was excluded because of its outstanding performance which would constitute a significant oulier for the regression analysis. Other regions located outside Europe as well as significant outliers (Murcia, Uusimaa) were excluded as well.

Derived variables: $y = 1000Y/POP$, $k = 1000K/POP$, $Pat = patents/L$, H=1000Lh.

8 The Technology Gap and European Regional Growth Dynamics

Lydia Greunz

Free University of Brussels, Belgium

8.1 Introduction

During recent decades, the process of convergence among European regions has slowed down considerably and according to a widening group of economists (Fagerberg and Verspagen 1996; Tondl 1997; Beine and Philippe 2000) it even ceased after 1980. However, important growth differentials still characterize the European landscape. This fact implies that some major assumptions underlying the neoclassical view of regional economic growth, such as constant returns to scale, perfect mobility of factors and immediate diffusion of knowledge may fail to match reality. At the other extreme, perfect immobility of knowledge as advocated by the circular and cumulative causation assumption (Myrdal 1957; Kaldor 1966), where regions build up their own technological capabilities and move along their own balanced growth path, can not reasonably be defended when one looks at the growing evidence for geographically and technologically mediated spillovers (Jaffe *et al.* 1993; Jaffe and Trajtenberg 1996; Maurseth and Verspagen 1999; Varga 1998; Caniëls 2000).

An alternative, somewhat intermediate way of investigating regional economic growth is the technology gap approach in the spirit of Abramovitz (1986). The main idea underlying the technology gap approach is that technological differences between countries open up the possibility that countries at a lower level of economic and technological development catch up by imitating the most productive technologies developed by the leading country (Fagerberg 1987). The basic foundation for such an "advantage in backwardness" is that imitating foreign best practices is supposed to be considerably less costly than innovation. However, technology is not assumed to be a free public good everywhere available to everybody. According to Abramovitz (1986, p. 390) "Countries that are technologically backward have a potentiality for generating growth more rapid than that of more advanced countries, provided their social capabilities are sufficiently developed to permit successful exploitation of technologies already employed by the technological leaders." To put it differently, a country's potential for rapid growth is strong not when it is backward without qualification, but rather when it is technologically backward but socially sufficiently advanced to understand, integrate and exploit new technologies developed by countries at the technological frontier. The concept of "social capability" refers to factors that reflect a country's technical competence, which is generally proxied by educational indicators such as number of years of education. If for Abramovitz (1986) a country's capacity to catch up depends on "social capability", it also requires a minimum level of "technological congruence". The latter refers to the fit between a country's factor supply and technological profile on the one hand, and the evolving technological frontier on the other.

Among the various studies investigating the technology gap approach (Fagerberg 1987, 1988a; Verspagen 1991; and others[1]), the relative recent studies of Amable (1993) and Fingleton (2000a) are especially interesting in the context of this Chapter since both of them adopt a simultaneous equation framework which is also our approach. Amable (1993) tested whether catching up took place during the period 1960-1985 by considering a cross section of 59 developed and developing countries and using variables expressed in terms of average values over the period. His system of equations can be summarized as follows: the labor productivity growth rate is assumed to depend on the initial technology gap, the rate of equipment investment, the percentage of the concerned age group engaged in primary education and the ratio of real government expenditures (minus defence and education) to real GDP. Equipment investment, innovative activity and the level of schooling are endogenised. The results suggest a general pattern of equilibrium divergence rather than convergence in productivity levels. While the reference to Amable (1993) is a fundamental one for our paper, it should be noted, as pointed out by Fingleton (2000a), that Amable's simultaneous system of equations fails to satisfy the order condition for identification.

Fingleton (2000a) takes up the idea investigated by Amable as well as his data set. With respect to Amable's study, the most important new hypothesis tested by Fingleton is that a country's level of technology influences the neighboring country's R&D activity and thus its productivity and output growth. The main channel of "technology transmission" between neighboring countries is assumed to be mutual trade, which is proxied by a geographic proximity matrix. The econometric findings support this hypothesis which, in a less restrictive manner, has already been put forward in the recent literature (Grossman and Helpman 1991, 1994; Coe and Helpman 1995; Verspagen 1997). Fingleton (2000a) comes to a conclusion similar to that of Amable (1993), finding that some countries will reach the frontier country's level of productivity in the steady state while others will never catch up.

While Amable (1993) and Fingleton (2000a) investigated the catching up hypothesis for a cross section sample of 59 developed and developing countries, the aim of this Chapter is to investigate whether catching up in the spirit of Abramovitz (1986) occurs on the European regional level. More precisely, an ex-

---

[1] For an overview of empirical work on catch up and growth, including its theoretical underpinnings, see Fagerberg (1994).

tended sample of 153 NUTS I and NUTS II regions[2] covering the entire European Union[3] is used in order to test the catching up hypothesis. The question of European regional convergence has largely been debated by means of absolute or conditional convergence models (Amstrong 1995; Capron 2000; Tondl 2001; and others[4]), but little attention has been devoted to the technology gap approach at the European regional level.

At this stage, one should consider the theoretical justification for stepping down the technological gap approach from the national to the regional level. Indeed, contrary to the neo-classical view where convergence is driven by labor and capital flows among regions until an equalization of factor returns in each region is achieved, for technology-gap theorists it is the technological difference between economies and their capacity to imitate which are the mechanisms for catching up. The major theoretical justification for applying the technology gap approach at the regional level relies on the increasingly commonly shared view that innovation is a systemic, territorial process. The degree of interdependence and proximity of economic, political, social and cultural regional factors determines the efficiency of the regional innovation system (De la Mothe and Paquet 1998). Currently, there is an abundant literature on this topic, which may be classified into different streams, namely the "innovation milieus" approach (Aydalot 1986, Camagni 1995; Camagni *et al.* 1999), the "industrial district" approach (Pyke and Sengenberger 1992), the "regional innovation systems" approach (Braczyk *et al.* 1998), the "learning regions" approach (Morgan 1997; Florida 1998), the "cluster" approach (Porter 2000) and the "social capital" approach (Putnam 1993; OECD 2001). All of them consider that regionally tied institutional and social variables are major determinants of the innovation dynamics and/or the capacity to imitate. If this is so, catching up or falling behind in the spirit of the technology gap approach should be examined within a regional framework rather than a national one, since the region's "social capability" and "technological congruence" which permits to catch up is largely dependent on regional specificity.

8.2 The Model and the Role of Variables

Similar to the approach of Fingleton (2000a), we adopt a "broad" Kaldorian view, but in this case our model specification is that (total) productivity growth depends on GDP per capita growth[5]. We refer to this as a dynamic quasi-Verdoorn Law be-

---

[2] Our sample is considerably larger than the ones applied in related studies investigating catching up among European Regions using R&D data (Clarysse and Muldur 2001; Fagerberg and Verspagen 1996; Cappelen *et al.* 1999; Fagerberg *et al.* 1997).

[3] However, the sample does not include Luxembourg and the Germany's new Länder since R&D data are not available for the latter.

[4] For an overview of recent studies of convergence at the European regional level, see Tondl (2001).

[5] This would be precisely equal to a Verdoorn Law (applied to the whole economy, not just manufacturing) if population growth was zero in all regions. Actually, during the period

low[6]. The dynamic quasi-Verdoorn Law together with innovative activity fed by "home" R&D investment and the "geographical neighborhood's innovations" influence the cumulative growth mechanism. The region's productivity growth, the structure of its productive system and its innovative activity are endogenised. Furthermore, the model accounts for both the region's endowment of "social capability" and its "technological congruence" with respect to leading regions. The structural system of equations can be formulated in the following way:

$$y = a_1 + a_2 G_{89} + a_3 p + a_4 S + \xi_1 \tag{8.1}$$

$$p = b_1 + b_2 y + b_3 I + \xi_2 \tag{8.2}$$

$$S = c_1 + c_2 PT + c_3 K + c_4 (1\text{-}TP) + c_5 GD + \xi_3 \tag{8.3}$$

$$PT = d_1 + d_2 BRD + d_3 PTGN + \xi_4 \tag{8.4}$$

$$BRD = e_1 + e_2 G_{89} + e_3 HQ + e_4 GRD + \xi_5 \tag{8.5}$$

where all variables[7] are expressed in terms of averages over the period 1989–1996 and where:

- y is the growth rate of real GDP per capita;
- G_{89} is the initial (1989) technology gap defined as $(1-(Y/Y^*))$ where Y/Y^* is the regional level of real GDP per capita relative to the one of the three best performing regions (Bruxelles, Hamburg, Île-de-France)[8];
- S represents the structure of the productive system and is proxied by the employment concentration in the service and industry sector;
- p is the productivity growth rate;

---

studied, the GDP growth rate was considerably higher than the rate of population growth and according to demographic forecasts (Eurostat 1999) it is reasonable to assume a similar situation in the future.

[6] The Verdoorn Law involves a relationship between manufacturing productivity growth (p) and manufacturing output growth (y) so that p=a+by where a is a constant autonomous growth of productivity. In our setting, this relation is modified in a twofold way. Firstly, it is applied to the entire economy rather than to the manufacturing sector. An overview of the theoretical underpinnings justifying the application of the Verdoorn Law to the aggregate economy is provided by Fingleton (2000). Secondly, we consider a relationship where GDP per capita growth rather than GDP growth influences productivity growth. The reason for this setting is that we wish to test for catching up in terms of per capita income and to derive the steady state, which is commonly expressed in terms of per capita values.

[7] For a detailed definition of variables see appendix 1.

[8] For reasons of non availability of more disaggregated data, the selection of the three best performing regions was based on the regional disaggregation level NUTS II. However, at a more disaggregated level, other regions such as München, Frankfurt and London perform better than Bruxelles and Hamburg.

I is manufacturing gross investment in tangible goods per manufacturing employee;
PT is the number of patents for 1000 inhabitants;
K is an indicator of physical infrastructure;
TP is Jaffe's measure of the technological proximity calculated on the basis of 118 different IPC classes;
GD is a measure of the geographical distance of a region with respect to the location of the technological leaders;
BRD is business sector R&D expenditure per capita;
PTGN is patenting activity of immediate geographical neighbors;
HQ represents high qualification;
GRD is government R&D expenditure;
ξ_i i=1..5 is the error term.

Eq. (8.1) states that the regional growth rate per capita depends on the initial technology gap (G_{89}), the growth rate of productivity (p) and the structure of the productive system (S). The technology gap variable (G_{89}) captures the idea that technological differences between regions allow the lagging ones to catch up with the leading ones by imitating their more productive technologies. A positive impact of this variable is therefore expected. As a measure of the technology gap, the level of real GDP per capita relative to that of the three best performing regions has been chosen[9]. The structure of the productive system (S) is considered to be one of the elements that allow a region to pick up the opportunity of backwardness. A high dependence on agriculture has been shown to be determinant of low regional growth (Fagerberg and Verspagen 1996), among other things because of low technological and market opportunities. Regarding the share of industry employment in total employment, the expectations are less clear-cut. Traditionally, manufacturing industry has been considered as the "engine of growth" (Kaldor 1966). However, in recent years technological progress is increasingly driven by the service sector (OECD 1996, 1997). To cover both arguments our variable, which should positively influence the GDP per capita growth, includes industry employment as well as service employment[10]. Finally, considering our dynamic quasi-Verdoorn relationship as simultaneous[11], productivity growth (p) determines, and is determined by, the growth of GDP per capita. This interrelationship

[9] In addition, several alternative measures of the technology gap were tested: technology gap based on the technological intensity (ratio of R&D expenditures on GDP), technology gap based on the technological base (R&D expenditure per capita) and technology gap based on the propensity to patent (the number of patents divided by R&D expenditure or alternatively per inhabitant). However since, in general, the time span is relatively long between the moment where R&D expenditures are realized and the moment where these expenditures have an impact on economic growth (European Commission 1997; Kline and Rosenberg 1986), the existence of alternative measures turned out not to be a significant issue in the context of our relatively short period of 8 years (1989-1996).

[10] For reasons of non availability of desaggragated data, the variable also includes public services which is however supposed to have little influence on technological progress.

[11] Simultaneity is confirmed by the Hausman specification test.

is picked up by Eqs. (8.1) and (8.2). Productivity growth (*p*) is also assumed to be positively influenced by manufacturing investment (*I*), a main channel of capital-embodied technology transfer.

In Eq. (8.3), the structure of the productive system (*S*) is supposed to be a positive function of the region's patenting activity (*PT*) as well as its communication infrastructure (*K*), but a negative function of its geographical distance (*GD*) and technological distance (*1-TP*) with respect to leading regions. As far as patenting activity is concerned (*PT*), it reflects the ability of a region to develop new technologies and to absorb technologies developed elsewhere. These factors largely determine the region's competitiveness, which in turn, permits to maintain or even to increase industrial and/or service employment. An efficient communication infrastructure (*K*) is a main component of a region's attractiveness and it is supposed to attract private investment (Quinet 1992). Moreover an efficient communication infrastructure should be considered as one of the channels of technology diffusion. Both mechanisms boost competitiveness upon which the employment level in industry and services has been shown to depend[12]. Technological proximity (*TP*) with respect to technological leaders aims at capturing "technological congruence", which is assumed to foster the lagging region's innovative activity and its potential for catching up. Indeed, the region's capacity to take productive advantage of another region's stock of knowledge is supposed to depend on the extent of technological similarity. The reason relies on the fact that each technology embodies a kind of unique language and concerns a specific set of applications which can only be shared when regions are technologically closely related. If technological proximity fosters the lagging region's innovative activity and therefore its competitiveness and capacity to maintain or increase the employment level in industry and services, the contrary is expected for technological distance (*1-TP*) with respect to technological leaders. Beside technological distance, it is also supposed that geographical distance (*GD*) with respect to technologically leading regions which may be assimilated to some kind of technological decision center, hampers the lagging region's potential for catching up.

Variables representing the level of innovative activity such as R&D expenditures should be viewed as prime candidates for capturing regional differences in "social capability". It has been shown that a certain level of R&D activity is a necessary condition for successful imitation (Fagerberg 1988a) in so far as it feeds the "knowledge base" upon which a country builds its technical competence (Dosi 1988). Having innovative activity confers an expertise that facilitates the assimilation of knowledge developed elsewhere, and thus magnifies R&D spillovers (OECD 1999). Eq. (8.4) takes into account this important catching up component. As far as business R&D expenditures (*BRD*) are concerned, this variable should positively influence the patenting activity (*PT*), a measure of R&D output. But not only does "home R&D expenditures" influence the "home patenting activity", neighborhood also matters since knowledge spills over, mostly between geo-

[12] While the density of the telecommunication network would have been a better candidate to capture the technology diffusion aspect, this variable is not available at the regional level.

graphical neighbors. The fact that geographical proximity matters for knowledge spillovers is largely supported by recent empirical literature. Knowledge spillovers occur between countries (Jaffe *et al.* 1993; Jaffe and Trajtenberg 1996; Sjöholm 1996, and others) but also between regions (Maurseth and Verspagen 1999; Bottazzi and Perri 1999; Caniëls 2000; Greunz 2001). Therefore, in our model, the patenting activity of immediate regional neighbors (*PTNG*) should positively influence home patenting activity (*PT*).

Finally, as stated by Eq. (8.5), the main determinants of business R&D investment (*BRD*) within a region are assumed to be the qualification level of the working age population (*HQ*), government R&D activities (*GRD*) as well as the extent of the initial technology gap (G_{89}). The qualification level of the working age population (*HQ*) is considered to be an important ingredient of "social capability". The technical competence of the labor force and the availability of various skills are essential to absorb technologies developed elsewhere as well as developing one's own. Contrary to university research, which is essentially oriented towards basic research without any particular commercial motivation[13], government R&D activities (*GRD*) mainly aim at supporting business enterprise needs in the field of R&D. For this reason, the former is expected to positively influence the latter. Finally, it seems reasonable to consider business R&D activities of a region as a negative function of its initial technology gap (G_{89}) since technological backwardness is associated with insufficiencies in terms of R&D efforts.

To summarize, our growth model contains five equations. All of them are over identified and the rank condition for identification is satisfied. Therefore unique estimates will be obtained for our coefficient for which we expect the following signs:

$a_2>0$	$a_3>0$	$a_4>0$		(see Eq. 8.1)
$b_2>0$	$b_3>0$			(see Eq. 8.2)
$c_2>0$	$c_3>0$	$c_4<0$	$c_5<0$	(see Eq. 8.3)
$d_2>0$	$d_3>0$			(see Eq. 8.4)
$e_2<0$	$e_3>0$	$e_4>0$		(see Eq. 8.5)

The growth rate of GDP per capita (y) is assumed to be a positive function of the initial technology gap (a_2), the rate of growth of productivity (a_3) and the structure of the productive system (a_4). The relation between the growth rate of GDP per capita (b_2) and productivity growth (p) is considered to be simultaneous. Moreover, the latter is assumed to depend on manufacturing investment (b_3). Patenting activity (c_2) and the endowment of an efficient communication infrastructure (c_3) should positively influence the structure of the productive system (S), while the contrary is assumed for geographical (c_5) and technological (c_4) distances with respect to technological leaders. "Home" R&D business expenditure

[13] Since it takes time for basic research to become commercially interesting for the private sector, in our model which covers only a period of 8 years, university R&D turns out not to significantly influence business R&D activity.

(d_2) should be a major determinant of "home" patenting activity (*PT*) but since knowledge spills over, the latter is assumed to depend also on the innovative activity of geographical neighbors (d_3). The main determinants of "home" business R&D expenditure are supposed to be the availability of highly qualified human capital (e_3), supportive government R&D activities (e_4) and the initial technology gap (e_2).

8.3 The Reduced Form and Dynamics

Coming back to the model described in the previous section it is useful to work out its reduced form and to derive the steady state levels of relative GDP per capita. In a reduced form, the per capita rate of growth is:

$$y = c + \Pi_0 G_{89} + \Pi_1 I + \Pi_2 HQ + \Pi_3 GRD + \Pi_4 PTGN + \Pi_5 K + \Pi_6 (1 - TP) + \Pi_7 GD \tag{8.6}$$

where

$$c = (a_1 + a_3 b_1 + a_4 c_1 + a_4 c_2 d_1 + a_4 c_2 d_2 e_1)(1 - a_3 b_2)^{-1} \tag{8.7}$$

$$\Pi_0 = (a_2 + a_4 c_2 d_2 e_2)(1 - a_3 b_2)^{-1} \tag{8.8}$$

$$\Pi_1 = (a_3 b_3)(1 - a_3 b_2)^{-1} \tag{8.9}$$

$$\Pi_2 = (a_4 c_2 d_2 e_3)(1 - a_3 b_2)^{-1} \tag{8.10}$$

$$\Pi_3 = (a_4 c_2 d_2 e_4)(1 - a_3 b_2)^{-1} \tag{8.11}$$

$$\Pi_4 = (a_4 c_2 d_3)(1 - a_3 b_2)^{-1} \tag{8.12}$$

$$\Pi_5 = (a_4 c_3)(1 - a_3 b_2)^{-1} \tag{8.13}$$

$$\Pi_6 = (a_4 c_4)(1 - a_3 b_2)^{-1} \tag{8.14}$$

$$\Pi_7 = (a_4 c_5)(1 - a_3 b_2)^{-1} \tag{8.15}$$

Given that the technology gap is defined as: $G=1-(Y/Y^*)=1-R$ where Y and Y^* are respectively the GDP per capita of the lagging region and of the three leading ones and since for the leading regions G and $(1-TP)$ must be zero, the rate of per capita GDP growth in leading regions y^* can be expressed in the following way:

$$y^* = c + \Pi_1 I^* + \Pi_2 HQ^* + \Pi_3 GRD^* + \Pi_4 PTGN^* + \Pi_5 K^* + \Pi_7 GD^* \tag{8.16}$$

where I^*, HQ^*, GRD^*, $PTGN^*$, K^* and GD^* are the values applying to the leading regions of the manufacturing investment in tangible goods per manufacturing employee, the highly qualified working age population, the government business R&D expenditures, the geographically mediated knowledge spillovers, the communication infrastructure and the geographical distance to technological leaders[14]. Subtracting Eq. (8.16) from Eq. (8.6) yields:

$$y - y^* = \Pi_0 G_{89} + \Pi_1(I - I^*) + \Pi_2(HQ - HQ^*) + \Pi_3(GRD - GRD^*) \\ + \Pi_4(PTGN - PNGT^*) + \Pi_5(K - K^*) + \Pi_6(1 - TP) + \Pi_7(GD - GD^*) \tag{8.17}$$

Since $R=Y/Y^*$, the rate of growth of the GDP per capita ratio is

$$\dot{R}/R = y - y^* \tag{8.18}$$

In replacing G_{89} by its alternative expression (1-R) in Eq. (8.17) and in taking account of the expression (8.18), it is possible to infer from Eq. (8.17) the dynamics of the ratio R:

$$\dot{R} = \alpha R - \beta R^2 \tag{8.19}$$

where

$$\alpha = \Pi_0 + \Pi_1(I - I^*) + \Pi_2(HQ - HQ^*) + \Pi_3(GRD - GRD^*) \\ + \Pi_4(PTGN - PNGT^*) + \Pi_5(K - K^*) + \Pi_6(1 - TP) + \Pi_7(GD - GD^*) \tag{8.20}$$

and

$$\beta = \Pi_0 \tag{8.21}$$

In steady state $\dot{R} = 0$ so that the equilibrium ratio of GDP per capita of lagging regions over GDP per capita of frontier regions, that we denote R_{SS}, can be expressed in the following way:

$$R_{SS} = \alpha / \beta \tag{8.22}$$

The above formulation indicates that the steady state ratio of GDP per capita of a region over GDP per capita of the frontier regions is a function of the region's current deficit/excess in terms of manufacturing investment, high qualification, government R&D expenditures, patenting activity of immediate neighbors, endowment of communication infrastructure as well as its technological and geographical distances with respect to technological leaders.

---

[14] For the calculation of the respective geographical distances among technological leaders see appendix 1.

8.4 Testing the Model

The model is tested on the European regional landscape over the period 1989-1996. A total of 153 European regions are covered by the sample which is composed of 120 NUTS II regions, 31 NUTS I regions and 2 NUTS 0 regions. For the latter - The Netherlands and Ireland - no regional data on R&D are available. This is also the case for Açores and Madeira. Belgian regions can only be covered at the NUTS I level. As far as Denmark is concerned which has no NUTS II regions, spatial aggregations of NUTS III regions have been performed on the basis of available R&D data[15]. Germany's new Länder as well as Luxembourg have been excluded given that R&D data do not exist for these regions. Finally, since the Greek regions Ionia Nisia and Voreio Aigai have no patenting activity, their technological profile could not be computed and we had to exclude them from the sample. This is also the case for the Spanish region Canarias, which has no geographical neighbor since this region was judged to be too far away from the European continent. Our final sample contains therefore 150 European regions including the three leading regions Bruxelles, Hamburg and Île-de-France.

Table 1 summarizes the main empirical findings. The system of simultaneous equations has been estimated by three different econometric methods: Full information maximum likelihood (FIML), two stage least squares (2SLS) and three stage least squares (3SLS). The Ordinary least squares method (OLS) is not appropriate in our case since some variables are endogenous.

Table 1 indicates, first of all, a high similarity between the estimates obtained by the three alternative methods. Statistically, the differences among the estimates are not significant. Secondly, one can observe that the coefficients show the expected signs such as described in the previous section. Thirdly, all coefficients are significant at the 5% level. Fourthly, even if a comparison with other studies seems difficult since neither the periods covered nor the model specifications are the same, it is interesting to note that our quasi-Verdoorn coefficient (b_2) is in line with the estimates of other studies. As stated by Fingleton (1999, p. 2, 2000b) the estimates of the Verdoorn coefficient turn quite consistently around 0.5 when the model is fitted to various data on manufacturing productivity growth and output growth. Finally, goodness of fit seems acceptable and does almost not vary among the estimation methods. Together these elements support our previous reflections about the functioning of the catching up mechanism, the role of variables and their interactions.

Since the estimated coefficients obtained by the three estimation methods are statistically similar, it is relatively difficult to choose a particular one. However, two arguments are in favor of the 2SLS method. Firstly, when deriving the steady state ratios of relative GDP per capita[16], and confronting them to the average value of these ratios prevailing during the period 1996-1998[17] one can verify whether or not the predicted steady states are coherent with respect to the 1996-1998 values.

---

[15] See appendix 2.

[16] This is explained in the next section.

[17] This period is not covered by our estimation.

To put it differently, when we observe an exclusively positively or negatively sloped growth path between the initial (1989) ratio of relative GDP per capita, the average value for this ratio over the period 1996-1998 and the steady state ratio, we conclude that the steady state outcome is in line with observed reality. This kind of "reality test" is slightly in favor of the 2SLS estimates for which about 95% of regions evolve in line with the predicted steady state ratio of relative GDP per capita.

Table 8.1 Estimates of the simultanous system of equations

	FIML		2SLS		3SLS	
	est.	t-stat.	est.	t-stat.	est.	t-stat.
GDP per capita growth (y)						
Constant	-0.26	-4.00	-0.29	-4.29	-0.21	-3.77
Initial technology gap *(G_{89})*	0.07	3.61	0.08	3.42	0.05	2.44
Productivity growth *(p)*	0.50	3.16	0.41	2.63	0.55	4.68
Structure of prod. system *(S)*	0.26	3.97	0.28	4.35	0.22	3.95
R^2_a	0.53		0.52		0.59	
SE of regression	0.02		0.02		0.02	
Productivity growth (p)						
Constant	-0.02	-3.31	-0.02	-5.21	-0.02	-4.90
GDP per capita growth *(y)*	0.60	5.71	0.55	7.33	0.60	8.35
Manufacturing investment *(I)*	0.06	3.96	0.07	5.68	0.06	5.35
R^2_a	0.68		0.68		0.68	
(SE of regression)	0.02		0.02		0.02	
Structure of productive system (S)						
Constant	0.92	32.56	0.93	32.45	0.94	44.75
Home patenting activity *(PT)*	0.47	3.28	0.39	3.69	0.39	4.56
Infrastructure endowment *(K)*	0.57	2.87	0.62	2.18	0.61	3.29
Technological distance *(1-TP)*	-0.07	-2.56	-0.10	-2.80	-0.08	-3.29
Geographical distance *(GD)*	-0.06	-4.10	-0.05	-3.27	-0.06	-5.86
R^2_a	0.47		0.50		0.49	
(SE of regression)	0.06		0.06		0.06	
"Home" patenting activity (PT)						
Constant	-0.01	-0.52	-0.01	-1.59	-0.01	-1.65
Business R&D exp. *(BRD)*	0.35	9.46	0.34	10.14	0.35	10.86
Patenting of neighbors *(PTGN)*	0.47	5.81	0.49	8.05	0.47	8.05
R^2_a	0.81		0.81		0.81	
(SE of regression)	0.04		0.03		0.03	
Business R&D expenditures (BRD)						
constant	0.26	5.97	0.23	5.38	0.26	6.52
Initial technology gap *(G_{89})*	0.62	-7.57	-0.57	-8.29	-0.62	-9.63
High qualification *(HQ)*	0.66	4.18	0.71	5.27	0.69	5.64
Government R&D exp. *(GRD)*	0.70	3.11	0.69	3.35	0.70	3.82
R^2_a	0.57		0.57		0.56	
(SE of regression)	0.10		0.10		0.10	

Note: number of observations: 147 regions.

For the second argument in favor of 2SLS we quote Klein (1974, p. 150) and Gujarati (1995, p. 679) who argues that "single equation methods, in the context of a simultaneous system, may be less sensitive to specification errors in the sense that those parts of the system that are correctly specified may not be affected appreciably by errors in specification in another part". In order to avoid that an eventual specification error in one equation being transmitted to the rest of the system it is more cautious to adopt 2SLS.

Having decided on the 2SLS estimates, it is useful to investigate the degree of stability of our estimated coefficients. The test procedure, in order to determine stability or instability, consisted to eliminate in turn each individual region from the sample and to perform 147 different regressions. Table 2 presents the result of such a parameter stability test and indicates that the full sample estimates are not significantly influenced by one individual region. The coefficients remain remarkably stable and the standard deviations are low. Furthermore we have investigated parameter stability by means of the cumulative sum of the recursive residuals test (CUSUM). Coefficient instability is rejected at the 5% level.

Table 8.2 Distribution of 2SLS coefficients

Equation	Distribution	Dependent variables			
GDP per capita growth (y)		G_{89}	*p*	*S*	
	minimum	0.073	0.371	0.257	
	maximum	0.095	0.475	0.321	
	mean	0.082	0.414	0.283	
	st. deviation	0.003	0.015	0.007	
Productivity growth (p)		*y*	*I*		
	minimum	0.506	0.063		
	maximum	0.567	0.074		
	mean	0.547	0.067		
	st. deviation	0.007	0.001		
Structure of productive system (S)		*PT*	*K*	*(1-TP)*	*GD*
	minimum	0.333	0.515	-0.126	-0.057
	maximum	0.425	0.777	-0.086	-0.042
	mean	0.393	0.620	-0.102	-0.051
	st. deviation	0.010	0.026	0.004	0.002
"Home" patenting activity (PT)		*BRD*	*PTGN*		
	minimum	0.311	0.458		
	maximum	0.349	0.519		
	mean	0.336	0.491		
	st. deviation	0.004	0.007		
Business R&D expenditures (BRD)		G_{89}	*HQ*	*GRD*	
	minimum	-0.612	0.639	0.581	
	maximum	-0.527	0.753	0.814	
	mean	-0.573	0.712	0.688	
	st. deviation	0.008	0.013	0.021	

8.5 Europe at Different Speeds?

The preceding estimation results can be used to get deeper insight into the respective growth dynamics followed by each individual region. In order to determine whether a region is currently engaged in a catching up process, or, on the contrary, in a process leading to relative underdevelopment, it is useful to evaluate for each individual region the steady state ratio of relative GDP per capita. This is possible by applying the theoretical considerations developed in section 8.3. The estimated 2SLS coefficients of the system of equations reported in Table 1 and Eq. (8.21), although valid on average over the period 1989–1996, allow one to compute the value of β which turns out to be 0.08. Eq. (8.20) permits one to evaluate the value of α for each individual region and the equilibrium ratio of relative GDP per capita (R_{SS}) is given by Eq. (8.22).

Since β turns out to be positive, three cases can be distinguished:

$$\text{case 1: } \alpha > 0,\ \beta > 0 \text{ and } \alpha \geq \beta,\ R_{SS} \geq 1 \tag{8.23}$$

$$\text{case 2: } \alpha > 0,\ \beta > 0 \text{ and } \alpha < \beta,\ 0 < R_{SS} < 1 \tag{8.24}$$

$$\text{case 3: } \alpha \leq 0,\ \beta > 0,\ R_{SS} \leq 0 \tag{8.25}$$

In case 1, no matter what the region's initial position is, it will converge to the three leading ones. This means, that taking into account the extent of the initial technology gap, "social capability" and "technological congruence" are strong enough to reach a steady state similar to, or even higher than, that of the technological leaders.

In the second case, there is an equilibrium level of the technology gap $(1 - R_{SS})$, so that the lagging region never converges to the three best performing European regions even if its level of development is initially close to that one of leading regions. In this case, catching up as well as falling back to a lower steady state ratio of relative GDP per capita can occur.

In case 3, "social capability" and/or "technological congruence" is too low to enable positive growth dynamics. Without an adequate set of policy measures, a reversal of the relative impoverishment process in which the region is currently engaged appears highly unlikely.

At this stage, it is worth remembering several facts that underlay our classification. Firstly, given that the technology gap is calculated with respect to the three best performing regions, the second case does not imply that convergence to the European level is an inaccessible objective. Secondly, the above classification and especially the levels of the steady state ratios of relative GDP per capita are only valid if no exogenous shock such as a new policy measure occurs. Thirdly, from the definition of case one, Eq. (8.23), it is straightforward that there is no constraint in our theoretical model that prevents current technology leaders to be overtaken. However, since such a fundamental change seems relatively unrealistic and probably requires a different modeling approach, we restricted the steady state ratio of relative GDP per capita to be equal to one for regions where this outcome occurred[18]. Similarly, as indicated by Eq. (8.25), there is nothing in our model that prevents a negative value for R_{SS}. Since only non-negative values make sense from an economic viewpoint, in Fig. 8.1 negative ratios of relative GDP per capita which occurred in some "declining regions" were restricted to equal zero[19].

The rows of Fig. 8.1 indicate which case, in the steady state, each region belongs to. Only 7 % of European regions are expected to converge to the development level of leading regions and the great majority of European regions, namely 82 %, move towards their own steady state. Since important differences characterize the regions belonging to case 2, it has been split up into three different groups according to the steady state ratio of relative GDP per capita. Under this more homogeneous representation, about 41% of European regions catch up to a steady state ratio of relative GDP per capita situated between 0.9 and 0.7 and about 25% evolve towards a steady state ratio of relative GDP per capita half of the one of technological leaders. Finally, for 26 % of European regions, relative backwardness seems to be a recurrent issue of their growth mechanism since their steady state ratios of relative GDP per capita are low, especially for the 11% of regions belonging to case 3.

In order to get a more detailed insight into the different regional growth paths, the columns of Fig. 8.1 indicate the ratios of relative GDP per capita prevailing in 1989. The juxtaposition of the initial situation and the steady state enables to establish a typology of regions based on their respective growth dynamics. Basically three types of regions can be distinguished.

The first type of region is moving towards a steady state ratio of relative GDP per capita that is higher than it is currently. These regions are situated above the diagonal cells of Fig. 8.1 and are darkly shaded. Converging and strongly converging regions exhibit an extremely high "social capability" especially in terms of qualification and innovative activity. Furthermore, their technological profiles are very close to the one of technological leaders and the low geographical distance to the latter favors knowledge spillovers. One can observe that this case only concerns German regions. Catching up regions are characterized by an above of average endowment of "social capability" and a relatively favorable position in terms of "technological congruence" with respect to leading regions. This is the case for most of Austrian, German, Danish and Belgian regions, the Netherlands but also for about half of the French regions and one third of the British regions. A relatively remarkable position is observed for Ireland and Burgenland. Although, for the latter, the initial ratio of relative GDP per capita is extremely low, they are the only "poor" regions that are able to catch up. This process could be accelerated in strengthening their respective R&D activities.

[18] This is the case for the German region DE12-Karlsruhe.

[19] This is the case for the Greek regions GR43-Kriti, GR13-Dytiki Makedonia, GR42-Notio Aigaio GR13-Dytiki Makedonia, GR11-Anatoliki Makedonia, GR21-Ipeiros, GR23-Dytiki Ellada, GR25-Peloponnisos, GR14-Thessalia and the Portuguese regions PT14-Alentejo, PT15-Algarve and PT11-Norte.

Steady state ratio of relative GDP per capita	Initial ratio of relative GDP per capita (1989): R_i=1	$R_i \in$]0.95 ; 0.65]	$R_i \in$]0.65 ; 0.45]	$R_i \in$]0.45 ; 0.05]	R_i=0
Case 1, R_{SS} = 1	Three leading regions	DE12-Karlsruhe DE13-Freiburg DE25-Mittelfranken DE3-Berlin DEA2-Köln DE14-Tübingen DE71-Darmstadt *Converging regions*	DEC-Saarland DE91-Braunschweig DE72-Gießen *Strongly converging regions*		
Case 2, $R_{SS} \in$]0.95 ; 0.65]		DE11-Stuttgart FR42-Alsace DK-1-Kobenhaven DEB3-Rheinhessen-Pfalz DEA1-Düsseldorf DE92-Hannover AT32-Salzburg SE05-Västsverige FR23-Haute-Normandie DE21-Oberbayern DE5-Bremen SE01-Stockholm UK5-South East AT13-Wien FI11-Uusimaa	DE22-Niederbayern DE73-Kassel FR82-Provence Alpes-Côte d'Azur NL-Netherlands DK-2 DEB1-Koblenz DE27-Schwaben DEA5-Arnsberg DE26-Unterfranken DE23-Oberpfalz DEB2-Trier DEA4-Detmold AT33-Tirol DE24-Oberfranken DEF-Schleswig-Holstein BE2-Vlaams Gewest DEA3-Münster DK-5-Arhus amt DK-3-Fyns amt BE3-Reg. Wallonne FR26-Bourgogne FR21-Champagne-Ardenne FR71-Rhône-Alpes AT34-Vorarlberg FR62-Midi-Pyrénées FR41-Lorraine FR3-Nord-Pas-de-Calais SE02-Östra Mellansverige DE94-Weser-Ems FR24-Centre DK-6-Nordjylland AT12-Nieder-öst AT31-Oberöst. FR81-Languedoc-Roussillon DK-4 UK6-South West FR43-Franche-Comté FR63-Limousin UK4-East Anglia AT22-Steiermark AT21-Kärnten UK3-East Midlands UK9-Pays de Galles DE93-Lüneburg SE04-Sydsverige FR61-Aquitaine *Catching up regions*		
Case 2, $R_{SS} \in$]0.65 ; 0.45]		IT2-Lombardia SE03-Smland med Öarna IT11-Piemonte IT12-Valle d'Aosta IT31-Trentino-Alto Adige IT6-Lazio IT51-Toscana IT4-Emilia-Romagna IT13-Liguria SE07-Mellersta Norrland IT32-Veneto IT33-Friuli-Venezia Giulia SE08-Övre Norrland *Strongly declining regions*	UKA-Ecosse SE06-Norra MellansVerige FI15-Pohjois-Suomi UK1-North IT71-Abruzzo ES21-Pais Vasco ES3-Comunidad de Madrid ES22-Comunidad Foral de Navarra IT52-Umbria ES23-La Rioja ES51-Cataluña UK8-North West FR22-Picardie FR72-Auvergne FR25-Basse-Normandie UK7-West Midlands UK2-Yorkshire Humberside FR51-Pays de la Loire FR53-Poitou-Charentes FR52-Bretagne FI12-Etelä-Suomi UKB-Northern Ireland	IE-Ireland AT11-Burgenland *Catching up regions*	
$R_{SS} \in$]0.45 ; 0.05]		FI2-Aland IT53-Marche	FI13-Itä-Suomi ES24-Aragón IT8-Campania IT72-Molise FI14-Väli-Suomi ES13-Cantabria ES52-Comunidad Valenciana ES12-Principado Asturias FR83-Corse PT13-Lisboa e Vale do Tejo IT91-Puglia ITB-Sardegna GR3-Attiki	IT92-Basilicata ES41-Castilla y León ES42-Castilla-la Mancha ES11-Galicia ES61-Andalucia ITA-Sicilia ES62-Murcia IT93-Calabria	
Case 3, R_{SS} = 0			ES53-Baleares GR43-Kriti GR13-Dytiki Makedonia GR42-Notio Aigaio GR13-Dytiki Makedonia *Declining regions*	GR12-Kentriki Makedonia PT12-Centro ES43-Extremadura GR11-Anatoliki Makedonia PT11-Norte GR21-Ipeiros GR23-Dytiki Ellada GR25-Peloponnisos PT14-Alentejo PT15-Algarve GR14-Thessalia *Declining regions*	

Fig. 8.1 Regional typology

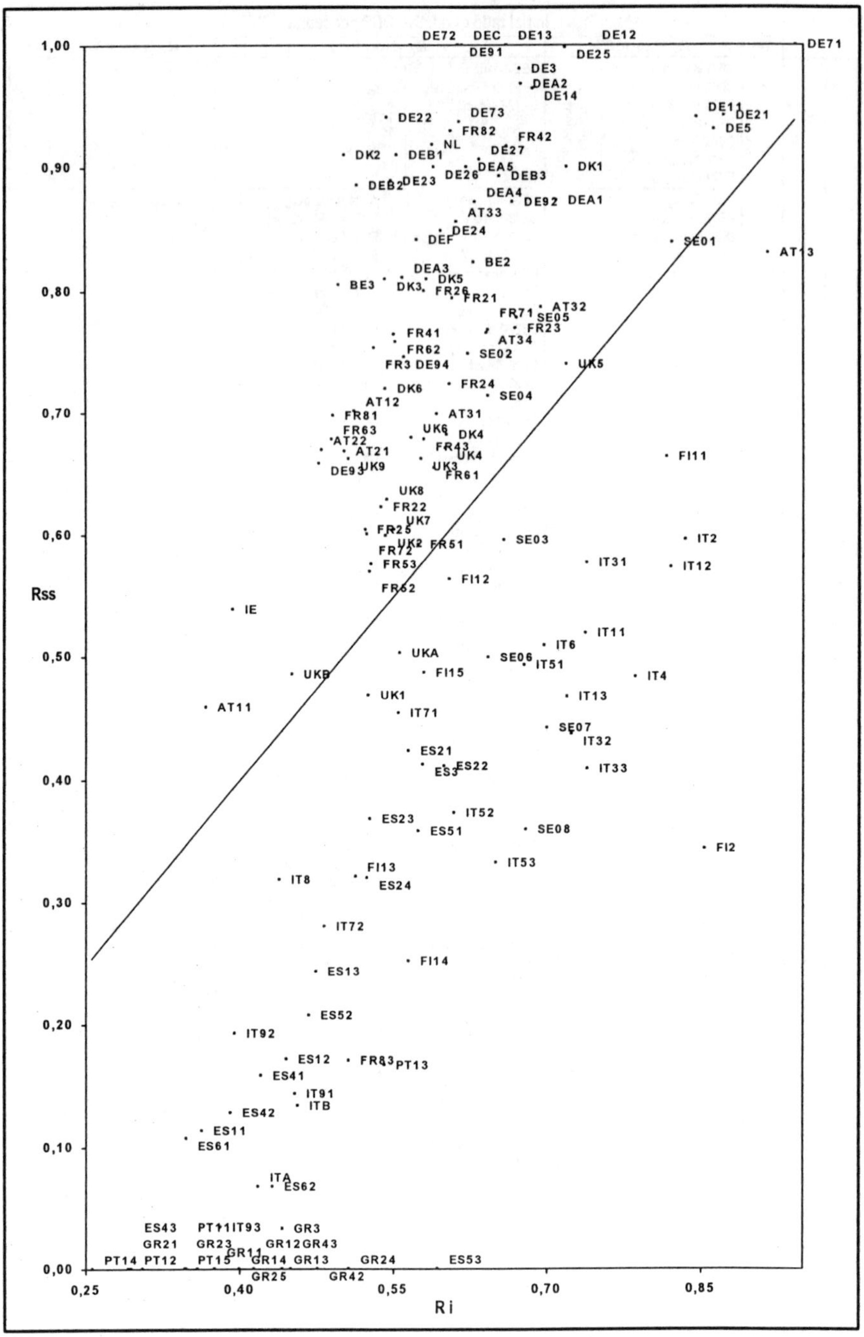

Fig. 8.2 Illustration of the different regional growth dynamics

The second type of region is characterized by a lower ratio of relative GDP per capita than is presently the case. These regions are situated below the diagonal cells of Fig. 8.1. A striking observation for this type of region is that most of them are currently "poor". Put it differently, "poor" regions are expected to get even "poorer". However, an exception to this prognosis is the case of the Italian regions. They are mostly relatively "rich", but are expected to decline strongly in terms of their respective ratios of relative GDP per capita. However, we think that this prediction should be tempered because of the "industrial district structure" of some Italian regions' productive system which is largely based on informal networks, co-operation and mutual trust (Pyke and Sengenberger 1992), as can be observed for Lombardia, Emilia-Romagna, Veneto and Toscana. In these regions endogenous factors considerably speed up technology diffusion, innovation adoption and knowledge creation and this may be insufficiently reflected by official data on R&D expenditure. As far as Greek regions, Portuguese regions and most of the Spanish regions are concerned, they evolve towards extremely low ratios of relative GDP per capita. An inversion of their negative growth dynamics appears difficult in view of their low endowment of "social capability" and the geographical as well as the technological distances that separates them from leading regions. This situation raises some important questions regarding the efficiency of their "Objective 1" development programs which are supposed to improve growth potential.

Finally, *the third case* concerns regions which are located on the lightly shaded diagonal cells of Fig. 8.1 and which remain in the same group throughout the period of transition. However, this does not mean that within their respective group catching up or falling behind cannot occur.

Since Fig. 8.1 does not report the values of the respective ratios of relative GDP per capita, it is useful to consider Fig. 8.2. The vertical axis of this figure reports the steady state values while the horizontal one indicates the initial situation. One can easily see that above the diagonal, regions are evolving towards a ratio of relative GDP per capita higher than their current one, whereas the contrary prevails for regions situated below. The orthogonal distance of a given point, i. e., a given region, to the diagonal may be interpreted as the importance or the strength of the growth dynamics.

One can observe form Fig. 8.2, that strongly positive growth dynamics are expected for the German regions Gießen (DE72), Niederbayern (DE22), and Trier (DEB2). It is appealing to notice that all German regions, without exception, follow a positive growth path, which, for the majority of them, is very important. This is also the case for Danish and Belgian regions. All of them are expected to catch up. Vestsjællans Amt and Storstøms Amt (DK2) display the strongest growth dynamics. As far as French regions are concerned, only Corse (FR83) follows a negative growth path. The others catch up and the highest growth performances are expected for Alsace (FR42) and Province Alpes-Côte d'Azur (FR82). A diametrically opposite expectation prevails for most regions belonging to the European periphery, especially the Southern one, but also for North Italian regions. Without exception, there is no Greek, no Portuguese, no Spanish, no Italian and no Finnish region that follows a positive growth path. The most alarming cases are the Greek and the Portuguese regions for which important underdevelopment can not be excluded.

Europe at different speeds? According to our analysis, the answer to this question is clearly affirmative. On average, "poor regions" get even "poorer" since, despite important policy efforts, their "social capability" remains underdeveloped. On the opposite, regions with initially "intermediate" technology gaps appear to be able to realize their catch up opportunity. The majority of them evolve towards high steady state ratios of relative GDP per capita. These observations open up some fundamental questions. Firstly, the numerous development programs undertaken in lagging regions, are they efficient? A real improvement of "social capability" in these regions probably cannot be achieved by adopting an ambivalent approach, but requires systemic policy intervention. Secondly, regarding the diverging growth paths followed by the European regions, what are the implications for the socio-economic cohesion within Europe? And, thirdly, since European regions are clearly evolving according to different speeds and are not likely to converge, is European enlargement a reasonable proposition from an economic perspective?

8.6 Conclusion

Over the last two decades, many studies have investigated the catching up hypothesis within the European regional landscape. These studies confirm the slow down of the convergence process after 1980. While most empirical research in this field is based on single absolute or conditional equation methods, our approach draws on the technology gap literature where "social capability" and "technological congruence" are considered as prerequisites for catching up. Compared to the traditional convergence approaches, the advantage of our framework relies on the explicit causal structure, which provides interesting insights not available through the estimation of reduced forms, and in the incorporation of "social capability" and "technological congruence".

Estimated on the basis of a total of 150 European NUTS I and NUTS II regions over the period 1989-1996, the simultaneous equation model of cumulative causation developed in this Chapter leads to the following comments. The average growth rate of real GDP per capita over the period 1989-1996 is shown to depend positively on the initial technology gap of a region with respect to the technology leaders, the average growth rate of productivity and the structure of the productive system. The relation between GDP per capita growth and productivity growth is treated as a simultaneous one and produces reasonable coefficient estimates that are in line with other studies. Patenting activity and an efficient communication infrastructure positively influence the structure of the productive system while increasing geographical and the technological distances from the technological leaders has a negative impact. R&D business investment, which is a positive function of government R&D expenses and of highly qualified human capital but a negative function of the initial technology gap, determines the patenting activity of a region which is also influenced by knowledge spillovers from geographical neighbors.

The estimated coefficients have been used to evaluate the different growth paths likely to be followed by the European regions. Considering Bruxelles, Hamburg and Île-de-France as references, we estimate that about 7% of European regions will converge to the level of development achieved by these three best performing European regions. Approximately 82% of European regions move towards their own steady states, which are relatively high for 41% of them. For 26% of European regions, relative backwardness seems to be the consequence of their inability to raise sufficient "social capability". These findings rely on the assumption that regional development policies do not change over time.

An obvious dichotomy exists between initially "poor" regions and regions with an initial "intermediate" technology gap. For the former, we expect an impoverishment process, but for the latter group there is a reasonable prospect that they will be able to realize their potential for catching up and for some of them the steady state ratio of relative GDP per capita will approach that of the technological leaders. European regions clearly evolve at different speeds. This fact could have important implications for European socio-economic cohesion and the economic sustainability of European enlargement.

Acknowledgements

The author wishes to thank Henri Capron and Bernard Fingleton for their useful comments and suggestions.

References

Abramovitz M. (1986) Catching up, forging ahead, and falling behind. *Journal of Economic History* 46: 385-406.

Abramovitz M. (1989) *Thinking about growth*. Cambridge University Press.

Amable B. (1993) Catch up and convergence: a model of cumulative growth. *International Review of Applied Economics* 7/1: 1-25.

Armstrong H. (1995) *Trends and Disparities in Regional GDP per Capita in the European Union, United States of America and Australia*. Final Report, 94/00/74/017, Commission of European Communities.

Aydalot P. (1986) *Milieux innovateurs en Europe*. Paris, GREMI.

Barro R. (1991) Economic growth in a cross section of countries. *Quarterly Journal of Economics* 106: 407-443.

Beine M. and Philippe J.-P. (2000) L'apport des tests de racine unitaire en panel à l'identification des clubs de convergence. In: Beine M. and Docquier F. (eds) *Croissance et convergence économiques des régions: Théorie, faites et déterminants*. Edn. De Boeck Université, Bruxelles.

Bottazzi L. and Perri G. (1999) Innovation, Demand and Knowledge Spillovers: Theory and Evidence from European Regions. *CEPR Discussion Paper*, N° 2279, London.

Braczyk H., Cooke P. and Heidenreich M. (1998) *Regional innovation systems.* University College London Press, London.

Camagni R. (1995) The Concept of Innovative Milieu and its Relevance for Policies in European Lagging Regions. *Papers in Regional Science* 1995/4:317:340.

Camagni R. *et al.* (1999) Le paradigme du milieu innovateur dans l'économie spatiale contemporaine. *Révue d'Economie Régionale et Urbaine* 3, 1999.

Caniëls M. (2000) *Knowledge Spillovers and Economic Growth.* Edward Elgar Publishing, Cheltenham.

Cappelen A., Fagerberg J. and Verspagen B. (1999) Lack of regional convergence. In: Fagerberg J. *et al.* (eds) *The Economic Challenge for Europe: Adapting to innovation based growth.* Edward Elgar Publishing, Cheltenham.

Capron H. (2000) Disparités de croissance entre régions européennes: résorption ou amplification? In: Beine M. and Docquier F. (eds) *Croissance et convergence économiques des régions: Théorie, faites et déterminants.* Edn. De Boeck Université, Bruxelles.

Clarysse B. and Muldur U. (2001) Regional cohesion in Europe? An analysis of how EU public RTD support influences the techno-economic regional landscape. *Research Policy* 30: 275-296.

Coe D. and Helpman E. (1995) International spillovers. *European Economic Review* 39: 859-887.

De la Mothe J. and Paquet G. (1998) *Local and regional systems of innovation.* Kluwer Academic Publishers, Boston, London.

Dosi G. (1988) Sources, procedures and microeconomic effects of innovation. *Journal of Economic Literature* 26: 1120-1171.

European Commission (1997) *Le deuxième rapport européen sur les indicateurs scientifiques et technologiques.* Office des publications officielles des Communautés européennes, Luxembourg.

European Commission (1999) *Sixième rapport périodique sur la situation et le développement économique et social des régions de l'Union Européenne.* Office des publications officielles des Communautés européennes, Luxembourg.

Eurostat (1999) Regional population decline in the EU: Recent trends and future perspectives. *Statistics in focus* 1-3/1999.

Fagerberg J. (1987) A technology gap approach to why growth rates differ. *Research Policy* 16: 87-89.

Fagerberg J. (1988a) Why growth rates differ. In: Dosi G. *et al.* (eds) *Technical Change and Economic Theory.* Printer Publishers, London, 432-457.

Fagerberg J. (1994) Technology and International Differences in Growth Rates. *Journal of Economic Literature* 32: 1147-1175.

Fagerberg J. and Verspagen B. (1996) Heading for Divergence? Regional Growth in Europe Reconsidered. *Journal of Common Market Studies* 34: 431-448.

Fagerberg J., Verspagen B. and Caniëls M. (1997) Technology, Growth and Unemployment across European Regions. *Regional Studies* 31: 457-466.

Fingleton B. (1999) Economic Geography with Spatial Econometrics: A "Third Way" to Analyse Economic Development and "Equilibrium", with application to the EU Regions. *EUI Working Paper ECO* No. 99/21.

Fingleton B. (2000a) Convergence: international comparisons based on a simultaneous equation model with regional effects. *International Review of Applied Economics* 14/3: 285-306.

Fingleton B. (2000b) Spatial econometrics, economic geography, dynamics and equilibrium: a third way? *Environment & Planning A* 32: 1481-1498.

Florida R. (1998) Calibrating the learning region. In: De la Mothe and Paquet G. (eds) *Local and regional systems of innovation*. Kluwer Academic Publishers, Boston, London.

Green W. (2000) *Econometric analysis*. Fourth Edn, Prentice-Hall, New Jersey.

Greunz L. (2001) A panel data analysis of geographically and technologically mediated knowledge spillovers between European regions. Paper presented at the 41st annual meeting of the Western Regional Science Association, Monterey, February 2002.

Grossman G. and Helpman E. (1991) *Innovation and growth in the global economy*. Cambridge MIT Press.

Grossman G. and Helpman E. (1994) Endogenous innovation in the theory of growth. *Journal of Economic Perspectives* 8: 23-44.

Gujarati D. (1995) *Basic Econometrics*, 3rd Edn., McGraw-Hill International Edns.

Jaffe A. and Trajtenberg M. (1996) Flows of Knowledge from Universities and Federal Labs: Modelling the Flow of Patent Citations over Time and across Institutional and Geographical Boundaries. *NBER Working Paper* 5712, Cambridge, MA.

Jaffe A., Trajtenberg M. and Henderson R. (1993) Geographic Localisation of Knowledge Spillovers as Evidence from Patent citations. *Quarterly Journal of Economics*, 108: 577-598.

Kaldor N. (1966) *Causes of the slow rate of growth in the United Kingdom*. Cambridge University Press.

Klein L. (1974) *A textbook of econometrics*, 2nd Edn., Prentice-Hall, Englewood Cliffs, NJ.,

Kline S. and Rosenberg N. (1986) An overview of Innovation. In: Landau R. and Rosenberg N. (eds) *The Positive Sum Strategy: Harnessing Technology for Economic Growth*. National Academy Press, Washington.

Mankiw N., Romer D. and Weil D. (1992) A contribution to the empirics of economic growth. *Quarterly Journal of Economics* 108: 407-437.

Maurseth P. and Verspagen B. (1999) Europe: one or several systems of innovation ? An analysis based on Patent citations. In Fagerberg *et al.* (eds) *The Economic Challenge for Europe – Adapting to Innovation based Growth*, Edward Elgar Publishing, Cheltenham, Glos.

Morgan K. (1997) The learning region: institutions innovation and regional renewal. *Regional Studies* 31/5: 491-503.

Myrdal G. (1957) *Economic theory and underdeveloped regions*. Duckworth, London.

OECD (1996) Measuring R&D in the Service. *STI working paper* 1996/7.

OECD (1997) Special Issue on New rationale and approaches in technology and innovation policy, *STI Review* 22.

OECD (1999) *Boosting Innovation, The cluster approach*. OECD, Paris.

OECD (2001) *The Well-being of Nations – The role of human and social capital*. OECD, Paris.

Porter M. (2000) Location, competition and economic development: local clusters in a global economy. *Economic Development Quarterly* 14: 15-34.

Pugno M. (1995) On competing theories of economic growth: gross-country evidence. *International Review of Applied Economics*, 9: 249-274.

Putnam R. (1993) *Making Democracy Work*. Princeton University Press, Princeton.

Pyke F. and Sengenberger W. (1992) *Industrial districts and local economic regeneration*. International Institute for Labour Studies, Geneva.
Quinet E. (1992) *Infrastructures de transport et croissance*. Economica, Paris.
Sjöhlom F. (1996) International transfer of knowledge: the role of international trade and geographic proximity. *Weltwirtschaftliches Archiv* 132:87-115.
Temple J. (1999) The New Growth Evidence. *Journal of Economic Literature* 36: 112-156.
Tondl G. (1997) The Ups and Downs of Regional Income Convergence in Europe. Evidence from a Cross-Section and a Cross-Section Time Series Analysis. *EUI working paper*, RSC 97/53.
Tondl G. (2001) *Convergence after Divergence? Regional Growth in Europe*. Springer-Verlag Wien New York.
Varga A. (1998) *University research and regional innovation: A Spatial Econometric Analysis of Academic Technology Transfers*, Kluwer Academic Publishers.
Verspagen B. (1991) A new statistical approach to catching up or falling behind. *Structural Change and Economic Dynamics* 2: 359-380.
Verspagen B. (1997) Estimating international technology spillovers using technology flow matrices. *Weltwirtschaftlichas Archiv* 133: 226-248.

Appendix 1: Definition of Variables for the Catching-Up Model

All variables are expressed in terms of average over the period 1989–1996 and row data come from Eurostat. The variables included in the catching up models are defined as follows:

y is the growth rate of real GDP per capita expressed in terms of Purchasing Power Standard (PPS) deflated by the GDP deflator with respect to the price level of 1990;

G_{89} is the technology gap in 1989 defined as $(1-(Y/Y^*))$ where Y/Y^* where is the regional level of real GDP per capita expressed in terms of PPS deflated by the GDP deflator with respect to the price level of 1990 relative to the one of the three best performing regions (Bruxelles, Hamburg, Île-de-France)[20];

S is the percentage of employment in industry and services with respect to total employment;

p is the growth rate of labor productivity defined as the ratio of real GDP per employee expressed in terms of Purchasing Power Standard (PPS) deflated by the GDP deflator with respect to the price level of 1990;

I is manufacturing gross investment in tangible goods per manufacturing employee expressed in terms of Purchasing Power Standard (PPS) deflated by the GDP deflator with respect to the price level of 1990;

---

[20] For reasons of non availability of more desaggregated data, the selection of the three best performing regions was based on the regional desaggregation level NUTS II. However, at a more desaggregated level, other regions such as München, Frankfurt and London perform better than Bruxelles and Hamburg.

PT is the number of patents for 1000 inhabitants;

K is an indicator of physical infrastructure defined as

$$K = \sqrt[4]{\frac{kmmot}{pop \times km^2}}$$

where *kmmot* is the number of kilometres of motorways, *pop* is population and km^2 is the surface of the region;

DT is Jaffe's measure of the technological distance. In our context it proxies the technological distance between the technological profile of a region and the average technology profile of the three leading regions. It is calculated in using patent application data from the European Patent Office (EPO) desagregated into 118 different classes of the International Patent Classification (IPC) and can be defined in the following way:

$$TD_{i,LR} = \frac{\sum_{k=1}^{118} f_{ik} f_{LRk}}{\sqrt{\left(\sum_{k=1}^{118} f_{ik}^2 \sum_{k=1}^{118} f_{LRk}^2\right)}}.$$

f_{ik} and f_{LR} are vectors of elements representing the share of a particular patent class k in the total of patents granted to region $_i$ and to the three technologically leading regions $_{LR}$. If the technological profiles of region $_i$ and the technological leaders $_{LR}$ are similar, $TD_{i,LR}$ is close to unity.

GD is a measure of the geographical distance of region $_i$ with respect to the location of the technological leaders and can be defined in the following way

$$GD_{i,LR} = mean(GD_{i,BE1}, GD_{i,DE6}, GD_{i,FR1})$$

where $GD_{i,BF1}$, $GD_{i,DE6}$, and $GD_{i,FR1}$ represent respectively the geographical distance[21] in terms of thousands of km between region $_i$ and the three leading regions Bruxelles, Hamburg and Île-de-France. The computation of the respective geographical distances among the three leading regions, is based on the same approach in considering that the distance of a given leading region with respect to its own is zero.

BRD is the business sector R&D expenditure per capita in PPS deflated by the GDP deflator with respect to the price level of 1990;

PTGN is the patenting activity of geographical neighbors and is obtained by applying a row standardized first order continuity regional neighborhood matrix to the regions' patent applications per 1000 inhabitants;

HQ is the proportion of highly qualified working age population (25 to 59 years) relative to total working age population;

GRD is government R&D expenditure per capita in PPS deflated by the GDP deflator with respect to the price level of 1990;

---

21 Distances have been evaluated using spherical trigonometry and taking regional capitals as regional references.

Appendix 2: List of Regions Included in the Sample

Code	Region	Code	Region
Austria			
AT11	Burgenland	AT31	Oberösterreich
AT12	Niederösterreich	AT32	Salzburg
AT13	Wien	AT33	Tirol
AT21	Kärnten	AT34	Vorarlberg
AT22	Steiermark		
Belgium			
BE1	Région Bruxelles-capitale	BE3	Région Wallonne
BE2	Vlaams Gewest		
Germany			
DE11	Stuttgart	DE73	Kassel
DE12	Karlsruhe	DE91	Braunschweig
DE13	Freiburg	DE92	Hannover
DE14	Tübingen	DE93	Lüneburg
DE21	Oberbayern	DE94	Weser-Ems
DE22	Niederbayern	DEA1	Düsseldorf
DE23	Oberpfalz	DEA2	Köln
DE24	Oberfranken	DEA3	Münster
DE25	Mittelfranken	DEA4	Detmold
DE26	Unterfranken	DEA5	Arnsberg
DE27	Schwaben	DEB1	Koblenz
DE3	Berlin	DEB2	Trier
DE5	Bremen	DEB3	Rheinhessen-Pfalz
DE6	Hamburg	DEC	Saarland
DE71	Darmstadt	DEF	Schleswig-Holstein
DE72	Gießen		
Denmark			
DK01	København og Frederiksberg Kom., Københavns-, Frederiksborg-, Roskilde amt	DK04	Sønderjyllands-, Ribe-, Vejle-, Ringkøbing-, Viborg amt
DK02	Vestsjællans-, Storstøms-, Bornhoms amt	DK05	Arhus amt
DK03	Fyns amt	DK06	Nordjyllands amt
Spain			
ES11	Galicia	ES41	Castilla y León
ES12	Principado de Asturias	ES42	Castilla-la Mancha
ES13	Cantabria	ES43	Extremadura
ES21	Pais Vasco	ES51	Cataluña
ES22	Comunidad Foral de Navarra	ES52	Comunidad Valenciana
ES23	La Rioja	ES53	Baleares
ES24	Aragón	ES61	Andalucia
ES3	Comunidad de Madrid	ES62	Murcia
Finland			
FI11	Uusimaa	FI14	Väli-Suomi
FI12	Etelä-Suomi	FI15	Pohjois-Suomi
FI13	Itä-Suomi	FI2	land

France			
FR1	Île de France	FR51	Pays de la Loire
FR21	Champagne-Ardenne	FR52	Bretagne
FR22	Picardie	FR53	Poitou-Charentes
FR23	Haute-Normandie	FR61	Aquitaine
FR24	Centre	FR62	Midi-Pyrénées
FR25	Basse-Normandie	FR63	Limousin
FR26	Bourgogne	FR71	Rhône-Alpes
FR3	Nord-Pas-de-Calais	FR72	Auvergne
FR41	Lorraine	FR81	Languedoc-Roussillon
FR42	Alsace	FR82	Provence-Alpes-Côte d`Azur
FR43	Franche-Comté	FR83	Corse
Greece			
GR11	Anatoliki Makedonia, Thraki	GR24	Sterea Ellada
GR12	Kentriki Makedonia	GR25	Peloponnisos
GR13	Dytiki Makedonia	GR3	Attiki
GR14	Thessalia	GR42	Notio Aigaio
GR21	Ipeiros	GR43	Kriti
GR23	Dytiki Ellada		
Ireland		IE	Irlande
Italy			
IT11	Piemonte	IT53	Marche
IT12	Valle d`Aosta	IT6	Lazio
IT13	Liguria	IT71	Abruzzo
IT2	Lombardia	IT72	Molise
IT31	Trentino-Alto Adige	IT8	Campania
IT32	Veneto	IT91	Puglia
IT33	Friuli-Venezia Giulia	IT92	Basilicata
IT4	Emilia-Romagna	IT93	Calabria
IT51	Toscana	ITA	Sicilia
IT52	Umbria	ITB	Sardegna
The Netherlands		NL	Pays-Bas
Portugal			
PT11	Norte	PT14	Alentejo
PT12	Centro (P)	PT15	Algarve
PT13	Lisboa e Vale do Tejo		
Sweden			
SE01	Stockholm	SE05	Västsverige
SE02	Östra Mellansverige	SE06	Norra Mellansverige
SE03	Smland med öarna	SE07	Mellersta Norrland
SE04	Sydsverige	SE08	Övre Norrland
United Kingdom			
UK1	North	UK7	West Midlands
UK2	Yorkshire and Humberside	UK8	North West
UK3	East Midlands	UK9	Pays de Galles
UK4	East Anglia	UKA	Ecosse
UK5	South East	UKB	Northern Ireland
UK6	South West		

9 Employment Growth of Small Computing Services Firms and the Role of Horizontal Clusters: Evidence from Great Britain 1991-2000

Bernard Fingleton, Danilo Camargo Igliori and Barry Moore

University of Cambridge, UK

9.1 Introduction

In recent years, two overlapping topics have received particular attention by governments and researchers throughout the world, particularly in Europe. The emergence of local economies based on high-technology clusters and the role of small and medium sized enterprises (SMEs) in the generation of employment.

The growing economic importance of some high-technology activities such as biotechnology, telecommunications, electronics, and computer services has motivated the British government to declare that knowledge is the most important driver sustaining competitive advantages and promoting economic growth (Department of Trade and Industry – DTI, "Our Competitive Future: building the knowledge driven economy", 1998). At the same time, it is currently recognized that many of these activities are geographically concentrated, forming so-called high technology clusters (see Baptista and Swann, 1998). It is also agreed that in the last two decades SMEs have been responsible for a large portion of the employment generation in many countries and policy support for them has been frequently claimed. The interface of these topics is clarified when one notices that some high-technology clusters are mainly formed by SMEs.

This Chapter examines the impact of horizontal clustering on SMEs employment growth using evidence from Computing Services over the period from 1991 to 2000. After an initial summary of theoretical considerations, the structure of the sector is analysed with regards to spatial distribution and employment performance over the 1990s. A new measure of horizontal clusters is then proposed and applied to map clustering in the sector. In the main section of the Chapter, spatial econometric models of employment growth in SMEs are estimated controlling for levels of horizontal clustering and a set of other initial conditions, particularly the presence of universities (again using a new measure which captures the quality and scale of the local research base). The models also capture the spillover effects across area boundaries. The regression results indicate that there is a significant association between the intensity of horizontal clustering and employment change. The Chapter concludes analyzing these econometric results.

9.2 High-Technology Clusters and Innovative Environments

Notwithstanding the importance of multinational corporations and global forms of production organisation in recent times, some regions have demonstrated a remarkable ability to sustain an above average economic performance, based mainly on the birth and growth of a large number of SMEs. As famous examples one could mention Silicon Valley and Route 128 in the US, the industrial districts in Italy and Germany, and the high technology clusters in Cambridge and Oxford in the UK. These regions are far from representing a homogeneous phenomenon regarding sectors, technologies and institutional environments. Nevertheless, they reveal some similarities, which have attracted the interest of many researchers. Among them is the existence of different kinds of relationships between the firms, from the traditional input/output supply chain linkages to a wide range of collaborative relationships, and the capacity to maintain considerable knowledge flows and innovation between organisations. Antonelli (1999), in analysing the emergence of the new knowledge industry identifies this mode of knowledge production as "scientific entrepreneurship". One aspect of this phenomenon of particular interest here is the incidence of clusters of SMEs in high-technology sectors and their role in supporting innovation, competitiveness, and growth. A fundamental idea of the "cluster" literature is a shift in focus from the firm to productive systems and an understanding of the phenomena of competitiveness as a collective result rather than the outcome of individual processes.

An early recognition of this phenomenon (although not focused on high-technology SMEs) is of course to be found in the work of Alfred Marshall (1920). For Marshall it was clear that specialisation as a result of an internal division of labour is one of the main drivers for an improvement in the efficiency and quality of the productive processes, and for the firm's growth (internal economies). However, these improvements could also be secured by geographical concentration of firms and external economies derived from integration among agents. Marshall identified three main factors related to the external economies, which could stimulate industrial concentration:

1. The existence of thick markets for specialized labour
2. The occurrence of knowledge and technology spillovers
3. The emergence of subsidiary trades

Following the approach suggested by Marshall, Krugman (1991a, 1991b, 1995) has attempted to explain the existence of geographical agglomeration through formal models, which associate increasing returns in the firm's production function with pecuniary external economies in a context of monopolistic competition. Two basic ideas characterize Krugman's approach. The first is that if increasing returns to scale exist at the firm level, then geographical concentration may occur as a result of the presence of pecuniary external economies. The second is that a cumulative process of concentration can be initiated by a particular combination of linkages between firms, transport costs, and the production under increasing returns. One element that is absent from Krugman's approach is consideration of technological externalities, which include spillovers due to information flows. These were considered to detract from the theoretical elegance of Krugman's formal models, but nevertheless they are accepted as necessary elements of a more complete explanation. The other strands in the literature reviewed below reflect their importance. We see empirical evidence for their existence in our econometric modelling.

Also inspired by the Marshallian approach is the discussion relating to modern European industrial districts. Usually a district specialises in one main product and includes many of its subsidiary activities in the region. Then firms, mainly SMEs, can exploit economies of scale by specialising in particular parts of the production process. However, these sorts of relationships are not sufficient to explain the dynamics of these districts. As many authors have emphasized (Pike et al., 1990, Pike and Sengenberger, 1992) cultural identity and different combinations of competition and co-operation among the firms are distinctive features of the new industrial districts.

A different perspective is offered by Porter (1990). Based on an extensive set of case studies he developed a theory of competitive advantage ("the diamond") founded on the interaction of factor conditions; demand conditions; related and supporting industries; and firm strategy, structure and rivalry. The idea is that the components of the diamond form a system, which determines the competitive environment of firms. According to Porter, this system is rooted nationally where the supporting facilities for the firm's performance such as information flows, institutions, infrastructure and competence formation are localized. However, Porter's later works emphasised sub-national concentrations, as the following cluster definition reveals (1998, p.197):

> "Geographic concentrations of interconnected companies, specialised suppliers, service providers, firms in related industries, and associated institutions (for example, universities, standard agencies, and trade associations) in particular fields that compete but also co-operate."

Although each of these approaches contain elements of relevance to high-technology clusters of SMEs, it was the work of the GREMI group (GREMI -

Groupe de Recherche Européen sur les Millieux Innovateurs) with their concept of innovative milieu (Camagni, 1991) that place innovative SMEs centre stage with obvious relevance for the high-technology sector.

A milieu becomes innovative as a result of the processes of interaction and learning carried out by its agents through a set of networks. Networking is a crucial feature of the local productive system, which contains an innovative milieu and can be internal to the milieu or involve external partners. Besides the firms, many different institutions can take part in the networks such as universities, centres of research, associations, and government agencies. On the other hand, spin-offs of new firms emerge as a side result of networking, information exchange, and innovation flows.

Another relevant idea is that the interaction between agents does not occur simply as a result of formal relations or material transfers, the informal links present in a local environment must also be taken into account. Related to this point is the work of Storper (1997) for who tacit knowledge flows and untraded interdependencies are key features of knowledge based clusters.

Different co-operative links between firms can emerge from these interactions. Combinations of elements can explain the networking outcomes, such as the same industry, cultural and social identity, location, business relations, and technological similarity. However, the crucial element is the presence of "trust" among the involved agents. In this context, economic clusters and SMEs become relevant to our understanding of inter-firm co-operation.

Two main types of co-operative relationships can be identified. The first relates to sharing related to particular assets or activities related to productive activity, such as the common use of equipment, or the sharing of a large order or joint marketing efforts. The second concerns information exchange. Here, information flows can include technical or economic information and can derive from a variety of informal or formal links. The rationale for the existence of co-operation is therefore based on the collective gains to related individuals in terms of their operational efficiency. What seems to be important for the firm's competitiveness is the combination of the stimulus due to competition together with adequate levels of inter-firm co-operation.

Regarding the high-technology clusters, the main sort of co-operative linkages are those related to knowledge exchange. While sharing productive activities remains potentially important, it appears that the generation and maintenance of innovation flows is particularly relevant for high technology clusters. In this sense, in addition to relationships between firms, the role of other institutions such as universities and research centres assumes a crucial importance.

The endogenous process of innovation promoted at the regional level through formal and informal networking can be summarised by the concept of collective learning, which was first adopted by the GREMI's researchers and which has recently been exploited in research on the Cambridge and Oxford high-technology concentrations (Keeble, Lawson, Moore and Wilkinson, 1999) and by the TSER European Network (Keeble and Wilkinson, 2000). Collective learning connotes a broad notion of the capacity of a particular regional innovative milieu to generate or facilitate innovative behaviour by its firms (Keeble, 2000: p.200). A particular aspect emphasized by Keeble is that collective learning processes can be seen as mechanisms of reducing uncertainty in a rapidly-changing technology context.

The capacity of a local environment to enhance innovation flows is also emphasised by the approach based on the notion of systems of innovation (Freeman, 1995). Inspired by the neo-Schumpeterian tradition, this approach states that the features of the institutions present in a region can positively impact on the innovative performance of firms, providing scientific, technical and organisational knowledge. The systems of innovation are therefore the results of the interaction between the firms and the respective set of institutions, including supplier/customer relationships, governmental incentive mechanisms, and educational systems. Notwithstanding the importance of growing international integration, it is recognised that national and regional systems of innovation remain as essential domains for the economic analysis of technical change.

Two additional points must be mentioned when discussing the systems of innovation approach: the role of multinational companies and the links between technical and organisational changes. Related to the former, it is recognised that large corporations can be crucial in maintaining the technological dynamism in regional systems. Their ability to transfer equipment and to organize training programmes and technical co-operation agreements may be an important aspect of the functioning of the system which helps avoid lock-in effects. Regarding the second point, interdependence between technical and organisational change mean that the latter acts as a constraint on the former and therefore if organisational development is slow, technical progress may be impeded.

Although, it is recognized that many other factors can influence the performance of SMEs, the specific features of clusters discussed above would seem to create an environment supporting their competitiveness. In addition to the possibilities of cost reduction through external economies, clusters and networks of SMEs can reduce uncertainty and foster innovation. These factors can be crucial in maintaining competitive advantage if one considers recent trends regarding forms of competition that are not based on price.

9.3 Empirical Evidence of Growth and Clustering of SMEs in Computing Services

The main data analysed in this Chapter are extracted from the Annual Employment Survey of the Office for National Statistics. They include employees in employment and number of establishments in Great Britain from 1991 to 2000. The spatial breakdown covers regions, counties, and unitary authorities and districts (UALAD). The SME classification adopted here is close to the one used by the UK government and the European Commission and is defined as firms with up to 199 employees. The industry classification follows the 1992 SIC provided in Table 9.1.

Table 9.1 Computing services

7210	Hardware consultancy
7220	Software consultancy and supply
7230	Data processing
7240	Data base activities
7250	Maintenance/repair of office machinery
7260	Other computer related activities

Source: 1992 SIC

9.3.1 Employment Growth Performance

Computing Services have experienced very intensive employment growth over the study period (from 169.41 thousand in 1991 to 490.6 in 2000), with the biggest increase coming from the smallest firms (between 0 and 10 employees). Figure 9.1 shows the employment performance by firm size over the 1990s.

In the period between 1991 and 2000, growth of employment in SMEs in Computing Services was concentrated in the London, the South East, and the Eastern regions, which together account for more than 62% of the total increase. Table 9.2 shows the distribution of absolute growth in SMEs by region. More detail about the precise location of the change in employment is given in Table 9.3 which identifies those UALADs (the top 20 out of 408 covering Great Britain) where employment change is largest over the period 1991 to 2000. This shows that employment growth in Computing Services is concentrated in selected parts of the major conurbations including London, Glasgow, Leeds, and Manchester. Figure 9.2 provides an overall picture of SMEs employment growth distribution over the period.

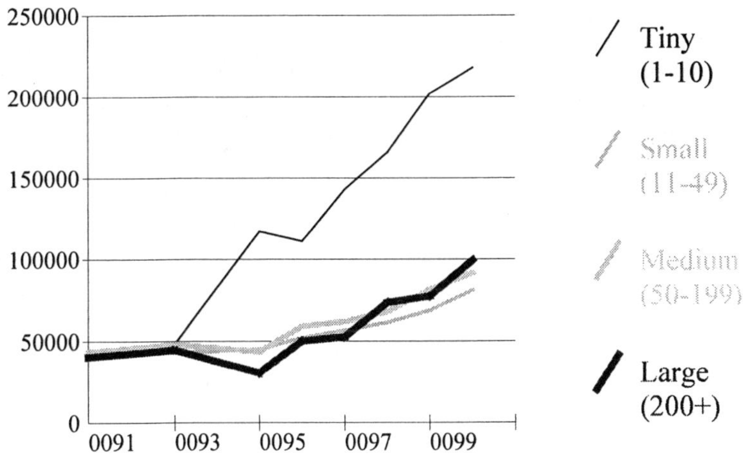

Fig. 9.1 Employees in employment by size of establishment, 1991-2000

Table 9.2 The absolute growth of employment by region in SMEs, 1991-2000

Computing Services		
	(000s)	%
Great Britain	262.11	100.00
Greater London	68.42	26.10
South East	63.29	24.15
East	30.93	11.80
North West	23.08	8.81
South West	17.31	6.60
West Midlands (GOR)	13.87	5.29
Scotland	12.91	4.93
East Midlands	12.22	4.66
Yorkshire and the Humberside	12.15	4.64
Wales	4.25	1.62
North East	3.67	1.40

Source: Annual Employment Survey / ONS

9.3.2 Geographical Concentration

During the last two decades, the high-technology sector has been consolidating a particular pattern of location in Great Britain. Since the late 1980s, a number of studies (Begg and Cameron, 1987; Keeble, 1988; Fingleton, 1992,1994; and Begg and Hodson, 2001) have analysed this locational pattern and the factors, which explain it. Generally speaking, these studies show that high-tech firms are likely to choose small towns in the South East and East Anglia with good transport facilities and proximity to London. However, more recent work by Begg and Hodson (2001), adopting a wider definition of high-technology than that based on Butchart (1987), show that such services are more likely to be found in larger cities. Again, the South is prominent and proximity to Heathrow airport seems to be very relevant. A final point relates to marketing. In order to build a positive image high-tech companies are likely to select regions where other successful high-tech firms have located, avoiding the old industrial areas. Although it is suggested that the features of high-tech business, in particular the large proportion of highly qualified employees, have contributed to a different locational pattern, the econometric results by Fingleton (1992) qualify this conclusion showing that in 1984, at least at the county level, the high-tech sectors were also strongly influenced by the same factors which explain the distribution of industry in general.

Table 9.3 The top 20 UALADS - Absolute employment growth in SMES, 1991-2000

Ranking	Computing Services	(000s)
1	Westminster, City of	8.4
2	City of London	4.5
3	Camden	4.1
4	Islington	3.8
5	Hammersmith and Fulham	3.5
6	Edinburgh, City of	3.4
7	Bracknell Forest	3.3
8	Leeds	3
9	Hounslow	3
10	Milton Keynes	2.9
11	Sefton	2.9
12	Wycombe	2.8
13	Tower Hamlets	2.7
14	Barnet	2.6
15	Wandsworth	2.4
16	Windsor and Maidenhead	2.4
17	St Albans	2.4
18	Southwark	2.3
19	Reading	2.3
20	Slough (UALAD)	2.3

Source: Annual Employment Survey / ONS

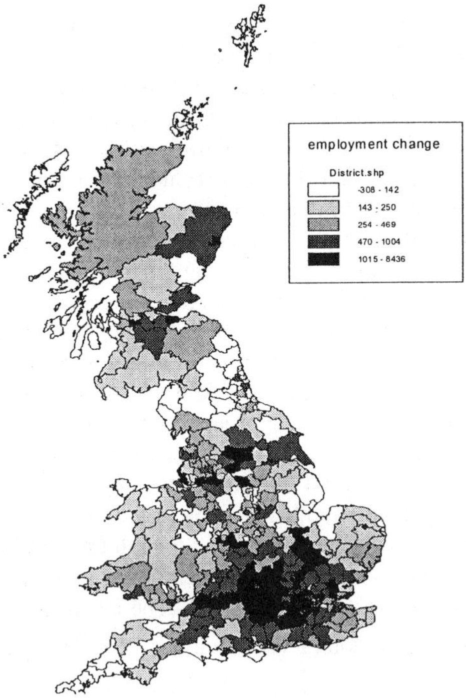

Fig. 9.2 SMEs employment growth in computing services 1991-2000 (UALAD Distribution)

Employment distribution data for Computing Services SMEs confirm the findings of previous studies of the location of high-technology industries. However, the employment distribution *per se* indicates the geographical concentration of the industry without controlling for the overall population and the size of each area's economy. One way to do this is by calculating the Location Quotient (LQ) for each industry. This is a measure of relative concentration and is defined as[1]

$$LQ = (E_{ij}/E_j)/(E_{in}/E_n) = (E_{ij}/E_{in})/(E_j/E_n) \qquad (9.1)$$

Where E_{ij} is employment in industry i in area (region, county or UALAD) j, E_j is total employment in area j, E_{in} is national employment in industry i and

[1] See DTI (2001a)

E_n is total national employment. Thus, an LQ greater than one indicates that there is an above average proportion of employment in a given industry in a given area. Although the LQ has been used in different studies, it has a relevant shortcoming in serving as a cluster measure. It only measures the local importance of an industry given the overall economic importance of the area. In other words, it captures whether an area is has a higher or lower share of a particular industry than the national share, but does not include any information related to the absolute size of the industry in that area. As a result, it is possible to obtain high LQs for very small local industries in absolute terms. If one assumes that scale matters for clustering results, the LQ is omitting "mass effects" that could exist when the area's industry is large, regardless of its relative concentration.

One way of dealing with this problem is to crosscheck looking, at the same time, both at LQ values and at a measure of size such as the local share of the national industry or simply number of jobs in the local industry. However, in this Chapter we adopt a different approach and propose an alternative measure for horizontal clustering (HC). This HC measure takes into account the relative local importance of an industry and the size of the agglomeration in terms of number of jobs. It is defined as the number of jobs in the local industry that exceeds its expected number. The expected number is then defined by the number of jobs in the industry that would occur if the share locally was equal to the share in the national economy, and therefore produce a LQ=1. The HC measure is explicitly calculated by first computing equation (9.1) but with E_{ij} replaced by $\hat{E}_{ij}$ to produce

$$LQ = (\hat{E}_{ij} / E_{in}) / (E_j / E_n) = 1 \tag{9.2}$$

where $\hat{E}_{ij}$ is the number of jobs that makes LQ = 1 given the other actual values. Finally, the measure of horizontal clustering is obtained by calculating

$$HC = E_{ij} - \hat{E}_{ij} \tag{9.3}$$

Table 9.4 presents the SMEs employment distribution, LQs and HCs in 1991 for Computing Services.

When calculated by region, the two key measures of geographical concentration show the same ranking, which is to be expected given the size of the regions (differences between HC and LQ only become significant at the county and UALAD levels of spatial aggregation). The greatest concentrations of employment in Computing Services are to be found in the South East, East, and London. The LQs for these regions are all greater than 1. The HC is positive in all regions indicating the presence of clusters at UALAD level within the regions[2].

[2] Note that the HC variable for the regions is here calculated as the sum of HCs for each UALAD in the region.

Table 9.4 Measures of employment concentration 1991

Employment (000s)		LQ		HC (000s)	
Great Britain	196.2	Great Britain	1	Great Britain	70.3
South East	52.9	South East	1.96	South East	32.1
Greater London	40.5	Greater London	1.38	Greater London	13.8
East	20.6	East	1.17	East	8.1
West Midlands (GOR)	18.1	West Midlands (GOR)	0.99	West Midlands (GOR)	5.2
Scotland	10.9	Scotland	0.62	Scotland	3.8
South West	14.3	South West	0.86	South West	3.7
North West	13.3	North West	0.58	North West	1.9
East Midlands	8.8	East Midlands	0.64	East Midlands	1.1
Wales	3.9	Wales	0.44	Wales	0.5
North East	4	North East	0.54	North East	0.2
Yorkshire and the Humber	8.8	Yorkshire and the Humber	0.53	Yorkshire and the Humber	0

Source: Annual Employment Survey / ONS

In what follows, we use the above-described HC as the measure of horizontal clustering for our empirical purposes. Table 9.5 present the top 20 UALADs in terms of HC in 1991. Figure 9.3 shows an overall picture of HC distribution in 1991 across the country.

Table 9.5 The top 20 HC UALADs, 1991 (000s)

Inverclyde	3.3
Havant	3.2
Wycombe	2.7
Portsmouth	2.6
Reading	2.4
Wokingham	2.2
Surrey Heath	2.2
Birmingham	2.1
Windsor and Maidenhead	1.6
Islington	1.6
Bracknell Forest	1.6
Hounslow	1.6
Hillingdon	1.5
Basingstoke and Deane	1.4
Slough (UALAD)	1.3
Bristol	1.2
West Berkshire (Newbury)	1.2
Hackney	1.1
Tower Hamlets	1.1
Warwick	1.1

Source: Annual Employment Survey / ONS

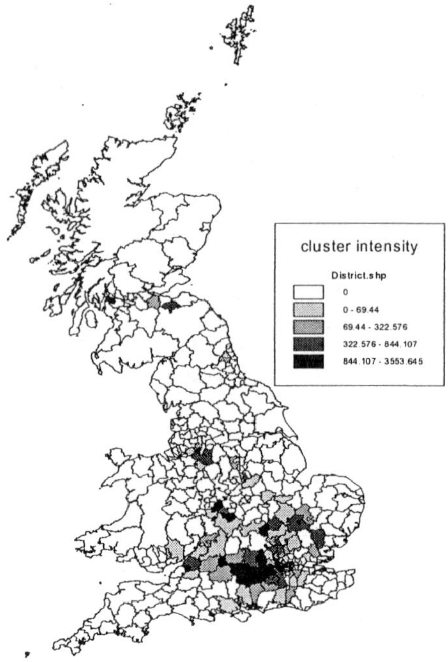

Fig. 9.3 Horizontal clustering in computing services 1991

The main hypothesis to be tested in this Chapter is that there is a positive correlation between employment growth in high-technology SMEs and geographical concentration or clustering. To provide an initial assessment of this relationship we have first calculated the correlation coefficients between the HC in 1991 and the absolute change in employees in employment in SMEs between 1991 and 2000 (SIEG), using data for UALAD and county levels. To allow for the interactions between large firms and SMEs, HCs include jobs in large firms. The coefficients indicate the existence of positive linear correlation between the variables, particularly at county level (0.76 at county level and 0.44 at UALAD level). However, as these correlation analyses do not control for other relevant variables which may also contribute to the employment performance of high-tech SMEs, a multivariate analysis is necessary.

9.4 Control Variables and Initial Conditions

To structure our econometric modelling approach we have selected a set of control and environmental variables that could have direct or indirect impacts on the employment performance of an industry's SMEs in a particular location. The basic idea is to test the relationship between employment growth and a set of 'initial conditions' present in the region, including the level of horizontal clustering. In selecting the control variables we have been guided by the literature reviewed above, suitably adapted to the characteristics of high-technology service industries.

The first set of control variables aims to reflect demand side conditions and also the supply of factors that are inputs to SME production. Three variables were selected. The first one is the total SME employment growth in the area (UALAD), including all industries. This is represented by the acronym STEG, which stands for Small Total Employment Growth. The idea here is that the variable captures the demand producing SME growth regardless of industrial sector. It also reflects the provision of inputs from SMEs in other sectors. The second variable is the large firm employment growth in the industry that occurred in the area (LIEG). This captures the role of large firms in creating opportunities for SMEs in the same industry. It also reflects inputs from large firms in the sector. The third variable is the total large firms' employment growth that occurred in the area (LTEG), which is treated as a proxy for the demand coming from large firms in other industries, as well as representing the provision of inputs from large firms in other sectors.

Given that computing services is a knowledge-based industry, the second set of controls focus on human capital as the key factor market variable. Two variables are included in this group. The first one is a simple indicator of the level of educational attainment in the area measured by the percentage of the population in 1991 with (two or more) A level qualifications or above (EDUC). The second variable aims to capture the impact of the presence of a science base in the area as a source of technology transfer but also as a source of highly qualified labour. The variable used is derived from the Research Assessment Exercise 1996 and measures the scale and quality local universities present in the area (UNIVIMP).

The quality of infrastructure provision is also hypothesised to contribute to the competitiveness and growth of firms in the area. Two variables are included in this group. Firstly, we include a general measure of urbanisation to reflect urbanization external economies, adopting population density as a proxy (POP DEN). This also reflects the greater differentiation or variety of suppliers in more densely populated areas, which would tend to enhance output. The second variable (NT_POP) concentrates on the so-called New Towns in which production conditions have been enhanced by government policy. In areas with New Towns, the variable takes a value equal to the area's population in 1991; otherwise its value is zero.

The fourth type of control variable focuses on market structure and firm rivalry. Arguably, a region with higher degree of competition may offer better incentives for innovation and productivity, contributing to the industry's performance and therefore its employment growth. The variable used is the number of establishments per employee in 1991 (MKTSTR).

9.5 Econometric Modelling

The strategy employed in the modeling exercise is to take the variables suggested by the preceding discussion and test whether they do actually appear to have a significant relationship with the growth of employment. The analysis is based on data for 408 UALADS. The first stage of the econometric analysis is to fit the simplest possible regression model estimated by OLS as a way of guiding us to a more correct specification and estimation method. Subsequently we give the results of estimation methods which are robust in the presence of endogenous variables, error heterogeneity, and non-normality. The existence of spatial effects leads to an autoregressive lag specification in which growth depends not only on local factors but also on factors in 'neighbouring' areas. In general, the model shows that local and neighbouring demand and supply conditions are important factors governing the growth of computing services, and while in essence the theory outlined above is largely supported, notably the effects of cluster intensity and university importance, with marginal significance for educational attainment, other variables (market structure, population density, New Town population size) turn out to be insignificant. Alternative specifications are considered as a means of deconstructing the significant spillover of effects from neighbouring regions. These are suggestive of a role for (horizontal) cluster intensity in neighbouring areas although this is not the only factor operating.

The simple regression model is summarized in Table 9.6 below. This suggest that the variables controlling for the effects of general economic growth in the area (small firms total employment growth - STEG, and large firms total employment growth - LTEG) and the growth of the sector in the area (as proxied by the growth of large firms in the sector- LIEG) are significantly related to the growth of small computing services firms in the area (SIEG). There is also an initial indication that cluster intensity (HC), educational attainment (EDUC) and population density (POPDEN) have significant effects. Overall, the levels of fit is very good (R-squared is 82.6%), but there is evidence and reason to believe that we cannot rely on these estimates.

Table 9.6 OLS regression model - computing services

Variables	Estimate	Standard Errors	t-ratio (398)
Constant	-393.6	60.1	-6.55
STEG	0.0743	0.0035	21.30
LIEG	0.1934	0.0299	6.46
LTEG	0.0105	0.0037	2.81
HC	0.3593	0.0416	8.63
UNIVIMP	0.0409	0.0213	1.91
EDUC	2366.	409.	5.78
MKTSTR	74.9	80.5	0.93
POPDEN	58.8	10.7	5.51
NT_POP	-0.279	0.406	-0.69

A priori, given that the data relate to 408 UALADs, it is highly likely that the observations will not be statistically independent as is assumed by the simple regression model. Area in close proximity in particular will tend to see similar growth in computing services employment not simply because they are similar in terms of the causal variable values. Other effects associated with their proximity are likely to come into play and therefore it is likely that the above simple regression model is misspecified since there is no attempt at this juncture to model any spatial effect. Moreover, with spatial data, it is commonly the case that we see error heterogeneity, so that for instance larger areas tend to show a greater variability than smaller areas. Applying some standard diagnostic tests to the regression model summarized in Table 9.6 does confirm these expectations. If we focus on a general (LM) test of misspecification from several sources as outlined by Anselin (1988), namely the joint test of error dependence (from all sources) plus error heterogeneity, taking the results at face value we reject the null in favour of spatial dependence and heterogeneity[3]. It is useful to observe that the main causes seem to be omitted spatial lag plus heterogeneity rather than spatial error autocorrelation. The one-directional tests (which do not allow for the other source of misspecification) indicate that error heterogeneity[4] is present (LM = 319.4, 9 degrees of freedom), spatial dependence (all sources) is evident (LM = 23.27, 2 degrees of freedom), and the test of the omitted endogenous lag gives LM = 22.45 which is highly significant in the χ^2 distribution with one degree of freedom. However, error autocorrelation, defined as spatially autoregressive or spatial moving average

---

[3] This is based on the Breusch-Pagan test for heteroscedasticity combined with the overall test for spatial dependence due to either error autocorrelation of omitted spatial lag. The test statistic is 342.7 which is very extreme when referred to the χ^2 distribution with 11 degree of freedom (9 for the variables thought to govern heterogeneity, plus 2 for the spatial dependence).

[4] This is reaffirmed by the Koenker-Bassett test using the dependent variable (SIEG) as the conditioning variable, which gives a test statistic of 12.94 which has a p-value in the relevant χ^2 null probability distribution (with 1 df) equal to 0.0003. This has the advantage of being robust to non-normality whereas the Breuch-Pagan test has low power with non-normality (although only with small samples).

residual autocorrelation, is absent (using the test due to Burridge, 1980, LM = 1.09, 1 degree of freedom, which is insignificant in the χ^2 distribution with 1 degree of freedom). Since the non-robust LM tests of error autocorrelation and spatial lag dependence clearly indicate the latter, we conclude that the spatial effect is due to lag dependence. However, there are further tests that could be carried out in a less clear-cut context. In general, with regard to error autocorrelation, testing the common factor hypothesis provides a way of testing the error autocorrelation hypothesis, since if the common factor restriction is rejected then the spatial effect is due to a combination of endogenous and exogenous lags. There is also other recent work (Anselin and Bera, 1998, Anselin, Bera, Florax and Yoon, 1996) dedicated to the question of the coexistence of spatial lag dependence and error autocorrelation, as for example in Case et. al. (1993), providing lag dependence tests robust to the presence of error autocorrelation and vice-versa. In this Chapter, while we provisionally accept lag dependence, we further examine a potential source of error autocorrelation due to omitted spatial variables, exploring this by testing whether the endogenous lag, which appears to be necessary from the above diagnostic tests, encompasses various regressor lags.

Evidently, we need to model spatial lag dependence and allow for error heterogeneity. A further indication of model misspecification due to omitted variables comes from Ramsey's reset test, which gives an F ratio of 3.408 which has a p-value of 0.03408 in the relevant F(2,396) distribution. However, while it is often used as a test for missing regressors, it also has power against misspecified functional form, and this weakens its appropriateness (no test can discriminate between missing regressors and nonlinearity, although the results are suggestive of the former). It should also be pointed out that the F test and other significance tests (t, LM tests) assume a normal distribution, and there is very strong evidence to gainsay that assumption. The Shapiro-Francia (1972) test of normality applied to regression residuals strongly rejects that assumption. Under normality the test statistic is N(0,1) (see Royston 1983), but for these data we obtain a z value of 6.62. Likewise, the Jarque-Bera test for normality of errors tells a similar story.

The tests of spatial dependence described above are based on a so-called W matrix, which in this case, with 408 UALADs, is a square matrix of dimension 408. The entries in the body of the W matrix reflect our hypothesis of spatial interaction. Therefore, typically the main diagonal of the W matrix contains zeros, and the off-diagonal elements contain values reflecting the spatial proximity of each pair of areas. In some studies the W matrix entries are binary 1, 0 values depending on whether the areas are contiguous or not, so then the hypothesis is that the spatial interaction is confined to the immediate neighbours of each area. However, this is rather a restricted and unreal assumption, since we would expect what happens to occur in more remote areas to play some part in the growth of employment locally, and areas that are not, because of the vagaries of local authority boundary definition, actually contiguous may in fact interact. Therefore it is standard practice to assume that interaction is a diminishing function of distance, with the effect decaying non-linearly as say an exponential or power function. In this case, we obtain the desired effect of stronger interaction locally rather than remotely, which is one of the attractive features of the binary matrix, but allow more remote interaction to also have some small effect. There are arguments from recent theoretical work (see Fingleton, 2001, for a discussion of this and estimates based on different assumptions about the structure of the W matrix) pointing to the negative exponential function as the natural outcome of an infinite number of choices that suggest it has the most solid theoretical basis, but since a power function is simply an exponential function with the argument in natural logs, it seems appropriate to assume a non-linear power function as the basis of the W matrix. We raise distance to the power 2 to give appropriate distance decay, and while this power is chosen a priori rather than estimated, the estimation of the parameter for the spatial lags does introduce a degree of freedom to offset this arbitrary choice. A further step in the construction of the W matrix is to scale it so that each row sums to 1. Hence

$$W_{ij}^{*} = 1 \big/ d_{ij}^{2} \qquad W_{ij} = W_{ij}^{*} \Big/ \sum_{j} W_{ij}^{*} \tag{9.4}$$

This has useful consequences for estimation, since the value for region *j* of the spatially lagged variable, defined as the *j*'th cell of the vector given by the matrix product *WX*, is then the weighted average of the values of the variable *X* in the areas that are 'neighbours' to *j*, and this helps in the interpretation of the spatial lag effects since the coefficients are more easily compared with the coefficient estimates for the initial variable. Also, with an endogenous lag, it usefully identifies a parameter value below[5] 1 as being consistent with a 'non-exploding' process while 1 and above leads to complex and little understood consequences for inference and estimation (the mathematical background to this and implications of spatial unit roots consistent with a parameter equal to 1 are discussed in Fingleton, 1999). However, the implication is standardising is that the interaction between areas is now not simply a function of absolute distance (see Anselin, 1988).

Another important consideration is the assumption above that the explanatory variables are exogenous. While this can be assumed to be the case for HC, UNIVIMP, EDUC, MKTSTR, POPDEN and NT_POP, partly because the data on which they are based does not lie within the period 1991-2000 and therefore can be assumed not to be determined by the growth of computing services, the same cannot be assumed for STEG, LIEG and LTEG. These indicators of local demand and supply will to some extent be a response to the growth of computing services as well as a cause of it, and we should allow for this endogeneity in our modelling. A further source of endogeneity is when there is an endogenous spatial lag, defined as the matrix product of *W* and the dependent variable SIEG, in the model. We noted above that our regression diagnostics had pointed to the need for a different specification involving spatial effects and that it appeared that this should be an endogenous lag rather than an autoregressive error process. This implies that we should use 2SLS estimation to achieve consistent estimates.

There are two issues that might concern us when considering 2SLS estimation in the spatial setting. The first issue is the problem posed whenever the estimate of

---

[5] Strictly, within the bounds equal to $1/e_S$ and $1/e_L$ where e_L and e_S are maximum and minimum eigenvalues of the *W* matrix. With a standardised *W* matrix, e_L is equal to 1.

the parameter on the endogenous lag enters that portion of the parameter space equal to or greater than the maximum eigenvalue reciprocal $1/e_L = 1$, or enters the zone below $1/e_S$, although in fact neither is problematic for the current estimation exercise. The second issue is the choice of instrumental variables, since in practice it tends to be somewhat difficult to find variables that are both independent of the errors and are correlated with the endogenous regressors. With regard to the selection of instruments, we use the 3 group method[6] (described in Kennedy 1992; Johnston 1984), which is commonly used to overcome measurement error problems, combined with the use of exogenous low order spatial lags[7]. Hence, the instruments comprise the three levels of STEG, LIEG and LTEG with levels -1, 0 and 1 according to whether the values of the variables were low, high or in between. The exogenous variables (HC, UNIVIMP etc) plus the lags of the exogenous variables (W multiplied by HC etc) provided additional instruments.

The resulting estimates are given in Table 9.7, together with robust Bootstrap estimates (which require neither normality nor homoscedasticity). It is noticeable from these that the evident heteroscedasticity and non-normality of the error distribution has had little impact on the point estimates and parameter standard errors. Note that when the endogenous regressor variables are lagged, then they too are endogenous. We therefore refer to the lags of the explanatory right hand side variables as regressor lags rather than endogenous lags, with the term endogenous lag reserved for the lag of the dependent variable. The letter I preceding the variable indicates that it is treated as being endogenous.

Table 9.7 Spatial lag model AR IV with endogenous lag

9.7a Spatial Lag Model – IV (2SLS) Estimation			
Variables	Estimate	t-value	p-value
IW_SIEG	0.348402	5.029947	0.000000
Constant	-358.309	-5.837827	0.000000
HC	0.322681	7.454679	0.000000
UNIVIMP	0.083158	3.232066	0.001229
EDUC	1252.25	2.444144	0.014520
MKTSTR	33.5139	0.410802	0.681217
POPDEN	22.3448	1.587851	0.112320
NT_POP	0.0160608	0.038053	0.969646
ISTEG	0.0606719	10.982471	0.000000
ILIEG	0.274316	5.259251	0.000000
ILTEG	0.00374262	0.669379	0.503254

$R^2 = 0.8077$ Sq. Corr. = 0.8294

$\sigma^2 = 119045.2099\ (345.0293\ )$

[6] This been shown in general to give consistent estimates (see Leser, 1966). We assume that the errors are not so large that they correlate with the groups, and that the instruments are correlated with the explanatory variables.

[7] In the spatial context, the available instruments extends to exogenous spatial lags, although these should be used cautiously because of the possibility of linear dependence in the matrix of instruments if higher spatial lags are used, as mentioned in Fingleton(2000).

9.7b Spatial Lag Model - Bootstrap Estimation, 999 replications			
Variables	Estimate	t-value	p-value
IW_SIEG	0.356728	5.515371	0.000000
Constant	-360.002	-6.149163	0.000000
HC	0.320528	7.846994	0.000000
UNIVIMP	0.0835085	3.502462	0.000461
EDUC	1241.59	2.456168	0.014043
MKTSTR	30.3487	0.405370	0.685206
POPDEN	21.8269	1.635093	0.102030
NT_POP	0.0296587	0.069160	0.944862
ISTEG	0.0605564	10.811572	0.000000
ILIEG	0.272524	5.404202	0.000000
ILTEG	0.00358549	0.658797	0.510026

R^2= 0.8147 Sq. Corr.= 0.8283

$\sigma^2 = 116511.3654$ (341.3376)

In what follows, we attempt to deconstruct the endogenous lag variable (W_SIEG) by introducing the significant regressors in lag form. The idea is that if the regressor lags are the dominant effect, then they will retain significance in the presence of the endogenous lag. On the other hand, if the endogenous lag encompasses the regressor lags, then they will tend to be insignificant in the presence of the endogenous lag, even though they may be significant when it is absent. We denote the regressor lags by WS_STEG, WS_LIEG, WS_HC, WS_UNIVIMP and WS_EDUC. Table 9.8a shows that the STEG and LIEG lags are evidently statistically significant, and this points to their relative importance because they retain significance[8] in the presence of the other (collinear) lags.

However, from Table 9.8b we see that the regressor lags are not significant in the presence of the endogenous lag W_SIEG, which is just significant in a one-tailed test with a 5% risk of Type I error. It appears that the endogenous lag encompasses the regressor lags, and contains significant additional explanatory information. Taken individually, regressor lags are significant, for example the HC lag in Table 9.8c. However, Table 9.8d indicates that the endogenous lag (W_SIEG) encompasses the HC lag, which loses significance in the presence of the endogenous lag. Tables 9.8e and 9.8f show a similar encompassing effect for the other important regressor lags.

[8] The regressor lags are quite highly correlated, and this has the effect of inflating standard errors thus reducing significance.

Table 9.8 Deconstructing the endogenous lag - IV (2SLS) estimates

	9.8a With Significant Regressor Lags Only		9.8b Regressor Lags plus Endogenous Lag	
Variables	Estimates	t-ratio (393)	Estimates	t-ratio (392)
Constant	-378.	-2.209338	-136.	-0.598024
ISTEG	0.06149	10.002592	0.05906	9.076447
ILIEG	0.2892	4.520346	0.2868	4.343104
ILTEG	0.00504	0.868944	0.00077	0.118891
HC	0.3226	6.198986	0.3364	6.192393
UNIVIMP	0.0788	2.833409	0.1009	3.195185
EDUC	967.	1.664431	1352.	2.106653
MKTSTR	35.	0.428512	37.	0.428665
POPDEN	21.2	1.409972	15.1	0.951250
NT_POP	-0.205	-0.477965	0.118	0.245362
IW_SIEG	-	-	0.822	1.675465
IWS_STEG	0.0383	2.472172	-0.0476	-0.887215
IWS_LIEG	0.389	2.209082	-0.099	-0.289299
WS_HC	-0.024	-0.166525	-0.235	-1.196707
WS_UNIVIMP	0.005	0.056232	0.104	0.981157
WS_EDUC	-315.	-0.186605	-1301.	-0.707410

	9.8c With HC Exogenous Lag Only		9.8d HC Lag plus Endogenous Lag	
Variables	Estimates	t-ratio (397)	Estimates	t-ratio (396)
Constant	-373.1	-6.054599	-360.5	-5.810133
ISTEG	0.06732	12.448704	0.05930	10.385749
ILIEG	0.2786	5.130689	0.00322	0.567795
ILTEG	0.00966	1.767893	0.2910	5.320082
HC	0.3206	6.831514	0.3437	7.239680
UNIVIMP	0.0430	1.772978	0.0848	3.256947
EDUC	2128.	4.420212	1276.	2.462546
MKTSTR	84.	1.035287	23.	0.281038
POPDEN	58.3	4.893289	17.1	1.147544
NT_POP	-0.151	-0.356795	0.011	0.024999
IW_SIEG	-	-	0.410	4.628300
WS_HC	0.266	2.170729	-0.178	-1.139652

	9.8e Endogenous Lag plus Regressor Lag for STEG		9.8f. Endogenous lag plus regressor lag for LIEG	
Variables	Estimates	t-ratio (396)	Estimates	t-ratio (396)
Constant	-309.	-2.979137	-326.5	-4.961067
ISTEG	0.06177	10.623340	0.06254	11.044394
ILIEG	0.2568	4.296167	0.2566	4.798418
ILTEG	0.00365	0.654315	0.00429	0.773307
HC	0.3106	6.509965	0.3038	6.693990
UNIVIMP	0.0871	3.286232	0.0871	3.397506
EDUC	1209.	2.341352	972.	1.754895
MKTSTR	40.	0.483797	41.	0.509841
POPDEN	24.3	1.684900	23.6	1.689750
NT_POP	0.034	0.081550	-0.088	-0.207407
IW_SIEG	0.467	2.204974	0.280	3.186884
IWS_STEG	-0.0189	-0.593993	-	-
IWS_LIEG	-	-	0.238	1.245406

Table 9.9 gives a final model with variables included on the basis of their evident significance in the foregoing analysis. Table 9.9a gives the IV estimates and 9.9b the robust Bootstrap estimates. These jointly reaffirm the importance of the intensity of the horizontal cluster to growth of employment in the small firms in the computing services sector, controlling for the effects of small firm employment growth (STEG), the growth of the computing services sector (LIEG), the importance of universities in the local economy (UNIVIMP) and the impact of human capital represented by educational attainment rates (EDUC). In addition, we see the significance of the endogenous lag, so that small firm computing services growth in 'neighbouring' areas influences (and is influenced by) growth locally. This we have attributed to the spillover of the effects of the causal variables across area boundaries, but in addition, there are other spillovers, which are not adequately represented by the causal variable spillovers. This has been shown by the encompassing of the regressor lags by the endogenous lag, and is explained by the fact that the endogenous lag captures the totality of external effects affecting employment growth, including the unexplained portion in the error. This could for instance include the transmission of knowledge across area boundaries and the fact that new spin-offs tend to be in the vicinity of the parent firm.

Table 9.9 Final model

9.9a IV (2SLS) Estimates

Variables	Estimates	t-ratio	P-value
Constant	-335.68	-6.240239	0.000000
IW_SIEG	0.392772	6.529738	0.000000
HC	0.331493	7.780907	0.000000
UNIVIMP	0.0953208	4.493164	0.000007
EDUC	1051.22	2.235487	0.025385
ISTEG	0.0619537	11.840220	0.000000
ILIEG	0.299613	5.642405	0.000000

$R^2 = 0.8028$ Sq. Corr. = 0.8252

$\sigma^2 = 120743.3599$ (347.4815)

9.9b Bootstrap Estimation, 999 replications

Variables	Estimates	t-ratio	P-value
Constant	-338.324	-6.484001	0.000000
IW_SIEG	0.396193	6.804595	0.000000
HC	0.330216	8.235498	0.000000
UNIVIMP	0.0948107	4.801954	0.000002
EDUC	1057.36	2.246699	0.024659
ISTEG	0.061929	11.757837	0.000000
ILIEG	0.298238	5.790016	0.000000

$R^2 = 0.8011$ Sq. Corr. = 0.8226

$\sigma^2 = 120513.7096$ (347.1508)

9.6 Conclusions

The spatial econometric model shows that local and neighbouring demand and supply conditions are important factors controlling the change in Computing Services employment, and supports the role attributed to both cluster intensity and human capital as represented by university importance (and to a lesser extent by educational attainment). The other variables (the impact of competition as reflected by market structure, the effect of the economic diversity and urban infrastructure as reflected by population density and New Town population size) turn out to be insignificant. It is also apparent that external effects spillover across area boundaries. Once these are 'deconstructed', it is apparent that while horizontal cluster intensity in 'neighbouring' areas is statistically significant, there is a multiplicity of across-area spillovers suggesting that this is not the only factor operating and that additional effects extending beyond the suite of variables considered are operating. This points to the importance of local spin-offs and knowledge flows creating technological externalities that transgress area boundaries.

Acknowledgements

The authors wish to thank PACEC - Public and Corporate Economic Consultants - for the data provision. Especial acknowledgement goes to Mr Nic Boyns for his suggestions and efficiency in preparing the variables. Usual disclaimer applies.

References

Anselin L. 1988 *Spatial Econometrics : Methods and Models*. Kluwer Academic Publications, Dordrecht.

Anselin L. and Bera A.K. (1998) Spatial Dependence In Linear Regression Models With An Introduction To Spatial Econometrics, in Ulah A. and Giles D.A. (eds) *Handbook of Applied Economic Statistics*, pp. 237-289. Marcel Dekker, New York .

Anselin L., Bera A.K., Florax R. and Yoon M.J. (1996) Simple Diagnostic Tests for Spatial Dependence, *Regional Science and Urban Economics*, 26: 77-104.

Antonelli C. (1999) The Evolution of the Industrial Organisation of the Production of Knowledge. *Cambridge Journal of Economics* 23: 243-260.

Baptista R. and Swann G. M. P. (1998) Clustering Dynamics in UK Computer Industries: A comparison with the USA, in Swann G.M.P., Prevezer M. and Stout D. (eds) *The Dynamics Of Industrial Clustering: International Comparisons In Computing And Biotechnology*. Oxford University Press, Oxford.

Begg I. G. and Cameron G. C. (1987) *High Technology Location and the Urban Areas of Great Britain*. Discussion Paper Series. Department of Land Economy. University of Cambridge, Cambridge.

Begg I. G. and Hodson D. (2001) *The Location of Jobs in High-technology Industries in Great Britain in the 1990s* mimeo, South Bank University: London.

Burridge P. (1980) On the Cliff-Ord test for spatial autocorrelation. *Journal of the Royal Statistical Society Series B* 42: 107-8.

Butchart R. L. (1987) A New UK Definition of the High Technology Industries. *Economic Trends* 400: 82-8.

Camagni R. (1991) *Innovation Networks: Spatial Perspectives*. Belhaven Press, London.

Case A.C., Rosen H.S. and Hines J.R. (1993) Budget Spillovers and Fiscal Policy Interdependence: Evidence from the States. *Journal of Public Economics* 52: 285-307.

DTI (1998) *Our Competitive Future: building the knowledge driven economy*. London.

DTI (2001) *Business Clusters in the UK - A First Assessment*. London.

European Commission (1996) Commission Recommendation of 3 April 1996 Concerning the Definition of Small and Medium-Sized Enterprises. *Official Journal* L 107 30/04/1996 0004 - 0009.

Fingleton B. (1992) The Location of Employment in High-technology Manufacturing in Great Britain. *Urban Studies* 29: 1265-1276.

Fingleton B. (1994) The location of high-technology manufacturing in Great Britain : changes in the late 1980s. *Urban Studies* 31: 47-57.

Fingleton B. (1999) Spurious spatial regression: some Monte-Carlo results with a spatial unit root and spatial cointegration. *Journal of Regional Science* 39: 1-19.

Fingleton B. (2000) Spatial econometrics, economic geography, dynamics and equilibrium : a third way? *Environment & Planning A* 32: 1481-1498.

Fingleton B. (2001) Theoretical economic geography and spatial econometrics : dynamic perspectives. *Journal of Economic Geography* 1: 201-225.

Freeman C. (1995) The 'National System of Innovation' in historical perspective. *Cambridge Journal of Economics* 19: 5-24.

Johnston J. (1984) *Econometric Methods.* McGraw-Hill, New York.

Keeble D., Lawson C., Moore B. and Wilkinson F. (1999) Collective Learning Processes, Networking and Institutional Thickness in the Cambridge Region. *Regional Studies* 33: 319-32.

Keeble D. (2000) Collective Learning Processes in European High-Technology Milieux, in Keeble D. and Wilkinson F. (eds.) (2000). *High-Technology Clusters, Networking and Collective Learning in Europe.* Ashgate, Aldershot.

Keeble D. and Wilkinson F. (eds.) (2000) *High-Technology Clusters, Networking and Collective Learning in Europe.* Ashgate, Aldershot.

Kennedy P. (1992) *A Guide to Econometrics.* Blackwell, Oxford.

Krugman P. R. (1991a) *Geography and Trade.* MIT Press, Cambridge, Massachusetts.

Krugman P. R.. (1991b) Increasing Returns and Economic Geography. *Journal of Political Economy* 99: 483 - 499.

Krugman P. R. (1995) *Development, Geography, and Economic Theory.* MIT Press, Cambridge, Massachusetts.

Leser C. (1966) *Econometric Techniques and Problems.* Griffin, London.

Marshall A. (1920) *Principles of Economics.* Macmillan, London.

Pike F. et al (1990) *Industrial Districts and Inter-firm Co-operation in Italy.* International Institute for Labour Studies, Geneva.

Pike F. and Sengenberger W. (1992) *Industrial Districts and Local Economic Regeneration.* International Institute for Labour Studies, Geneva.

Porter M. E. (1990) *The Competitive Advantage of Nations.* Macmillan Press, London.

Porter M. E. (1998) Clusters and the New Economics of Competition *Harvard Business Review*, November/December 77-90.

Royston J. P. (1983) A simple method for evaluating the Shapiro-Francia W test of non-normality. *The Statistician* 32: 297-300.

Shapiro S. S. And Francia R. S. (1972) An approximate analysis of variance test for normality. *Journal of the American Statistical Association* 67: 215.

Storper M. (1997) *The Regional World: Territorial Development in a Global Economy.* Guilford, New York.

Wilkinson F. and Moore B. (2000) Concluding Reflections: Some Policy Implications, in Keeble D. and Wilkinson F. (eds.) (2000) *High-Technology Clusters, Networking and Collective Learning in Europe.* Ashgate, Aldershot.

Appendix - The University Impact Variable (UNIVIMP)

In order to build a variable that could serve as a proxy for the impact of universities on SMEs employment growth in Computing Services two aspects were considered: research quality and staff size. The data source was the 1996 Research Assessment Exercise and the departments considered were the following: Agriculture; Anatomy; Applied Mathematics; Biochemistry; Biological Sciences; Chemical Engineering; Chemistry; Civil Engineering; Clinical Dentistry; Clinical Laboratory Sciences; Community Based Clinical Subjects; Computer Science; Earth Sciences; Electrical and Electronic Engineering; Environmental Sciences; Food Science and Technology; General Engineering; Hospital Based Clinical Subjects; Mechanical, Aeronautical and Manufacturing Engineering; Metallurgy and Materials; Mineral and Mining Engineering; Nursing; Other Studies and Professions Allied To Medicine; Pharmacology; Pharmacy; Physics; Physiology; Pre-Clinical Studies; Psychology; Pure Mathematics; Statistics and Operational Research; Veterinary Science

The first step consists of deriving a number for each university. In order to do that we initially multiply the rate of each department by its Category A Research Active Staff. Then, we sum up the department numbers obtaining the university impact measure. In the second step, we construct the variable for each area (UALAD) by adding up the numbers pertaining to the universities located within each area.

10 Externalities and Local Economic Growth in Manufacturing Industries

Stefano Usai[1] and Raffaele Paci[2]

[1] *University of Sassari,* [2] *University of Cagliari, Italy*

10.1 Introduction*

The recent revival of interest in issues relating to economic geography has put new vigour into research efforts intending to explain economic phenomena without neglecting space. Such a revival has two main causes as pointed out by Fujita, Krugman and Venables (1999): on the one hand the real-world concerns arising with the ongoing processes of regional integration, especially in Europe; on the other hand, from the theoretical point of view, the availability of the analytical tools necessary to deal with increasing returns bounded in space which characterise agglomeration forces.

Theory and empirics have, however, developed quite apart (see surveys by Ottaviano and Puga, 1998 and Bruhlart, 1998). Most importantly, empirical studies have become increasingly interested in testing for the effectiveness of externalities, even though lacking an appropriate investigation of the different mechanisms which are behind such phenomena (Breschi and Lissoni, 2001). Moreover, such empirical testing has too often been carried out with quite a partial perspective despite the fact that local economic growth is a complex and multifaceted issue.

This Chapter has a two objective. Firstly, we outline, as a synthesis of previous contributions, a general eclectic model of local economic growth to provide the theoretical background which should allow us to be as explicit as possible about what can be really inferred from our results and under what assumptions. The model includes a general taxonomy of different factors which may explain economic growth, measured by employment dynamics, in a specific industry and location. Secondly, we intend to assess within a wide-ranging framework the role of

* We would like to thank Barbara Dettori and Giuliana Caruso for the excellent research assistance. We thank the editor of this volume and an anonymous referee for very helpful comment. We have also benefited from comments by participants to the ERSA Conference in Zagreb, AISRE Conference in Venice, Cost Action 17 in Cambridge and IREM seminar in Naples.

a large set of potential determinants of the process of local agglomeration of economic activity.

In particular, the crucial mechanism which explains why firms tend to concentrate in a certain local industry is represented by the technological milieu in that area. The idea is that technology is not completely exogenous and freely available in the whole economy but it is spatially bounded (Jaffe *et al.*, 1993; Lawson and Lorenz, 1999). As a result, its level and growth rate depend on several elements which are specific to each industry and location. Firms take advantage of the external economies created by the presence in certain local industries of a fertile technological environment which increases firms' productivity.

The crucial issue we address in the Chapter is therefore the identification of a variety of factors which determines technology in the local industry and consequently influences local economic growth.[1] We introduce a useful classification of determinants of local technology in order to provide a general setting for testing different scenarios. Such a classification includes the usual distinction among several types of externalities, such as specialization, urbanisation and network economies. In this respect, it is more difficult to discriminate among knowledge and rent externalities, that is those which are due to pure diffusion of information and those which are mediated by the market, respectively. Our general framework allows us to consider also other elements which do not enter directly as inputs in the production function of firms but which certainly influence local technology and therefore firms' productivity: the availability of human capital, the characteristics of the social environment, and the accessibility to public infrastructures.

This Chapter is very much intended as a further step in the research path started by Glaeser *et al.* (1992) with their seminal contribution on the economic growth of US cities and followed by many studies on other countries' experiences. Henderson *et al.* (1995), for example, extended the study on a number of sectors of US Metropolitan areas. Other interesting contributions are Combes (2000b) and Funke and Niebuhr (2001) who study employment zones' dynamics in France and Germany respectively. As for Italy there have already been some important contributions along this line of research, such as those by Cainelli and Leoncini (1999), Cunat and Peri (2000), Forni and Paba (2001).

Our original contribution rests mainly on the attempt to widen and generalise the framework of analysis and to assess the relative importance of explanatory factors in different settings. In order to do this, we apply our model to the case of Italy which probably represents the most interesting experience of multifaceted economic development at the local level: the spatial dualism between the rich North and the underdeveloped South, the industrial districts based on a network of small and medium sized firms, and the development poles built around large companies in the heavy industries. Our empirical analysis makes use of a very ample database on socio-economic indicators for the Italian Local Labour Systems

[1] An alternative way is to use a direct measure of technology and therefore to investigate on the association between production and technology at the local level. A similar approach has been followed by Feldman and Audretsch (1999) Kelly and Hageman (1999) and Paci and Usai (2000a, 2000b).

(LLS). LLS are 784 groupings of municipalities, which cover the whole national territory, characterised by a high degree of self-contained flows of commuting workers (Sforzi, 1997). This high level of geographical breakdown appears particularly fruitful for the analysis of local growth since the production activities have, by construction, a high degree of self containment that makes it easier the identification of the explanatory factors at the local level. The information on local labour systems is also disaggregated with respect to 97 manufacturing sectors. The data are limited to the two census data 1991 and 1996 where the number of LLS is fixed, while it is problematic to extend the analysis to previous census data since the number of LLS is considerably different[2]. This limited temporal extension obviously introduces a limit to our empirical analysis, which is restricted in scope to the short-term.

All in all, the available databank consists of 76,048 observations. Thanks to such a large set we can show that local economic growth is not a homogeneous phenomenon, especially with respect to the geographical dimension. In other words, across the Italian macro regions several patterns of territorial development emerge in the nineties, characterized by significant differences in the relevance of the explanatory factors.

The outline is as follows. The second section presents the main stylised facts which come about from the descriptive analysis of manufacturing employment growth. In the third section we present a simple eclectic model to orientate the empirical analysis. The econometric estimations are presented and interpreted in section 10.4. In section 10.5 we directly address the issue of spatial autocorrelation in the growth process. Section 10.6 concludes.

10.2 Descriptive Analysis

The performance of the Italian manufacturing employment in the nineties has been dismal as pointed out in several studies (Cunat and Peri, 2000). In particular, manufacturing industries have lost, from 1991 to 1996, some 370,000 jobs out of around 5,200,000, with an average fall per year of 1.47%. However, the most striking feature of this industrial decline has been its considerable variety in terms of spatial distribution. As a result, geographical heterogeneity is one of our main focus.

First of all, it is worth noting that the distribution of workers in the local labour systems has become less polarised along time: concentration indexes show less dispersion in 1996 with respect to 1991. Nonetheless, employment dynamics follows the usual North-South pattern, even though some important qualifications emerge from the data especially among northern regions, as it is shown in Figure

---

[2] For instance, 944 LLS were identified in Italy in the 1981 Census.

10.1. As a matter of fact, it is useful to distinguish between the following six macro areas (with the included NUTS 2 regions between parentheses):

North-West	(Lombardia, Piemonte, Val d'Aosta, Liguria)
North-East	(Trentino, Friuli, Veneto, Emilia)
Center-North	(Toscana, Umbria, Marche)
Center-South	(Lazio, Abruzzo, Molise)
South	(Campania, Puglia, Basilicata, Calabria)
Islands	(Sicilia, Sardegna).

Table 10.1 Manufacturing employment in macro regions

	Employees (000)		Annual average variation
	1991	1996	
North West	2069	1896	-1.75
North East	1406	1359	-0.69
Centre North	683	649	-1.04
Centre South	364	331	-1.89
South	511	457	-2.24
Islands	194	166	-3.14
Italy	5228	4857	-1.47

From Table 10.1 it is therefore possible to note that there is a dualism within the North itself: the North-East shows a relatively good performance of manufacturing employment growth with a fall of "only" –0.69% per year, whilst the North-West stays below the national average due to a fall of 1.78% per year.

Such a variation is mainly the outcome of different stories of development. The one of the North-West is very much the development history of the Italian industrial system of large heavy industries with Turin, Milan and Genoa as main metropolitan centres, giving rise to the so called "industrial triangle". The one of the North-East is, on the contrary, a more recent story of industrial and service development based on local networks of small and medium dynamic firms and plants scattered throughout the area. This is the widely studied development model of the "industrial districts" (see, among many others, Brusco, 1982; Piore and Sabel, 1984). The regions of the Centre-North have a similar performance suggesting that the Italian model of small and medium enterprises agglomeration systems, typical of these two areas, have been rather successful in going through such a troublesome period.

At the other extreme, the South and the Islands show the worst performance with a loss of almost 3% of employees per year during the period from 1991 to 1996. Again this may be interpreted, as the result of the path followed so far by such regions as the outcome of externally induced process of industrialisation and the crowding out of weak domestic firms. The structural crisis of such heavy industries and the slow process of recovery and growth of a renewed structure of endogenous firms are behind such negative results.

It should be, however, remembered that we are just focusing on manufacturing and that the dynamics of total employment may be different. In particular the analysis of manufacturing sectors may be partial anytime the process of deverticalisation and outsourcing is carried out by involving the service sector. An analysis of the interaction of the dynamics of the two macro sectors is beyond the objectives of this study but it is certainly worth doing in the future.

Table 10.2. Manufacturing employment in selected areas

Local Labour System	Region	Employees		Annual average variation
		1991	1996	
A. Top 10 LLS				
Melfi	Basilicata	2659	9713	25.9
Lungro	Calabria	77	151	13.5
Palomonte	Campania	100	196	13.5
Belvedere Marittimo	Calabria	494	903	12.1
Taverna	Calabria	44	80	12.0
Custonaci	Sicilia	305	505	10.1
Sant'Agata de' Goti	Campania	669	1105	10.0
Frasso Telesino	Campania	169	265	9.0
Morcone	Campania	256	391	8.5
Apice	Campania	96	143	8.0
B. Worst 10 LLS				
La Maddalena	Sardegna	763	187	-28.1
Samugheo	Sardegna	416	140	-21.8
Naro	Sicilia	242	110	-15.8
Montebello Ionico	Calabria	640	304	-14.9
Tarquinia	Lazio	1506	748	-14.0
Manfredonia	Puglia	2237	1180	-12.8
Santa Teresa di Gallura	Sardegna	230	123	-12.5
Pisticci	Basilicata	4149	2257	-12.2
Accadia	Puglia	172	96	-11.7
Tricarico	Basilicata	258	144	-11.7

Table 10.3. Manufacturing employment in macro sectors

Sectors	Employees (000)		Annual average variation
	1991	1996	
Food, beverages and tobacco	476	446	-1.34
Textiles and clothing	823	692	-3.48
Leather	244	230	-1.12
Wood	186	170	-1.79
Paper and printing	284	260	-1.76
Coke and petroleum	29	24	-3.71
Chemical	239	209	-2.67
Rubber and plastic	179	198	1.96
Non metals	276	252	-1.85
Metals	785	761	-0.62
Machine and Instruments	541	553	0.45
Electrics and Electronics	491	457	-1.45
Transport	358	287	-4.48
Other manufacturing	315	318	0.22
All sectors - Italy	5228	4857	-1.47

It is also interesting to look at individual cases to spot "miracles" as much as "disasters" as reported in Table 10.2. Miracles seem to be concentrated in the South. They refer mostly to small areas where the location of new firms has a very strong impact. The most remarkable success is the one of Melfi associated to Fiat, which is the site of a plant for the production of vehicles, thanks to the financial and fiscal incentives available to the Objective 1 regions of the EU. Among the disasters we may find some cases which are due to structural crises both in the South (i.e., La Maddalena which has paid the price of the crisis involving a military arsenal) and in the North (Ivrea which has followed the decline of Olivetti). Obviously miracles and disasters are often the result of idiosyncratic shocks affecting specific sectors which are more prevalent in certain areas rather than others. As a matter of fact, Table 10.3 shows that there is as much variability from sector to sector as from one area to another one. The most successful sector is the Rubber and Plastic

Table 10.4 Moran test on spatial autocorrelation of manufacturing employment growth among LLS

Contiguity order	Italy	North-West	North-East	Centre-North	Centre-South	South	Islands
First order contiguity							
standardized Z values	6.43	4.22	4.76	1.11	0.98	1.79	0.74
probability level	0.00	0.00	0.00	0.14	0.17	0.04	0.23
Second order contiguity							
standardized Z values	6.49	2.93	3.94	0.61	1.03	2.34	-0.23
probability level	0.00	0.00	0.00	0.26	0.15	0.01	0.41
Third order contiguity							
standardized Z values	5.36	1.68	2.50	0.76	0.63	1.05	-0.06
probability level	0.00	0.05	0.01	0.23	0.27	0.16	0.48

sector with almost 2% of annual growth. Only two more sectors (Machine and Instruments and Other manufacturing) display a positive growth rate. The worst performance is shown by the Transport sector which loses 4% of its employment each year from 1991 to 1996. In this Chapter the focus is on the geographical rather than on industrial heterogeneity, the latter being a subject which deserves an analysis on its own[3].

Finally, as for the problem of spatial dependence, we find contrasting evidence. At the global level, the Moran index for the whole country, reported in Table 10.4, indicates the presence of spatial dependence among nearby local labour systems.[4] However, when one looks at the local level by means of the LISA (local indicator of spatial autocorrelation) reported in Figure 10.2, one discovers that there are only 211 LLS out of 784 which are spatially associated in a statistically significant way at the 10% level. Moreover, Figure 10.2 shows clearly that a cluster of sig-

[3] Interesting examples of such analyses are Combes (2000b), who directly investigates the different performance of the manufacturing and the service sectors in France and Pagnini (2001), who focuses on manufacturing sectors in Italy at the provincial level. Moreover, it might prove interesting to focus on specific sectors as it is done in Maggioni (2000) who studies the dynamics of high tech sectors.

[4] The spatial analysis presented throughout the Chapter is based on a row standardised contiguity matrix.

nificant associations is polarised in the North-East. One possible reading is that spatial correlation is not necessarily a uniform phenomenon and that it may depend on several geographical factors. As a matter of fact, in Table 10.4 the Moran statistic for the first order contiguity proves statistically significant at the 1% level only for the macro-areas in the North.

This evidence is a further hint of the necessity for an appropriate division of the national territory in some sub-regions to enable a careful look at geographical specificities.

10.3 The Model

The main hypothesis under examination may be summarised in a simple economic model based on previous contributions proposed by the literature on local growth; in particular, the model shares some features with Glaeser *et al.* (1992), Henderson *et al.* (1995), Cunat and Peri (2000). Suppose that a firm in industry j and location i follows a production function given by:

$$Y_{ij} = A_{ij} (L_{ij})^{1-\alpha} \tag{10.1}$$

where $0 < \alpha < 1$, Y is the level of output, A represents the state of technology and L are units of labour. This production function does not include firm specific capital as input and therefore is not able to take into account capital accumulation by the firm which may be labour saving. This simplification is mainly due to the lack of data on capital stocks (and even on investment flows) at the local industry level. As in all previous studies which have introduced such hypothesis (see the seminal work by Glaeser *et al.*, 1992), one assumes that this does not perturb too much the interpretation of the main results. For the case of our empirical analysis, the period of time examined is quite short (1991-1996) and therefore the absence of the capital stock accumulation process can be considered as a more reasonable assumption.

Another limiting factor for studies on local industry growth is the lack of output data at the sectoral and territorial level; this implies that we are not able to examine the growth process in terms of labour productivity but we have to focus on the employment dynamics. For each local industry ij the equilibrium employment level requires the value of labour marginal product to be equal to the wage rate, that is:

$$A_{ij} [(1-\alpha)(L_{ij})^{-\alpha}]P_{ij} = W_{ij} \tag{10.2}$$

where P_{ij} and W_{ij} are, respectively, the price and the nominal wage rate of sector j in location i. Let us focus on the dynamics by introducing the temporal dimension t and rewriting (10.2) in terms of growth rate of employment thanks to the transformation in logarithms:

$$\alpha \log(L_{ijt+1}/L_{ijt}) = \log(A_{ijt+1}/A_{ijt}) + \log(P_{ijt+1}/P_{ijt}) - \log(W_{ijt+1}/W_{ijt}) \tag{10.3}$$

We now concentrate on the determination of the technology growth rate in a certain local industry ij. It is important to remark that this variable stands for several elements which, although not directly included as inputs in the production function (10.1), still play a crucial role in determining the external environment and therefore firms productivity.

It is well known that the interpretation of the term A and the definition of its determinants have been one of the most controversial issues in the economic literature. Our specific interest in this Chapter is simply to outline a very general framework which enables us to evaluate empirically the role played by several factors often indicated by the literature on local growth. In general, the growth rate of technology in a local industry A_{ij} can be decomposed into three components: i) specific to the local industry, $\overline{A}_{ij}$; ii) specific to the area, A_i; iii) specific to the industry, A_j. In other words we can write:

$$\log(A_{ijt+1}/A_{ijt}) = \log(\overline{A}_{ijt+1}/\overline{A}_{ijt}) + \log(A_{it+1}/A_{it}) + \log(A_{jt+1}/A_{jt}) \tag{10.4}$$

Let us now examine in details the determinants of each channel of the overall growth rate of technology.

First of all, technological change has to be attributed to the idiosyncratic factors which are specific to each industry and to each location. Among these variables we assume that the key role is played by the specialisation externalities (SE) and by the scale effects and the degree of competition (SC), which can be summarised in a linear function such as:

$$\log(\overline{A}_{ijt+1}/\overline{A}_{ijt}) = \chi_0 + \chi_1 SE_{ijt} + \chi_2 SC_{ijt} \tag{10.5}$$

Following a long standing tradition, we start considering Marshallian externalities measured by an index of production specialisation, SE.[5] This variable measures pecuniary and localisation externalities such as a suitable supply of labour

[5] The index of relative specialisation for sector j and LLS i based on employment is computed as:

$$SE_{ij} = (E_{ij} / \sum_i E_i)/(\sum_j E_j / \sum E)$$

. The index has been standardized within the range –1 : +1.

and primary and intermediate goods (Ellison and Glaeser, 1999), the provision of specific goods and services (Bartelsman *et al.* 1994), and the availability of specific infrastructures and networks. However, this specialization index takes also into account the intra-industry flows of knowledge which occurs among similar firms located in the same area (Henderson *et al.*, 1995, Maskell and Malmberg, 1999). In general, the specialisation or Marshallian externalities capture the advantages gained by firms producing similar products within a bounded geographical location.

The second variable is the average dimension of plants *(SC)* which has been included in previous studies to consider two effects. First, the inverse of *SC*, that is the number of firms over workers, is interpreted by Glaeser *et al.* (1992) as a direct measure of the degree of local competition. Secondly, the number of employees per firm can be seen, more straightforwardly, as a proxy for economies of scale which may affect labour productivity (O' hUallachàin and Satterthwaite, 1992). In principle, it would be better to distinguish between the two effects defining two different indicators and including both of them in the estimated equation. Unfortunately, the lack of data on employment of individual firms does not allow the construction of a concentration ratio as a more appropriate indicator of local competition. The interpretation of this indicator remains, therefore, ambiguous.

Secondly, the changes of technology in a certain location *i,* common to all industries, may be due to some features which characterise the whole local labour system or even a bigger territorial unit which contains it, such as the province or the region. Local factors refer to a large set of externalities and other indicators of market conditions which are distinctive to each area and affect in the same way all industries located in that area. Among them, diversity externalities *(DE)*, network externalities (*NE*), human capital (*HK*), social capital (*SK*) and public capital (*PK*). This translates in the following specification:

$$\log(A_{it+1}/A_{it}) = \beta_0 + \beta_1 DE_{it} + \beta_2 NE_{it} + \beta_3 HK_{it} + \beta_4 SK_{it} + \beta_5 PK_{it} \quad (10.6)$$

Diversity externalities (DE) in the production activities (also known in the literature as Jacobs or urbanisation externalities; Jacobs, 1969) are measured by the inverse of the Herfindal index applied to employment.[6] They are expected to positively influence local growth under the hypothesis that a firm located in a certain area can benefit from the presence in the same area of a wide range of other firms operating in different sectors since it can enjoy inter-industries cross fertilisation.

---

[6] The diversity externalities for LLS *i* has been computed as the inverse of the Herfindal index based on the employment shares in all industries.

The index has been also normalised by the same index computed for the Italian case as follows, where the summations are over j:

$$DE_i = 1/\sum_j (E_{ij}/E_i)^2 \Big/ 1/\sum_j (E_j/E)^2$$

Network externalities (NE) are intended to take into account the influence of the size of the economic system, measured by the density of the resident population in each LLS, where a firm is located (Ciccone and Hall, 1996). We expect a positive effect on local growth when a larger population density means a higher local demand and the availability of a wider supply of local public services. At the same time, we may have a negative effect if congestion effects prevail giving rise to pollution and higher rents[7]. We have also included a second proxy for network externalities which focuses on the supply side taking into account the presence of small firms within the local economy. The idea is that a larger share of small firms is helpful for local growth since they often have to operate outside their optimal production scale through cooperation and integration with other firms at the local level. This stimulates the creation of local externalities. The opposite happens with large firms which are more vertically integrated and therefore are less involved in local networks.

Technological change in a certain area may be also affected by the stock of human capital available in that area. The literature has emphasized the positive role of human capital on growth (Mankiw *et al.*, 1992) although the correlation between economic development and education attainment has often turned out to be weak (Benhabib and Spiegel, 1994). In our case we have included two proxies to measure the availability in the local area of labour forces with different levels of education: the share of population with just the primary education and the share of population with a university education. The hypothesis is that the former variable measures low level education and therefore should negatively affect local growth, while a higher availability of well educated labour represents an advantage for the localization of firms thus fostering local growth.

Another important element often indicated by the literature as a key determinant of local growth is social capital.[8] Also in this case it is not an easy task to find the proper indicator for this very complex and intangible phenomenon. To measure the degree of trust in the local society we include an index of the propensity to cooperate among firms based on the number of inter-firms agreement and participations in consortia surveyed by the industrial census at the provincial level. The idea is that a higher propensity to cooperate among firms in a certain area helps local growth since it facilitates knowledge diffusion, decreases transaction costs, and enables firms to take advantage of local externalities. We also introduce a second indicator to capture the characteristics of the social environment: an index of the existence of organised crime at the provincial level. In this case, the hy-

[7] This variable is also meant as a control variable to take into account for the high heterogeneity in the dimension of local labour systems which include both urban and rural areas. It is interesting to notice that we have found similar result using the absolute size of the resident population. For a comprehensive debate on the importance of controlling for the size of the area (instead of the size of the local industry) one may refer to Combes (2000a).

[8] The classical references on the role of social capital are Coleman (1990) and Putnam (1993). See also Temple and Johnson (1998).

pothesis is that a high level of crime is detrimental to local development since it increases firms' costs and reduces expected revenues.[9]

The last locally specific variable we consider is public capital, given by a synthetic indicator of physical infrastructures at the provincial level. Following a large body of literature, we assume that a higher availability of infrastructures in a specific area encourages the localization of firms and thus positively influences local growth.[10]

The third component of technological growth rate of the local industry is specific to each industrial sector while it is common to all areas. It can be attributed to technological progress and opportunities within each industry at the national level (*STP*). The specification is as follows:

$$\log(A_{jt+1}/A_{jt}) = \gamma_0 + \gamma_1 STP_{jt} \tag{10.7}$$

To capture these nationwide effects we include the sectoral differences in technological progress and opportunities which are assumed to be identical across the country. The index is a stock measure of patenting activity per sector at the European Patent Office (EPO).

As for the last two elements of equation (10.3), price changes depend on the conditions prevailing in the output market (*OMC*):

$$log(P_{ijt+1}/P_{ijt}) = \phi_0 + \phi_1 OMC_{ijt} \tag{10.8}$$

Due to the lack of data at the product and LLS level, we use price changes for nine macro-sectors at the regional level. From the profit maximizing condition (2), we expect them to produce a positive influence on growth. Finally, wage changes are a function of labour market conditions (*LMC*):

$$log(W_{ijt+1}/W_{ijt}) = \delta_0 + \delta_1 LMC_{ijt} \tag{10.9}$$

Also in this case, we use wage changes for macro-sectors at the regional level. From the model, they are assumed to affect local growth negatively.

Concerning the latter two variables – prices and wages - we believe that the regional level may prove to be more appropriate than local labour systems as the proper geographical unit to approximate output and labour market conditions. The macro-sector level, on the contrary, is believed to be too large and, therefore, its

---

[9] The existence of a negative relationship between crime and growth has been suggested by Murphy *et al.* (1993).

[10] Several papers have highlighted the positive effect of infrastructures on local growth; see, among many others, Aschauer (1989), Eberts (1990) and Munnell (1990).

average may conceal some relevant information. Consequently, more research on this point is required.

The combination of (3) and (4), taking into account (5) through (9), gives rise to the following specification of employment growth which is suitable for estimation:

$$\log(L_{ijt+1}/L_{ijt}) = \eta + \chi_1 SE_{ijt} + \chi_2 SC_{ijt} + \beta_1 DE_{it} + \beta_2 NE_{it} + \beta_3 HK_{it} + \beta_4 SK_{it} + \beta_5 PK_{it}\ \gamma_1 STP_{jt} + \delta_1 OMC_{ijt} - \phi_1 LMC_{ijt} \qquad (10.10)$$

where η is a combination of the intercepts in equations 5-9.

In conclusion, from the model we can derive a taxonomy of five types of determinants of local employment growth:

1) Local and industry specific factors
2) Local specific factors
3) Industry specific factors
4) Output market conditions
5) Labour market conditions

As we have already pointed out, this very simple theoretical model has several shortcomings due to the lack of data on output and capital stock at the local industry level. However we believe it is still useful as a guide for the empirical analysis of the local growth process.

10.4 Econometric Estimation

Unlike previous studies, in this Chapter we can attempt to simultaneously consider different factors which are bound to affect local economic growth thanks to the broad perspective suggested by our theoretical framework. Actually, in the search of the best specification we do not apply the usual general to specific approach which consists of a sequence of deletions of variables which are found not significant from a statistical point of view. On the contrary, we carry out an analysis of parameter stability with respect to different subsections of our main sample. In other words we apply the same general specification to sub samples identified with respect to geographic features to establish if there is any difference in the value, sign and significance of the estimated coefficients.

The dependent variable in the econometric analysis is given by the average annual growth rate of employment over the period 1991-96 for each 97 three-digit manufacturing industries in 784 LLS. This indicator is normalised by the value it takes at the national-industry level so that we can directly investigate the relative performance of each local industry with respect to the national average. A similar normalisation procedure with respect to the national values has been followed for the exogenous variables expressed by an index; that is, all indicators except the stock of patents. A table with the source of the data and the definition of each indicator is reported in Table A.1 in the appendix.

It is important to remark that all our regressors are exogenous to the local industry employment growth rate. This is unambiguous for the variables which refer to the beginning of the period considered, but this is also the case for the two growth rates - prices and wages - included in the regression. As we have already pointed out, both variables are defined for a very broader territorial (20 regions) and sectoral (9 macro-sectors) level with respect to the one used for the dependent variables (784 and 97, respectively). In such a case it is unlikely that the growth rate of a local industry can exert any influence on the growth rate of its macro-sector at the regional level.

We have estimated our model using Feasible Generalised Least Squares (FGLS) to correct for the presence of heteroskedasticity[11]. To take into account the risk of variables omission with respect to the industry dimension we include sectoral fixed effects. We have also tried to control for local fixed effects but they turn out to generate problems of multicollinearity given the simultaneous presence of several explanatory variables specific to each area.

10.4.1 Aggregate Results

Results of the FGLS estimation of equation (10.10) at the national level (with 76,048 observations given by 97 manufacturing sectors and 784 LLS) are reported in the first column of Table 10.5. Most explanatory variables are significant at the 1% level and show a plausible sign.

The first interesting outcome is the absence of externalities due to productive specialisation: the coefficient of SE is negative and highly significant. Such a result is, to some extent, not surprising given that it has been found in the US (Glaeser *et al.*, 1992), in France (Combes, 2000b) and, most importantly, in Italy at both the provincial (Forni and Paba, 2000) and regional level (Cainelli *et al.*, 2001). Nonetheless, the absence of Marshallian externalities also at the local labour system level is a notable and rather unexpected result. On this point it is worth remembering that our analysis covers a short time period characterised by a bad economic crises which may have induced stronger reorganization processes in those local productive systems which were highly specialised. Such systems are, as a matter of fact, characterised by a relatively more pervasive and rigid structure which is more costly to modify and transform (Combes, 2000). An this point, we also note that most highly specialised local production systems in Italy operate in traditional and mature sectors and that the negative relationship between initial specialisation and employment growth can also be linked to a product cycle mechanism.

---

[11] Heteroskedasticity is also taken into account by using White-robust standard errors.

Table 10.5 Econometric results

Variables			Italy	North-West	North-East, Centre-North	Centre-South, South and Islands
Local and industry specific variables	SE	specialisation externalities	-0.243 [a]	-0.188 [a]	-0.169 [a]	-0.335 [a]
	SC	scale effect - competition	0.000 [a]	0.000	-0.003 [a]	0.001 [b]
Local specific variables	DE	diversity externalities	0.067 [a]	0.026 [a]	0.030 [a]	0.091 [a]
	NE1	population density	-0.001	-0.003 [a]	-0.001	0.000
	NE2	small firms	0.015 [a]	0.022 [a]	0.010 [a]	0.004
	HK1	primary education	-0.017 [a]	-0.034 [c]	-0.023 [a]	-0.009
	HK2	university education	0.024 [a]	0.002	0.006 [c]	0.026 [a]
	SK1	cooperation propensity	-0.181 [c]	-0.168	0.306 [a]	-0.247
	SK2	crime rate	-0.001 [b]	0.001	-0.004	-0.001 [b]
	PK	infrastructure	0.000	-0.003	0.002	0.013 [c]
Industry specific variables	STP	sectoral technological progress	0.012	-0.003	-0.008	0.024
Output market cond's	OMC	price growth rate	-0.003 [b]	-0.001 [b]	0.006 [b]	-0.001
Labour market cond's	LMC	wage growth rate	0.000	0.003 [a]	0.000	0.000
		n. of observations	76048	13580	24444	38024
		Adj. R-squared	0.15	0.12	0.15	0.25
		S.E. of regression	0.39	0.41	0.40	0.37

Dependent variable: employment growth in the local sector, annual average 1991-1996 (LG)

Level of significance: a=1%, b=5%, c=10%

As for the average firm size, this is found to be positively related to local growth. This result can be interpreted as evidence in favour of either economies of scale enhancing local growth or of competition hindering it. In the following section we attempt to provide some interpretative elements to disentangle this ambiguity.

As far as local specific determinants are concerned, it has to be noted that positive diversity externalities appear strong and stable, substantiating most of the evidence collected by previous studies.[12] In particular, the coexistence of a positive effect due to diversity and a negative one due to specialisation, confirms previous results (see Paci and Usai, 1999, 2000b). Most importantly, this reveals the need for appropriate indicators to evaluate such effects independently. As for network externalities, the size of the local system measured by population density is negatively linked to employment dynamics but it is not significant. A positive and significant influence on local growth is, on the contrary, associated to the presence of a relatively higher quota of small firms in the area.

All these results reinforce the idea that - especially in a period of negative business cycle like the one being considered - it may prove crucial to rely on a production system based on a diversified network of small flexible firms.

The indicators referring to the different qualities of capital (human, social and public) show interesting composite results. First, both primary and university education emerge as relevant determinants of local growth: the former with a negative impact and the latter with a positive one. However, this relationship proves more complex (as it is also shown in Lodde, 2000 and Di Liberto, 2001) when one moves to a more detailed geographical analysis (see section 10.4.2). Secondly, the importance of social capital is not signalled by the coefficient of the variable which measures cooperation among firms, which is negative but significant only at the 10% level[13]; on the contrary, the relevance of social environment appears in the negative and significant sign of the crime index[14]. Thirdly, the indicator of infrastructures, that is public capital, is positively related to local economic growth but it is not significant.[15]

Finally, the last three indices prove rather fragile. In particular, technological progress specific to a certain industry either does not seem to diffuse nationwide in any homogenous way or simply does not affect employment growth. A possible explanation is that our indicator for national technology – i.e. the stock of Italian patents at the European Patent Office - may be inadequate in taking into account the complex nature of technological progress. Moreover, we may have multicollinearity problems with the sectoral fixed effects included in the regressions. As for output and labour market conditions, both indicators show coefficients with the wrong sign – negative and positive, respectively. However, only the price growth

---

[12] Forni and Paba (2001) suggest the relevance not just of variety per se but also of a specific type of variety defined in terms of input-output linkages. In a different setting the same result is obtained by Cunat and Peri (2000). Such specification, particularly problematic at this level of disaggregation, is an objective for future research.

[13] On the importance of social capital in the regional growth process in Italy see Helliwell and Putnam (1995). A complete array of interesting findings at the provincial and city level can be found in the works by Peri (1997) and Forni and Paba (2000).

[14] The complex relationship between criminal and economic activity has been thoroughly analysed in Marselli and Vannini (1999). In particular, empirical evidence on the link between crime and economic growth in Italy can be found in Attanasio and Padoa-Schioppa (1991) and in Forni and Paba (2000).

[15] The positive effect of public infrastructures on economic growth for the Italian regions has been shown by Paci and Pigliaru (1995) and Picci (1999).

rate is significant but just at the 5% level. As we have previously remarked this may depends on an inappropriate measurement of these variables at the sectoral level.

10.4.2 Results for Macro Regions

The descriptive analysis in section 10.2 and evidence in Table 10.1 suggest the presence of diversified patterns of development in Italian macro regions which hints at parameter variability with respect to geographical location[16].

In Table 10.5, columns 2 to 4 summarise the results of FGLS estimations for three macro regions which have been singled out thanks to some preliminary regressions applied to the macro-areas included in Table 10.1. Such preliminary regressions have allowed us to ascertain that the North-East and the Centre-North have homogenous patterns of employment dynamics with similar sets of coefficients and therefore they have been aggregated into the macro region NEC. The same happens for the Centre-South, the South and the Islands which are aggregated into the macro area CSI. The third region we consider is the North-West which presents a rather peculiar growth pattern. It is immediately clear that there appear significantly different impacts, in qualitative and quantitative terms, with respect to this territorial division.

In particular, we have the confirmation that productive specialisation accounts for negative effects on growth in all areas. Nonetheless, the impact on labour dynamics is relatively stronger in the CSI region (-0.335) with respect to the North (-0.188 for NW and –0.169 for NEC). This result may have several interpretations. On the one hand, it is possible that the histories of specialisation of these areas are affecting the current dynamics in a path-dependent fashion. In other words, it can be argued that areas in the North have selected their specialisation path following their competitive advantage, whilst in the South most local labour systems have been imposed by the industrial specialization pattern through public intervention[17]. Once the public intervention is eliminated, the exogenously imposed productive specialisation no longer has a positive influence on local growth. On the other hand, we might recall the short-run nature of our data. On this aspect, we have argued above, following Combes (2000b), that there may be asymmetric effects associated with specialisation: it enhances local growth during expansions but favours employment declines during recession, such as in the nineties, due to

---

[16] Cunat and Peri (2000) analyse aggregate economic growth in LLS introducing differences among macro-areas. However, such differences are assumed to be caught just by considering changes in the coefficient of macro-regional dummies with respect to different specifications of the estimated model. Consequently, no structural differences in parameters are directly taken into account.

[17] In the two decades from 1971 to 1991 it has been shown that the productive structure of the South has become more and more similar to the one of the North mainly as a result of public intervention (Rombaldoni and Zazzaro, 1997)

inflexibilities and rigidities. According to Brusco (1982), firms of the industrial districts in the North-East and in the Centre-North are more apt at reacting to the downturns of business cycles thanks to their high flexibility. Quite the contrary applies to the local systems located in the South and the Islands, where the productive structure is either based on medium and large plants which depend on subcontracting with firms located in the North; or, alternatively, on a high number of small and medium firms unable to constitute a real systemic network.

On this line of interpretation, it is worth noting that the average firm size in the NEC, contrary to all the other areas, has a negative impact on employment dynamics. This is probably because in such a macro-area economies of scale are not achieved within the plant but within a network of several plants and firms located in the area. At the same time, this result can be interpreted as a sign of the positive outcome of competition which enhances efficiency and innovation. In this light, one may also note that the presence of small firms is a strong positive determinant of growth in the NEC and in the NW. By contrast, small firms have a neutral role in the CSI (positive coefficient but not significant) confirming the well known fact that small firms in the South are often isolated entities which are relatively unable to exploit external economies available in the local system thanks to inter-firm interactions. As for the other network externalities, that is population density, it is worth noting that distinction among macro-regions allows one to detect a negative and significant coefficient for the North-west where some congestion effects are plausible. No effect is at work in the other two macro-areas.

Other interesting considerations come about from the analysis of the effects of diversity, that is urbanization economies. They have a positive effect on local growth in all macro-areas. In particular, it is worth noting that diversity proves particularly effective in improving local growth perspectives in the South (0.91) with respect to the North (0.24 in the NW and 0.29 in the NECN).

The presence of high rates of primary education has a negative impact in the NEC, whilst it is not effective in the NW and in the macro-areas of the South. In this latter area university education proves especially important. Infrastructures are a positive discriminatory factor only in the CSI even though the level of significance is only 10%[18]. This indicator remains insignificant in the regions of the North.

One other notable difference concerns the significance and sign of social capital, when measured by the cooperation index. Interestingly, cooperation proves extremely important in the NEC whilst it is neutral in the NW and in the CSI (negative and not significant). This brings further support to the role played by the favourable social settings and institutions in the success of industrial districts in the NEC (Becattini, 1991). As for the other indicator of social environment - the degree of diffusion of the organised crime in the area - this proves important in

[18] A higher effectiveness of public infrastructures in Central and Southern Italian regions with respect to Northern ones has been found also by La Ferrara and Marcellino (2000).

discriminating local system growth only in the South where it has the expected negative sign and significance at 5% level.[19]

Finally, it is to be noted that price growth rate is positive and significant at the 5% level in the NEC and that wage growth rate is positive and significant, at the 1% level, in the NW.

10.5 Spatial Dynamic Model: an Extension

In section 10.2, looking at the evidence provided by Table 10.4 and Figure 10.2, we acknowledge the presence of spatial interdependence at the aggregate level: i.e. there seems to be an interaction in the dynamics of manufacturing employment among contiguous local labour systems. However, the high number of observations in our dataset does not allow the implementation of the spatial econometric analysis to the whole Italy. Consequently, we opt to focus on one macro-area: the North-East and Centre-North (NEC). Our choice is based on the fact that the spatial dependence process is not homogenous nationwide (Table 10.4). In particular, from Figure 10.2 it clearly emerges that the significant LISAs are clustered particularly in the North-East and the Centre-North. Moreover, the interest in the NEC area is justified by the fact that this region represents a success story for local economic growth in Italy based on a diffused network of industrial districts. As a matter of fact, the Moran index applied to the residuals obtained in the FGLS regression reported in Table 10.5 for the NEC (Z=29.7) clearly shows the presence of spatial autocorrelation[20].

When autocorrelation is present two alternative solutions are available. First of all, one can apply a spatial lag model, which implies the inclusion of the spatially lagged dependent variable in the specification. This solution is interpreted as a proper substantive spatial model because it directly incorporates spatial interaction. Alternatively, one can implement a spatial error model, where the spatial dependence is referred to the disturbance term correcting what is called nuisance dependence.

We have implemented the former model because it allows a direct assessment of the existence, nature and strength of spatial interaction. The spatial lag model cannot be estimated with the usual OLS (or FGLS), given that the spatial lagged dependent variable is endogenous, which makes the estimated coefficients biased and inconsistent. Therefore, we have estimated our spatial lag model making use

[19] Other indexes of criminal activity have been used, such as number of homicides and crimes against property. Results are almost identical.

[20] It should be remembered that our data set has two dimensions (geographical and sectoral) and that autocorrelation of residuals may have not just a spatial component but also a sectoral one. In this analysis we focus on just spatial autocorrelation.

of the generalised method of moments (GMM) [21]. The spatial lagged variables have been previously generated with the software *Spacestat* on the basis of a row-standardised contiguity matrix. Following Keleijan and Robinson (1993), we have employed the spatial lags of explanatory variables[22] as instrumental variables.

Table 10.6 Spatial dynamic models: North East and Centre North

Variables			Regr 1	Regr 2	Regr 3
Local and industry specific variables	SE	specialisation externalities	-0.225 [a]	-0.224 [a]	-0.225 [a]
	SC	scale effect - competition	-0.003 [b]	-0.003 [b]	-0.003 [b]
Local specific variables	DE	diversity externalities	0.071 [a]	0.073 [a]	0.074 [a]
	NE1	population density	0.014 [a]	0.013 [a]	0.009 [c]
	NE2	small firms	0.016 [b]	0.014	0.018 [c]
	HK1	primary education	-0.148 [a]	-0.145 [a]	-0.147 [a]
	HK2	university education	-0.002	0.003	0.000
	SK1	cooperation propensity	1.303 [a]	1.237 [a]	1.107 [a]
	SK2	crime rate	-0.011	-0.010	-0.012
	PK	infrastructure	-0.017	-0.014	-0.018
Industry specific variables	STP	sectoral technological progress	-0.011	-0.008	-0.013
Output market cond's	OMC	price growth rate	0.027 [a]	0.023 [a]	0.025 [a]
Labour market cond's	LMC	wage growth rate	-0.002 [b]	-0.002 [b]	-0.002 [b]
Spatial lag variables	W_LG	employment growth rate, 1° contiguity	-0.524 [a]	-0.543 [a]	-0.544 [a]
	W2_LG	employment growth rate, 2° contiguity		0.349	0.240
	W_DEN	contiguous LLS population density			0.011

Dependent variable: employment growth in the local industry (LG), annual average 1991-1996. Estimation method: GMM with spatial lag variables and industry fixed effects, White covariance.
Instruments: spatial lag explanatory variables with 1st order contiguity
Level of significance: a=1%, b=5%, c=10%. Number of observations: 24444

[21] An alternative estimation technique is the maximum likelihood (ML) estimation. However, as Anselin (1999, pp. 22-23) notes, this is not a feasible option with very large data sets due to the memory constraints in the estimation procedures.

[22] We do not insert too many instruments (spatial lags higher than one) because this is likely to create problems of multicollinearity (see Fingleton, 2001).

The main results of a selection of GMM specifications are reported in Table 10.6. Most results confirm FGLS estimations which prove, in general, rather robust. Among robust coefficients we find those associated with specialization and diversity externalities, scale economies and competition effects, primary education and cooperation propensity. On the contrary, it is interesting to notice a change in the sign and the significance of the coefficient for population density. In the spatial lag model it turns out to be always positive and statistically significant. There are no congestion effects at work in the NEC area and this result is more consistent with a production process based on industrial districts which are spread around in the territory, usually away from big cities.

Moreover, there are some indicators which became statistically significant in these specifications while they were not in the FGLS estimation: the indicators of output and labour market conditions become both significant and with the correct sign.

With regard to the spatial dynamics in the three regressions, some interesting results appear which deserve special consideration. In all the regressions the first-level spatial lag of the dependent variable comes out negative and significant, whereas it is positive and insignificant at the second level. This outcome, albeit peculiar, may be interpreted as a consequence of the bi-dimensional nature of our data, which are both local and sector specific. In other words, when each growth rate is referred to a local industry, and most importantly homogeneous territorial units are correctly identified by means of employment flows, polarization phenomena are likely to emerge. Such phenomena may give rise to a negative relationship between the dynamics of one sector in a local system and the dynamics of the same sector in contiguous areas. In regression three the insertion of the population density of the contiguous area proves to be insignificant, while other results remain unchanged.

10.6 Conclusions

This Chapter contributes to the literature which aims to unravel the complex issue of local economic growth. We present a theoretical model which is meant to put our empirical analysis on the right track, in terms of hypotheses to be tested and implied assumptions. The analysis refers to the situation of Italy in the nineties, which proves to be an extremely interesting and revealing case study.

Our results prove quite interesting particularly because they suggest that local growth may be a process characterized by significant differences across macro regions. First of all, we find robust evidence on the negative influence of specialisation externalities on labour dynamics at the local industry level, especially for the Southern regions. A possible explanation for the absence of Marshallian externalities in the local industry is the short period considered, which is characterised by a process of reorganisation of local manufacturing systems. Moreover, most highly specialised local production systems in Italy operate in traditional and mature industries and therefore the negative relationship between initial specialisation and employment growth can also be linked to product cycle mechanisms. As for the average firm size, this is found to be positively related to local growth. This result can be interpreted as evidence in favour of either economies of scale enhancing local growth or of competition hindering it.

Among the most robust determinants of local industry growth, it is worth mentioning the positive role of diversity externalities. The presence in a certain location of a production system characterised by a wide range of other firms operating in different sectors produces positive effects on labour dynamics since it favours inter-industries cross fertilisation and guards against idiosyncratic shocks.

As for network externalities, the size of the local system measured by population density has no significant effect on the growth of LLS. A positive and significant influence on local growth is also associated to the presence of a relatively higher quota of small firms in the area. The other determinants of local industry growth exert the expected effect: a negative impact on labour dynamics is found for low education and crime rate, while a positive one is detected for university education, cooperation propensity, infrastructures.

However, it is important to remark that the significance of these factors possesses a high variation across macro regions. For instance, the average firm size in the North-East and Centre-North, contrary to all the other areas, has a negative impact on employment dynamics which can be interpreted as evidence of external economies of scale and of a growth-enhancing effect of competition. Moreover the presence of small firms retains its positive effect in the two macro areas of the North but loses it in the South. Another notable feature of the North-East macro area is the relevance of social capital, measured by the level of cooperation among firms. This confirms the importance of favourable social settings and institutions in the success of industrial districts in this area.

As for the spatial dynamics, the analysis is carried out only for the North-East and Centre-North where the employment dynamics of the nineties has been relatively successful and where most industrial districts are located. The GMM estimation, while confirming most results of the FGLS estimation, reveals some interesting, and unexpected, circumstances. The estimated coefficient of the spatial lag of the dependent variable is consistently negative. This implies that at the local industrial level there are polarisation phenomena at work and that positive externalities are self-contained within the boundaries of local labour systems once we have controlled for a large set of local determinants.

This work clearly suggests that it is difficult to find a general explanation of local economic growth suitable for the entire country and that it is crucial to test for the robustness of the determinants especially with respect to the geographical setting. As a result, policy implications are rather complex. Nonetheless, there are remarkably robust results indicating a clear positive role to diversity and a negative one to specialisation at least during a recession. These results together with the one regarding the positive role of small firms should be borne in mind when shaping policy intervention in fragile areas. Other important indications, albeit more standard, are those which signal the importance of social and human capital.

As for future research, in this Chapter we have neglected a detailed investigation of industrial heterogeneity which deserves a thorough study on its own. In this vein it is crucial that the analysis be extended to the service sector in order to analyse the cross effects among industries. At the same time it may prove rather intriguing to implement the same analytical and empirical framework to specific sub samples of industries, such as, for example, high tech sectors.

References

Anselin L. (1999), Spatial Econometrics, *mimeo*, University of Texas at Dallas.

Aschauer D.A.(1989), Is Public Expenditure Productive?, *Journal of Monetary Economics*, 23, 177-200.

Attanasio O. and Padoa Schioppa F. (1991), Regional Inequalities, Migration and Mismatch in Italy, 1960-1986, in Padoa Schioppa F. (ed.) *Mismatch and Labor Mobility*. Cambridge: Cambridge University Press.

Bartelsman E. Caballero R. and Lyons R. (1994), Customer- and Supplier- Driven Externalities, *American Economic Review*, 84, 1075-1084.

Becattini G. (1991), The Marshallian Industrial Districts as a Socio-economic Notion, in Pike F., Becattini G and Sengenberger W (eds.), *Industrial Districts and Inter-Firm Cooperation in Italy*. Geneva: International Institute for Labour Studies.

Benhabib J. and Spiegel M. (1994), The role of Human Capital in Economic Development: evidence from aggregate cross-country data, *Journal of Monetary Economics*, 34, 143-174.

Breschi S. and Lissoni F. (2001), Knowledge Spillovers and Local Innovation Systems: a Critical Survey, *Industrial and Corporate Change*, 10, 975-1005.

Brulhart M. (1998), Economic Geography, Industry Location and Trade: The Evidence, *World Economy*, 21, 775-801.

Brusco S. (1982), The Emilian Model: Productive Decentralisation and Social Integration, *Cambridge Journal of Economics*, 6, 167-184.

Cainelli G. and R. Leoncini (1999), Externalities and Long-Term Local Industrial Development: Some Empirical Evidence for Italy, *Revue d'Economie Industrielle*, 90, 25-39

Cainelli G., R. Leoncini and A. Montini (2001), Struttura produttiva e sviluppo regionale di lungo periodo in Italia", *Moneta e Credito*, 215, 1-25.

Ciccone A. and Hall R. (1996), Productivity and the density of economic activity, *American Economic Review*, 86, 54-70.

Coleman J. (1990), *Foundations of Social Theory*. Cambridge MA: Harvard University Press.

Combes P.F. (2000a), Marshall-Arrow-Romer Externalities and City Growth, *CERAS working papers* 99/06.

Combes P.F. (2000b), Economic Structure and Local Growth: France, 1984-1993, *Journal of Urban Economics,* 47, 329-355.

Cunat A. and Peri G. (2000), Job creation: Geography, Technology and Infrastructures, *IGIER wp*175, Bocconi University.

Di Liberto A. (2001), Stock di capitale umano e crescita delle regioni italiane. Un approccio panel, *Politica Economica*, 17, 159-184.

Eberts R.(1990), Public Infrastructure and Regional Economic Development, *Economic Review, Cleveland Federal Reserve Bank*, 1, 15-27.

Ellison G. and Glaeser E. (1999), The Geographic Concentration of Industry: Does Natural Advantage Explain Agglomeration?, *American Economic Review Papers and Proceedings*, 89, 301-316.

Feldman M. and Audretsch D. (1999), Innovation in Cities: Science-Based Diversity, Specialization and Localized Competition, *European Economic Review*, 43, 409-429.

Fingleton, B. (2001), Theoretical economic geography and spatial econometrics: dynamic perspective, *Journal of Economic Geography*, 1, 201-225.

Forni M. and Paba S. (2000), The sources of local growth: evidence from Italy, *Giornale degli Economisti e Annali di Economia*, 59, 1-49.

Forni M. and Paba S. (2001), Knowledge Spillovers and the Growth of Local Industries, *CEPR Discussion Papers* n. 2934.

Fujita M., Krugman P.and Venables A.J. (1999), *The Spatial Economy*. Cambridge MA: MIT Press.

Funke M. and Niebuhr A. (2001), Regional Geographic R&D spillovers and economic growth – evidence from West Germany, forthcoming in *Papers in Regional Science*.

Glaeser E.L., Kallal H.D., Sheinkman J.A.and Shleifer A. (1992), Growth in Cities, *Journal of Political Economics*, 100, 1126-1152.

Helliwell J. and Putnam R. (1995), Economic growth and social capital in Italy, *Eastern Economic Journal*, 21, 295-307.

Henderson V., Kunkoro A. and Turner M. (1995), Industrial Development in Cities, *Journal of Political Economics*, 103, 1067-1090.

Jacobs J. (1969), *The Economy of Cities*. London: Jonathan Cape.

Jaffe A., Trajtenberg M. and Henderson R. (1993), Geographic Localisation of Knowledge Spillovers as Evidenced by Patent Citations, *Quarterly Journal of Economics*, 108, 577-598.

Kelejian H.H. and Robinson D.P. (1993), A Suggested Method of Estimation for Spatial Interdependent Models with Autocorrelated Erros, and an Application to a County Expenditure Model, *Papers in Regional Science*, 72, 297-312.

Kelly M. and Hageman A. (1999), Marshallian Externalities in Innovation, *Journal of Economic Growth*, 4, 39-54.

La Ferrara E. and Marcellino M. (2000), TFP, Costs, and Public Infrastructure: An Equivocal Relationship, *IGIER wp 176*, Bocconi University.

Lawson C. and Lorenz E. (1999), Collective Learning, Tacit Knowledge and Regional Innovative Capacity, *Regional Studies*, 33, 305-317.

Lodde S. (2000), Education and Growth: Some Disaggregate Evidence from the Italian Regions, *Rivista Internazionale di Scienze Economiche e Commerciali*, 47, 267-293.

Maggioni M. (2000), Structure and dynamics of high-tech clusters: an econometric exercise. *Quaderni del dipartimento, Dipartimento di Economia, Istituzioni e Territorio*, Università di Ferrara.

Mankiw N.G., Romer D. and Weil D. (1992), A contribution to the empirics of economic growth, *Quarterly Journal of Economics*, 107, 407-437.

Marselli R. and Vannini M. (1999), *Economia della criminalità*. Torino: Utet.

Maskell P. and Malmberg A. (1999), Localised Learning and Industrial Competitiveness, *Cambridge Journal of Economics*, 23, 167-185.

Munnell Alicia H. (1990), How does Public Infrastructure Affect Regional Economic Performance?, *New England Economic Review*, Federal Reserve Bank of Boston, 11-33.

Murphy K., Shleifer A. and Vishny R. (1993), Why is rent-seeking so costly to growth?, *American Economic Review*, 83, 409-414.

O' hUallachàin B. and M. Satterthwaite (1992), Sectoral growth patterns at the metropolitan level: an evaluation of economic development incentives, *Journal of Urban Economics*, 31, 25-58.

Ottaviano I. P. G. and Puga D. (1998), Agglomeration in the Global Economy: A Survey of the "New Economic Geography", *World Economy*, 21, 707-731.

Paci R. and Pigliaru F. (1995), Differenziali di crescita nelle regioni italiane: un'analisi cross-section, *Rivista di Politica Economica*, 85, 3-34.

Paci R. and Usai S. (1999), Externalities, Knowledge Spillovers and the Spatial Distribution of Innovation, *GeoJournal*, 49, 381-390.

Paci R. and Usai S. (2000a), Technological Enclaves and Industrial Districts. An Analysis of the Regional Distribution of Innovative Activity in Europe, *Regional Studies* 34, 97-104.

Paci R. and Usai S. (2000b), The Role of Specialisation and Diversity Externalities in the Agglomeration of Innovative Activities *Rivista Italiana degli Economisti*, 5, 237-268.

Pagnini M. (2001), L'agglomerazione spaziale nei comparti manifatturieri in Italia: aspetti metodologici, interpretazioni ed evidenza empirica, *mimeo*, Banca d'Italia.

Peri G. (1997), Do civic spirit and economic diversity help local technological accumulation? Evidence from Italian cities and provinces 1961-1991, *CESPRI wp* 101, Bocconi University.

Picci L. (1999), Productivity and Infrastructure in the Italian Regions, in *Giornale degli Economisti e Annuali di Economia*, 58, 329-353.

Piore M. and Sabel C. (1984), The second industrial divide: Possibilities for prosperity. New York: Basic Book.

Putnam, R. (1993), *Making Democracy Work: Civic Traditions in Modern Italy*. Princeton: Princeton University Press.

Rombaldoni R and Zazzaro A. (1997), Localizzazione delle imprese manifatturiere e specializzazione regionale in Italia: 1970-1990, *Rassegna dei lavori dell'ISCO*, XIV, 151-181.

Sforzi F. (1997) (ed.), *I sistemi locali del lavoro 1991*, Argomenti n.10. Roma: ISTAT.

Temple J. and Johnson P.A. (1998), Social Capability and Economic Growth, *Quarterly Journal of Economics*, 113, 965-990.

Appendix

Table A.1 Variables description and sources

Variables	Index	Level of aggregation area *	Level of aggregation sector **	Sources
Dependent variable				
- Local industry growth	annual average growth rate of employment (S)	LLS	3-digit ateco91	1991 - 1996 Industrial Census
1. Local and industry specific variables				
- Specialisation externalities	index of employment relative specialisation (S)	LLS	3-digit ateco91	1991 Industrial Census
- Scale effects, competition	Number of employees over number of plants (S)	LLS	3-digit ateco91	1991 Industrial Census
2. Local specific variables				
- Diversity externalities	inverse of Herfindal index for employment (S)	LLS	-	1991 Industrial Census
- Network externalities				
Population density	resident population over Km^2 (S)	LLS	-	1991 Population Census
Small firms	quota of workers in firms with less than 50 employees (S)	LLS	-	1991 Industrial Census
- Human capital				
Primary education	population with primary education / pop > 9 (S)	LLS	-	1991 Population Census
University education	population with university education / pop > 24 (S)	LLS	-	1991 Population Census
- Social capital				
Cooperation propensity	quota of firms with inter-firms agreements (S)	province	-	Industrial Census Long Form
Crime rate	index of organised crimes (S)	province	-	Censis
- Public capital	general index of phisical infrastructure (S)	province	-	UnionCamere
3. Industry specific variables				
- Sectoral technological progress	stock of sectoral patents, cumulative value 1978-1991	-	3-digit ateco91	CRENoS databank on EPO data

4. Output market conditions				
- Price growth rate	value added deflator, annual variation, 1991-96	region	macrosectors	CRENoS REGIO-IT data-bank
5. Labour market conditions				
- Wage growth rate	gross wages, annual variation, 1991-96	region	macrosectors	CRENoS REGIO-IT data-bank

(S) means that the indicator has been standardised to the national value

* Local Labour System=784; Province= 92; Region=20.

** 3-digit ateco91= 97; macrosectors= 9: food and beverages, textile and clothing, wood and plastic, paper and printing, chemical products, non metals, metals, metal products and machinery, transport

Fig. 10.1 Manufacturing employment dynamics in the local labour systems in Italy (1991-1996)

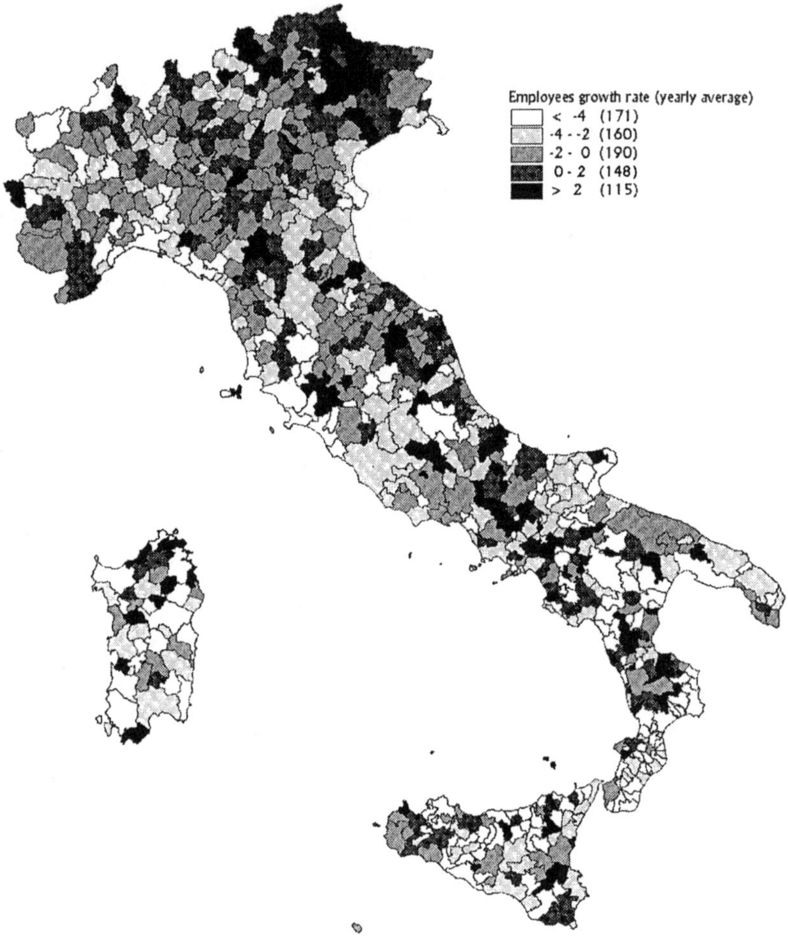

Fig. 10.2 Local indicators of spatial association (significant at 10%) of employment dynamics in the local labour systems in Italy (1991-1996)

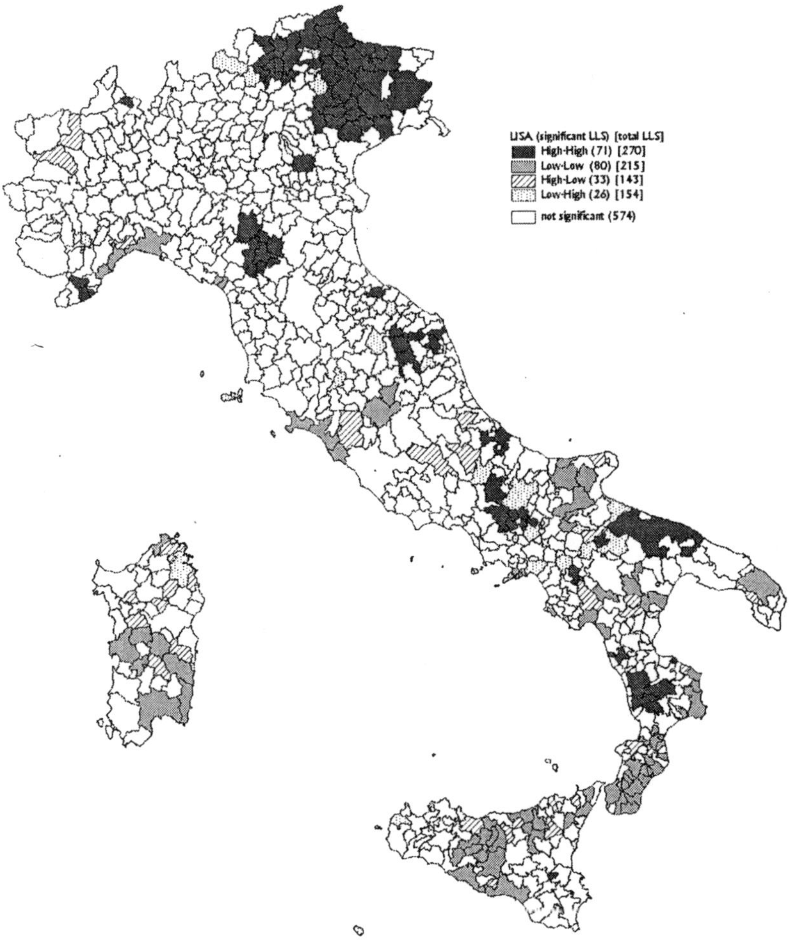

11 Regional Disparities in Income and Unemployment in Europe

Ådne Cappelen[1], Fulvio Castellacci[2], Jan Fagerberg[2] and Bart Verspagen[3]

1Research Department, Statistics Norway, 2 Center for Technology, Innovation and Culture (TIK), University of Oslo, 3 Eindhoven Center for Innovation Studies (ECIS)

11.1 Introduction

Greater equality across Europe in productivity and per capita income has been one of the central goals for the European Community since the early days of European economic integration. Various policy measures have been introduced to help achieve this goal (the so-called "Structural Funds"). For a long time the regions of Europe were on a converging path and, hence, the existing set of policies seemed to have the desired effect (Molle 1980). More recent evidence has, however, challenged these previous findings by showing that the tendency towards convergence came to a halt in the beginning of the 1980s (Fagerberg and Verspagen 1996, Cappelen, Fagerberg and Verspagen 1999). This evidence, based on data for 12 EU countries up 1997 and presented in section 2 of this Chapter, confirms that there is very little regional convergence going on within EU member states. To the extent that there has been any convergence, it appears to be at the country level mainly due to strong growth in Portugal and Spain.

These findings beg new questions about the effectiveness of existing regional policies. The EU Structural Funds were reformed and their funding increased significantly from 1989. The purpose was to make them more effective in reducing the gap between advanced and less-advanced regions and strengthening economic and social cohesion in the European Community. For an analysis of regional policy in the EU, including its rationale and the need for reform see Begg and Mayes (1993), Begg (1997). The reorientation of European regional policy, the increase of the budget, and the recent slowdown of convergence all underline the need for an assessment of the outcomes of regional policies. A likely enlargement of the European Union and the possible role that regional policies may play in an enlarged union, further underlines the need for an improved understanding of how these policies work. The distribution of regional support is the subject of section 11.3.

Earlier studies show that differences in economic growth between European regions can be reasonably well explained by innovation-activities in the region, the potential for exploiting technologies developed elsewhere and complementary factors affecting the exploitation of this potential (Fagerberg and Verspagen 1996, Fagerberg, Verspagen and Caniëls 1997, Cappelen, Fagerberg and Verspagen 1999). As will be argued in section 11.4 below, regions that are initially backward have a high potential for growth due to the backlog of technological knowledge developed in advanced regions that they may exploit. However, because the backward regions are also the regions that receive most support from European sources, it is not straightforward to separate the effect of 'catching-up' and regional policy. In other words, even if regions that receive European regional support grew rapidly, how could we be sure that realized growth was higher than what would have been realized without support? This is the subject of section 11.4.

We show in our descriptive section 11.2 below that the dispersion of regional unemployment within countries has not declined over time. This is perhaps surprising given the fact that much of the regional support is directed to regions with high unemployment. In section 11.5 we show that the estimated effect of regional support on unemployment is at best only marginally beneficial. Thus regional support has different effects on the dispersion of income and unemployment. Section 11.6 concludes and we suggest some policy lessons.

11.2 Regional Convergence?

It is by now well established that the distribution of regional incomes per capita in Europe became more equal after World War II (Molle 1980, Molle and Cappellin 1988). However, this convergence in regional incomes seems to have slowed down or come to halt after 1980 (Fagerberg and Verspagen 1996, Cappelen, Fagerberg and Verspagen 1999). This is in particular the case for the group of nine EU countries (the six founding members and the three countries that became members in 1973). This finding does not depend on how one measures dispersion of incomes. During the 1980s three relatively poor southern European countries joined the Union. These countries had at the time of entry roughly two thirds of the EU income per capita (with Portugal as the laggard). The "catch up" in incomes of these three countries compared to the nine "original" EU-members, has meant that convergence has continued for these twelve EU members as a group. During the 1990s the EU has been enlarged by three relatively rich countries, Austria, Finland and Sweden in addition to a relatively poor one (DDR). The inclusion of these four countries has also contributed to convergence (for the fifteen countries as a group), in particular because the relative income of Sweden has declined towards the EU-average. In addition the enlarged Germany with the Neue Ländern has contributed to a relative decline in incomes and brought Germany much closer to the EU average. However the inclusion of the Neue Ländern has obviously increased the dispersion of regional incomes within Germany as well as the EU as a whole.

This short description tells us that when studying dispersion of regional incomes in the EU over time, it is important to adjust for significant changes in the number of regions within the EU. Our study is confined to the twelve EU countries that comprised the union before the new countries entered in the 1990s and with a definition of Germany that is nearly identical to former West-Germany. Our descriptive analysis in this section is based on data for 163 NUTS II regions (see the appendix for a detailed presentation of the regions included and our data source). In our econometric analysis the sample is reduced even further and down to nine countries (Denmark, Ireland and Luxembourg are excluded as each consists of only one NUTS II regions). For some countries we had to use only NUTS I regional data as data were not available at the NUTS II level and the sample then comprises 105 regions. Incomes are made comparable by using purchasing power parities (PPS). In recent regional databases from EUROSTAT, regional GDP figures are based on the European System of Accounts (ESA95) available from 1995. For earlier years, the data are based on previous national accounts standards. These figures are linked to the most recent ones by taking growth rates from the old data and calculating comparable data using 1995 as the bridge year, as shown in more detail in the appendix. Comparing these revised figures with the earlier data (Cappelen *et al.* 1999) shows that standard measures of dispersion are not much affected by the change to ESA95.

Table 11.1 shows various measures of dispersion of GDP per capita in the European Union from 1980 to 1997. The focus is on standard deviation of the logs of GDP per capita relative to the relevant average. This choice is quite common in the literature. It is often called sigma convergence with reference to the neoclassical literature but can simply be interpreted as a common measure of dispersion without any theoretical connotation. We have used unweighted regional data and not weighted by population. The reason is that we regard each region as an "individual" or relevant observational unit. We would like to understand why dispersion of incomes between units (regions) differs over time irrespective of their size. The fact that migration from the poorest region in a country to a region with medium income may reduce dispersion simply due to changes in relative sizes is not so relevant for an understanding of regional income differences as such. The figures in Table 11.1 show that regional dispersion has changed very little since 1980. In particular, looking only at those countries that became members of the EU before the 1980s (EU 9), there is no convergence in incomes at all since 1980 at the regional level, see also Figure 11.1. However, if we include the three countries that joined the EU during the 1980s (EU 12), there is regional convergence in incomes. In our analysis in section 11.3 below, we have to aggregate the regions somewhat due to limited data availability at the NUTS II-level. Our sample consists then of 105 regions and is shown as (EU 12-3) in Table 11.1. Looking at the dispersion of regional incomes for this level of aggregation gives very much the same result as for the large sample. The standard deviation is slightly lower by 1997 compared to the first half of the 1980s. The level of the standard deviation is clearly higher for this level of aggregation which is mainly due to the fact that several regions with medium income levels have been aggregated. Dropping Greece, Portugal and Spain from this sample (EU 9-3) result in a minor increase in the standard deviation of regional incomes. This again confirms that convergence in incomes within the EU in the last two decades is mainly due to relatively fast growth in these three southern countries.

As an additional measure of dispersion we have estimated the Gini-coefficient of regional GDP per capita. The Gini-coefficient gives less weight to extreme observations compared with the more commonly used standard deviation. As regional incomes are far from obeying a normal distribution, but have typically a "twin peaks" distribution, (Quah 1996), the standard deviation has a less intuitive interpretation. The Gini-coefficient shows the same pattern as the standard deviation of the log of GDP per capita. However, the estimated standard deviations of the Gini-coefficient (in parenthesis in Table 11.1 and dotted lines in Figure 11.2) show that the decline in dispersion is not significant even for EU12. Our conclusion is therefore that although there is some tendency towards convergence in regional incomes among the 12 EU-countries since 1980, it is not significant.

Table 11.1. Dispersion of national and regional GDP per capita in Europe, 1980-1997

	1980	1985	1990	1997
Standard deviation EU9 - 127 regions	0.21	0.22	0.21	0.21
Standard deviation EU9 - 9 countries	0.22	0.23	0.20	0.17
Standard deviation EU12 - 163 regions	0.28	0.28	0.27	0.26
Standard deviation EU12 - 12 countries	0.29	0.31	0.29	0.25
Standard deviation EU(9-3)- 74 regions	0.22	0.21	0.22	0.23
Standard deviation EU(12-3)-105 reg.	0.31	0.31	0.30	0.27
Gini-coefficient EU12 - 163 regions	0.151	0.153	0.147	0.140
	(0.011)	(0.011)	(0.010)	(0.009)

Note: GDP figures based on PPS.

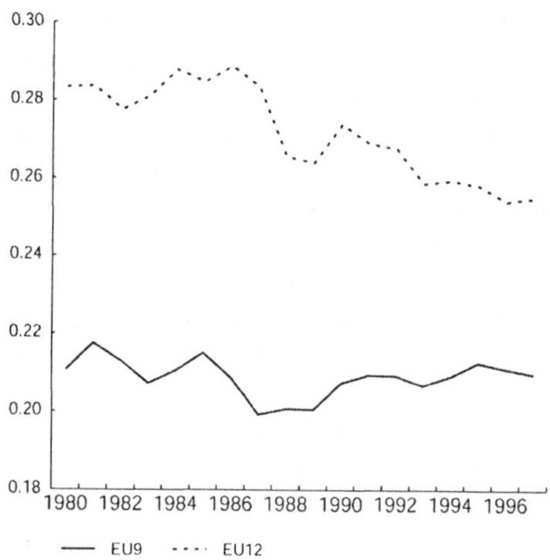

Fig. 11.1. Standard deviation of log GDP per capita in EU-regions

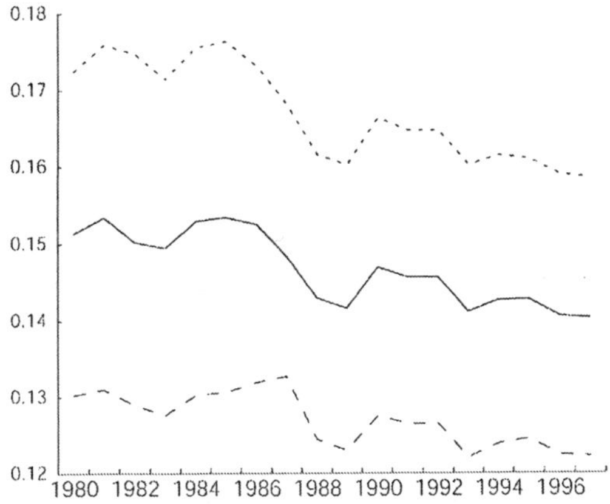

Sources: Statistics Norway and Eurostat.

Fig. 11.2. Regional inequality of GDP per capita in the EU. Gini coefficient and estimated confidence interval

There is, however, another interesting feature in the data in Table 11.1. There is a much clearer tendency for convergence in national incomes than in regional incomes. Looking at the standard deviation for national incomes for either the EU9 or EU12, there is a stronger decline at the country level than at the regional level. Thus, countries are converging but regions within countries are not, implying that there are elements of divergence within countries. This is indeed what we observe when looking at the standard deviation of regional incomes for each of the nine countries for which we have regionally disaggregated data (EU12 except Denmark, Ireland and Luxembourg), see Figures 11.3 and 11.4. Within Portugal and Spain there are clear signs of less dispersion. For the Netherlands this is also the case but data are highly influenced by the fall of oil prices between 1985 and 1986 that affects incomes strongly in Groningen. For many other countries dispersion either shows a small increase or there are cycles in dispersion with no clear trend.

This short descriptive analysis shows that the process of regional convergence in Europe is complex, especially because the differences in growth rates strongly depend on the period considered and on the countries included in the analysis. The concern of the EU with preventing any increasing disparities in regional economies is justified, and its recent effort in trying to reduce differences in GDP per capita among different areas of the Union will be considered in section 11.3 below.

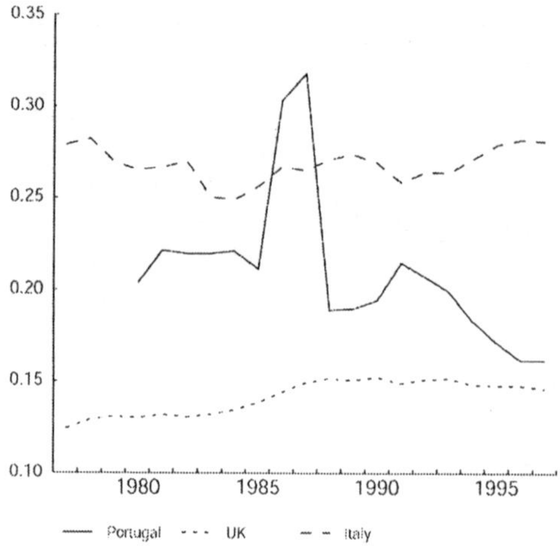

Fig. 11.3. Standard deviation of log GDP per capita in regions by country

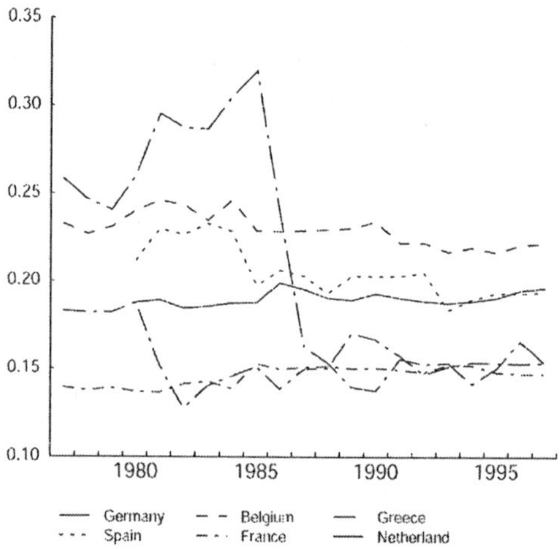

Fig. 11.4. Standard deviation of log GDP per capita in regions by country

From a welfare policy perspective it is interesting to study the dispersion of regional unemployment. For this variable we have (more or less) data from 1983 to 1998 for the 163 regions specified in the appendix. As shown in (EC 1997) the level of unemployment for the EU15 is closely and positively correlated with the regional dispersion of unemployment as measured by the standard deviation over the cycle. If this indicates a relationship in a structural sense, policies that would bringing down the level of unemployment in the EU should also reduce regional disparities in labour market outcomes and improve social cohesion. However, our data at the regional level seem to indicate a rather different story and in this case the level of regional aggregation seems to matter much. We use unemployment figures at the NUTS II level while (EC 1997) use NUTS I data. Dispersion of unemployment within each country is shown on figures 11.5 and 11.6. We see that dispersion clearly has increased in Italy and Belgium and has been reduced in Germany. There is also some indication of increased dispersion over time in Greece, France and the Netherlands, while Portugal Spain and the UK show no clear trends.

The level of unemployment seems to be inversely related to the dispersion within countries, the opposite of the finding for EU15 as a whole in (EC 1997). In Figure 11.7 we show the standard deviation of unemployment rates for EU6 and EU9 (corresponding to EU9 and EU12 above but now excluding Denmark, Ireland and Luxembourg) and the level of total unemployment in the EU. Nearly all European countries experienced a cyclical peak in economic activity around 1990, which led to low unemployment levels. These countries also experienced a peak in the regional dispersion of unemployment at that time and irrespectively of whether or not there was a positive or negative trend in the dispersion (with Greece as an exception). Furthermore, it is interesting to note that as unemployment gradually came down during the late 1990s from its peak around 1994, the regional dispersion of unemployment is again increasing in most cases (with Greece and Germany as the exceptions). Thus, as economic activity is picking up, it seems that unemployment is reduced relatively more in the centrally located regions within each country in the EU.

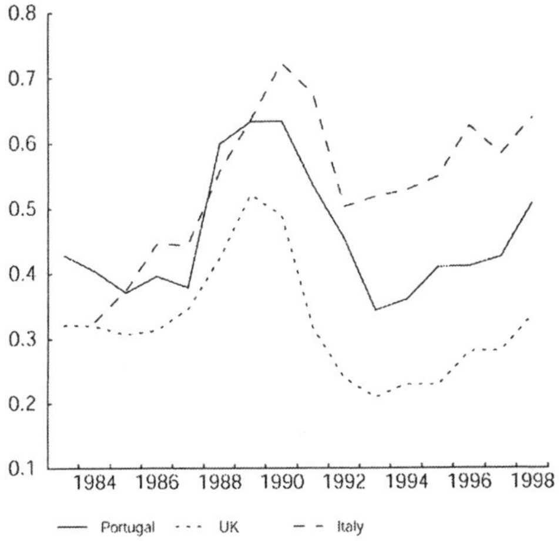

Fig. 11.5. Standard deviation of regional unemployment rates by country

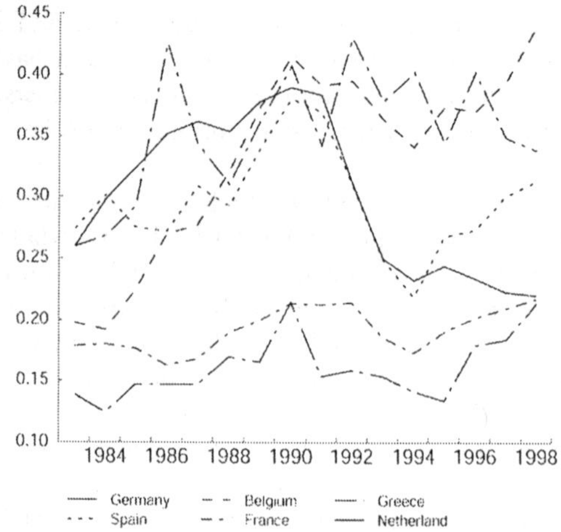

Fig. 11.6. Standard deviation of regional unemployment rates by country

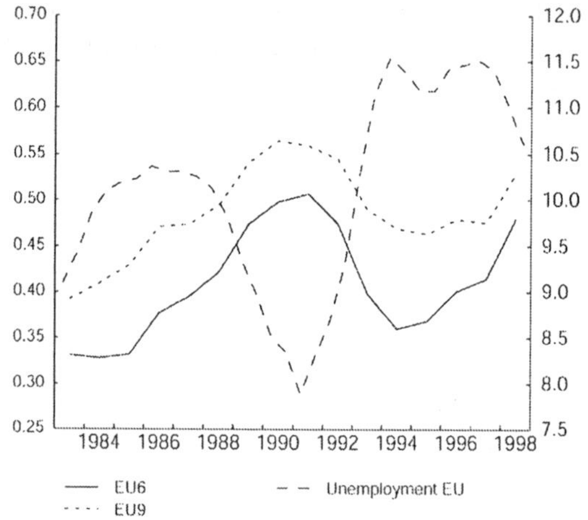

Fig. 11.7. Standard deviation of log unemployment EU-regions compared to rate of unemployment in EU (axis right)

Although we do not attempt to analyse the cyclical behaviour of GDP and unemployment in this Chapter, it is worth considering for a moment why the pattern emerges. We think one possible explanatory factor is industrial structure and how regional labour markets adjust. In sparsely populated areas with agriculture as the main activity, labour markets will have a relatively high share of self-employed people who do not register as unemployed in recessions. Thus there is less cyclical variation in unemployment in such regions. On the other hand there are also few job opportunities in these regions and so the level of unemployment may be relatively high permanently. Although migration could modify this so that unemployment acts as a permanent "push factor" there is ample evidence that labour mobility is not high in many countries in Europe, not to mention mobility between the European countries. In more centrally located regions with less agriculture and with wage labour dominating, there is cyclical hiring and firing that makes unemployment more cyclical but perhaps less permanent or structural. In a cyclical upswing, activity in the centrally located regions is generally more cyclically sensitive. Increased hiring brings down unemployment while this does not happen to the same extent in peripheral regions where activity generally is less cyclically sensitive due to the industrial structure. Thus the dispersion of unemployment increases while at the same time the national unemployment level goes down. In recessions, the opposite takes place. Another possible explanation is that labour markets in core regions have changed during the 1990s with higher turnover and more short-run unemployment due to job changes. Thus dispersion between regions has been reduced because of higher short run unemployment in core regions combined with the traditionally high long-term unemployment in the periphery. However, the observed pattern is not specific to the 1990s so the latter explanation cannot explain both the 1980s and the 1990s.

Looking at Figure 11.2 and 11.7 together, there is no simple relation between dispersion of GDP per capita and dispersion of unemployment for the EU as a whole. The two measures of dispersion are clearly capturing different aspects of regional imbalances that deserve separate analysis and perhaps policies. This result holds also at the country level. We will return to these issues later in this Chapter.

11.3 Regional Support in the European Union

Regional support is one of the key policy areas in the European Union. The idea driving this set of policies is the notion of social and economic 'cohesion', i.e., the desire to reduce differences in welfare between regions in the Union. In this section we present the data we use later in our analysis on the effect of regional support on economic growth and unemployment.

The first official regional policy initiative was the creation of the European Regional Development Fund (ERDF) in 1975, see (Corvers 1995) for a historical description of EU regional policy. Later on the European Social Fund (ESF, mostly concerned with employment), the European Agricultural Guidance and Guarantee Fund (EAGGF, aimed at developing agriculture), as well as several smaller measures were added (we will refer to the complete group of funds as 'regional funds' or 'structural funds'). Allocation of funds was initially done by fixed national quota.

The structural funds went through several reforms (1979, 1984), until in 1989 a completely new system was in place. In the new system, several 'objectives' were formulated, at which the regional funds were to be aimed. For the purposes of this Chapter, three of these objectives are of crucial importance:

- Objective 1, aimed at regions lagging behind in terms of GDP per capita, defined as regions with GDP per capita lower than 75% of the Community average.
- Objective 2, aimed at regions in industrial decline, as indicated by (high) unemployment and (low) employment growth.
- Objective 5b, aimed at rural and agricultural regions, as indicated by the share of employment in agriculture and GDP per capita.

The other objectives cannot easily be attributed to individual regions, and hence we will not take these into the analysis. In addition to the re-orientation of the allocation of regional funds according to these objectives, the 1988 reform increased the budget for regional policy at the European level significantly.

Table 11.2 gives an indication of the magnitude of regional support before and after the reform. During the period 1980-1984, which we take as a reference for the period before the reform, the average region in our sample received European regional support equal to around 0.18% of its GDP. Note that this value is influenced by the fact that Spain and Portugal were not members of the European Community at the time, and hence did not receive any support. Without these two countries, the mean value is 0.23% of GDP. During the period 1989 - 1993, the mean increases to 0.53% (column 'All Objectives'). The data are generally available for every region but for some countries or periods they are only given by objective. We have then spread the support to the eligible regions within the country using population shares. Some regional support is categorised as multi-regional by the EU and is consequently not taken into account and included in Table 11.2. According to our figures 90% of the regions receive some support, but only Greek and Portuguese regions, some Spanish and a few Italian regions receive support in excess of one percent of their GDP.

During this most recent period, Objective 1 support is by far the most important in terms of the total budget (0.49%-points of the total 0.53%) while Objective 5b support is by far the smallest part of the total budget. The countries that receive the largest amount of support (relative to GDP) are Portugal and Greece. In fact, these countries only receive Objective 1 support, up to almost 2% of regional GDP on average. Spain (0.6%) follows at some distance, then Italy (0.25%). Thus, it is clear that the largest amount of the total budget for European regional support goes to Southern European regions.

For the period 1989-1993, we also have data on national public and private matching funds. The provision of these funds is in fact a prerequisite for obtaining structural funds at all. On average, national public and private matching funds are about as large (in terms of budget) as the European funding. Public matching funds are about two-thirds of total matching funds. Although we will not explicitly take into account the role of the national public and private matching funds in the present Chapter, it is worth noticing that they are indeed important for the recent EU regional policy. One of the main purposes of the 1988 reform was to strengthen the coordination between the regional policy of the member states and the EU structural funds on long term plans and objectives (see Armstrong, Taylor and Williams, 1994, for a description of the 1988 reform).

Table 11.2. Regional support in per cent of GDP, average over regions in our sample

	1980-84	1989 - 1993						
	Total	Obj. 1	Obj. 2	Obj. 5b	All Object.	National	Private	Total
Belgium	0.015	0.000	0.023	0.004	0.027	0.033	0.010	0.070
Germany	0.026	0.000	0.013	0.008	0.022	0.031	0.012	0.065
Greece	1.024	1.759	0.000	0.000	1.759	0.897	0.151	2.808
Spain	0.000	0.518	0.058	0.022	0.598	0.474	0.280	1.352
France	0.070	0.000	0.035	0.036	0.070	0.100	0.063	0.233
Italy	0.269	0.218	0.013	0.013	0.243	0.218	0.107	0.568
Netherlands	0.030	0.000	0.025	0.007	0.032	0.062	0.022	0.116
Portugal	0.000	1.872	0.000	0.000	1.872	1.102	0.880	3.854
UK	0.167	0.077	0.070	0.004	0.151	0.158	0.080	0.389
Mean	0.178	0.494	0.026	0.010	0.531	0.342	0.178	1.051

Source: calculations on data taken from EUROSTAT regional yearbooks (1980-1984 support data) and Commission (1997) (1989-1993 support data).

11.4. EU Structural Funds and Regional Convergence in Incomes 1980-1997

When studying regional convergence in incomes the following accounting framework is often useful. Income per worker in a region (y_i) can be decomposed into the product sum of labour productivity by industry (y_{ij}) multiplied by the employment share of each industry (s_{ij}) of that region. In the case of two industries we simply have

$$y_i = y_{i1}\, s_{i1} + y_{i2}\,(1 - s_{i1}) \qquad (11.1)$$

We may now distinguish between three channels of convergence. First we may have within industry convergence between regions as labour productivity in any industry becomes more equal between different regions. The causes for this process may be the creation of a single internal market in Europe by removing trade barriers and increase competitive pressures. Also technological catch-up and diffusion will stimulate this form of convergence. Some studies have studied this process for European regions (de la Fuente, 1996) who shows that there is little sigma convergence within industries in Spain. Second, between industries (within regions) convergence may take place implying that labour productivities between industries become more equal. Why should that occur? Again diffusion of technologies is a relevant factor as different industries take advantage of technological changes and inventions in advanced industries. Also changes in the skill structure of the labour force may lead to this. If low-skilled labour becomes scarcer, low productivity industries may have to adjust in order to attract labour either by technological change or simply by increasing relative prices to pay for higher costs. This lead us two one explanation for the third channel of convergence which is a change in the industry structure as the share of labour in low-productivity industries falls relative to the average. Changes in relative prices of products may cause this, but so will changes in income levels. A falling share of agriculture in total GDP is often explained partly by reference to the falling budget share for food in consumption.

Any explanation of growth differences and convergence needs theoretical underpinning and not only statistical decomposition. In previous studies we have relied on the "technology gap theory" of growth. The hypothesis that technological catch-up requires substantial efforts and capabilities in the receiving country, is discussed and tested in Fagerberg (1987,1988). Verspagen (1991) and Amable (1993) analyse the possibility that countries may end up in a low-growth trap without the necessary assets. For an overview of empirical work on catch-up and growth, including its theoretical underpinnings, see Fagerberg (1994). Technology is not assumed to be a public good. On the contrary, it is argued that successful adoption of new technology is costly and requires a host of complementary factors of the sort that Abramovitz (1994) classifies under the terms 'social capability' and 'technological congruence'. Hence, following this perspective regional growth may be seen as the outcome of three sets of factors:

- Innovation activities in the region,
- Potential for exploiting technologies developed elsewhere (diffusion), and
- Complementary factors affecting the ability to realise this potential.

There are two major challenges when applying this perspective. The first has to do with finding indicators of innovation and the potential for diffusion, the second with identifying and measuring the 'complementary factors'. For innovation we use R&D intensity, defined as business enterprise R&D personnel as a percentage of total employment[1]. For diffusion potential we use, as customary in the literature, the initial level of GDP per capita in the region (log-form). The higher this level, the smaller the scope for imitating more advanced technologies developed elsewhere. Regarding complementary factors, there are many candidates that can be defended theoretically, from variables related to various types of investments (education, infrastructure and physical capital) to structural factors of various sorts. However, data are scarce, especially among the former. The 'complementary' variables that we have taken into account include:

- Physical infrastructure (kilometers of motorways per square kilometer),
- Population density (the number of inhabitants per square kilometer),
- Industrial structure (employment shares in agriculture and industry), and the
- Long-term unemployment rate (duration of more than one year).

---

[1] In some cases R&D data were missing at our level of aggregation and we had to use interpolation or regional breakdown using data from later years.

Among these, we would expect the first two to have a positive impact on technology diffusion, since both a more developed infrastructure and a higher population density increase the profitability/reduce the cost of introducing new technology. Regarding industrial structure, it is one of the standard results in the existing empirical literature on regional growth that industrial structure matters. In particular, a high reliance on agriculture has been shown to be detrimental to regional growth (Fagerberg and Verspagen, 1996), among other things because of low technological opportunities, and slow growth of the market. On the share of 'industry' in total employment the expectations are less clear. Traditionally this sector – particularly manufacturing – has been regarded as the 'engine of growth' (Kaldor, 1967). However, technological progress in recent decades has been more geared towards services than industry and many traditional industries have been characterized by slow growth. Finally, the level of unemployment is included as a possible complementary factor. We interpret this as a measure of the cohesion of the broader social and economic system in the region. The higher the share of the labour force that is excluded from work on a long-term basis, the less well this system works. Hence it is an indicator of institutional failure, and as such it might be expected to have a negative impact on growth. For instance, it may hamper inflows of qualified people and encourage outflows, as empirical research in this area indeed suggests (Fagerberg, Caniëls and Verspagen, 1997). Long-term unemployment also leads to deprecation of skills and lack of learning by doing in parts of the workforce.

To this framework we have added regional support from the EU as a possible growth-enhancing factor. As was shown in the previous section, the most important form of support (Objective 1 support) is allocated to regions on the basis of GDP per capita. In addition, Objective 2 support is allocated partly on the basis of unemployment rates, while Objective 5b support is allocated partly on the basis of the share of employment in agriculture. We have performed a cluster analysis for our 105 region sample with the above mentioned variables as the inputs. European regional support was broken down by Objective (1, 2, 5b) in this analysis. We arbitrarily fix the number of clusters to five, and apply a so-called K-means clustering algorithm. All variables were standardized before entering in the clustering algorithm.

We obtained one cluster of only two regions, and four larger clusters. The cluster of two regions consists of highly urbanized small regions (Brussels in Belgium and Cueta y Mililla in Spain) and is not shown in the table. The nature of the other four clusters has a clear interpretation from the perspective of European regional policy. The cluster characteristics are documented in Table 11.3. Note that because the data were standardized, a value of zero corresponds to the sample mean, and plus (minus) one corresponds to one standard deviation above (below) the mean. Cluster 1, which we label the 'little support' cluster, is a cluster of 38 advanced regions that receive little support. They score below average on all three support-variables. Their score on GDP per capita, R&D and infrastructure is positive. The second cluster consists of regions that receive a lot of Objective 1 support. This cluster comes with a positive score agriculture, and a negative score on GDP per capita and R&D. In a sense, these regions are the opposite of those in the

Table 11.3. A cluster analysis of European regions, 1989-97

Variable	Clusters			
	1 - 'little support'	2 - 'Objective 1'	3 - 'Objective 2'	4 - 'Objective 5b'
No. of regions	38	36	9	20
Agriculture	-0.65	0.99	-0.51	-0.22
Manufacturing	0.38	-0.55	1.02	0.05
Long-term un-employment	-0.50	0.45	0.37	-0.27
Infrastructure	0.78	-0.59	-0.05	-0.48
Obj. 1 support	-0.64	1.15	-0.52	-0.64
Obj. 2 support	-0.17	-0.49	2.67	0.06
Obj 5b support	-0.32	-0.49	0.41	1.35
Pop. density	0.11	-0.25	-0.21	-0.30
GDP per cap. 1988	0.82	-0.99	-0.08	0.17
R&D	0.85	-0.80	-0.23	-0.06
Growth* rate	4.23	4.80	4.26	3.64

* Variable not included as input in the clustering algorithm.

previous cluster. The next cluster is a small one, characterized as 'Objective 2' regions. Score on this form of support is very high. As could be expected by the nature of Objective 2 support, this cluster scores high on the share of manufacturing in employment. Scores on other variables are not very high in absolute value. The final cluster is the Objective 5b cluster, with a high score on this form of support. The bottom line of the table has been included for illustrative purpose and gives the mean of the annual growth rate of GDP per capita over the period 1989 - 1997. It shows that the regions differ somewhat in terms of growth. The Objective 1 regions show the highest growth rate, followed by the 'little support' and 'Objective 2' clusters. The 'Objective 5b' cluster lags behind in terms of growth.

The conclusion of the cluster analysis is that the three forms of European regional support that we distinguish after the 1989 reform are indeed aimed at relatively homogenous groups of regions. One can indeed speak of a 'typical Objective 1 region', and the same holds for the two other objectives. Hence, as illustrated in Table 11.4 below, the three forms of European regional support are closely correlated with various structural characteristics of regions, among which are the main variables of interest in our theoretical model set out above. As Table 11.4 shows, there is a close relation between European structural funds on the one hand, and GDP per capita and the share of agriculture in employment on the other hand.

This correlation might pose problems when estimating our equation for regional economic growth because a possible multicollinearity between some of our explanatory variables could weaken the precision of the single estimated coefficients, so that we would be unable to separate the effect on regional growth of some of the technology gap variables from the effect of the EU structural funds. In order to deal with this problem, we pooled the data for the period 1989-1997 (after the reform) with the data for the previous period 1980-1989.

Table 11.4. Correlation coefficients between selected variables, 1989-1997

	European support (% of GDP)	GDP per capita, 1988	Long-term unemployment, 1990
GDP per capita, 1988	-0.83		
Long-term unemployment, 1990	0.08	-0.31	
Share of agriculture, 1990	0.80	-0.73	0.04

We have also introduced as additional explanatory variables a "time slope dummy" (TSD) for each independent variable of the model. TSD is one in the first period and zero in the second. The "time slope dummies"[2] test the hypothesis of stability between the first and the second period of each explanatory variable. In other words, they give us information not only on whether or not the model is stable over time, but also on what are the sources of the (possible) structural break. Although we started out with TSDs for all variables, only the significant ones are reported. We have estimated the model both without country dummies (testing the hypothesis of convergence "between regions in Europe as a whole", irrespective of country borders) and with country dummies[3]. The results are presented in Table 11.5.

Let us start making some general considerations on the validity of the estimated models. First, all the conditional factors taken into account have the expected signs. The degree of significance varies but is generally quite satisfactory. The estimated parameters have signs that are quite robust with regards to model specification, that is the inclusion of dummies and whether or nor we include the three most recent southern countries. In this sense the model results are qualitatively robust. The size of the estimated parameters varies somewhat but not that much considering the estimated standard errors. Second, the negative effect on regional growth of structural (agriculture and manufacturing) and institutional factors (unemployment); and the positive effect on regional growth of infrastructures and the innovative activity internally developed (proxied by R&D) are supported by data. Furthermore, even the EU structural funds (variable EU) seem to have affected growth in a positive and significant way, supporting the hypothesis that the regional funds have been a complementary factor in fostering regional convergence in Europe in the whole period considered. On the whole, the significance and robustness of the estimated coefficients for all these conditional factors support the idea of technology-gap to explain regional growth in Europe.

Regarding the changes over time, as already explained above, we test the hypothesis of a structural change between the first (1980-1988) and the second pe-

[2] Also known as "differential slope coefficients"

[3] Testing the hypothesis of convergence "between regions within each country", as the between countries differences are taken into account by the dummies.

riod (1989-1997) by including as explanatory variables in our model time slope dummy variables (TSDs). The inclusion of such variables help us not only to test whether or not a structural break has taken place between the two periods, but also which are the factors that have determined such change. Then, focusing on the magnitude and the significance of the estimated TSDs included in each of the four regressions, we observe the following. First, the hypothesis of a structural change between the two periods is confirmed, as clearly shown by the high degree of significance of the TSDs in all the regressions.

In our preferred model for the large sample we can note that the positive sign of the estimated coefficient of the TSD relative to initial GDP per capita indicates a faster conditional convergence in the second period than in the first. This is in line with the observation in section two that convergence was slow during the 1980s compared to the 1990s. The positive sign of the TSD relative to unemployment suggests that the effect on regional growth of the long-term unemployment is stronger in the second period than in the first, in fact the effect is roughly zero in the first period. The negative sign of the TSD relative to infrastructure indicates that the positive contribution of this variable to growth is stronger in the second than in the first period, again the effect in the first period is close to zero. It is also worth noticing that the reform of regional support policies has increased the effect on economic growth. Although being positive also during the early 1980s the effect has clearly increased in the 1990s. The reform itself, in addition to the increase in the total budget, is clearly part of the explanation of why there is stronger (absolute) convergence among the regions in the European Union during the 1990s than during the 1980s.

Regarding the conditional factors of the technology-gap approach, notice that when we exclude the southern states that became members during the 1980s, both the infrastructure variable and the R&D variable become slightly larger and more significant. On the other hand the industrial structure variables become less significant but with the expected sign at least in the last period. However, unemployment which is not of any importance in the first period (the sum of the two coefficients is zero) turns out to be highly important in the 1990s. This could indicate that with closer integration in Europe following the establishment of the internal market, regions with bad job prospects loose skilled people that migrate to other regions. Consequently these regions have relatively slow growth.

One dimension to take into account when interpreting the results of the estimations is the change of the model after including country dummies. If we do not include country dummies as explanatory variables in the model, we test convergence "between regions in Europe as a whole" irrespective of country borders; while if we include country dummies, we test convergence "between regions within each country", as the between countries differences are taken into account by the dummies.

Table 11.5. Explaining regional growth, 1980-1997

	Large sample	Large sample with dummies	Small sample with dummies
Constant	0,032 (4,07)		
Initial GDP per capita	-0,006 (-2,22)	-0,008 (-2,78)	-0,017 (-3,62)
Initial GDP per capita - TSD		0,003 (3,52)	
Agriculture	-0,049 (-3,65)	-0,025 (-3,35)	-0,045 (-1,46)
Agriculture-TSD	0,035 (2,49)		0,083 (3,13)
Manufacturing	-0,009 (1,04)	-0,025 (-3,01)	-0,020 (-1,75)
Infrastructure	0,001 (2,27)	0,0007 (1,56)	0,0007 (1,81)
Infrastructure - TSD	-0,001 (-1,94)	-0,001 (-1,94)	
Unemployment	-0,0005 (-2,28)	-0,0007 (-3,00)	-0,0013 (-3,87)
Unemployment - TSD	0,0011 (5,11)	0,0008 (3,75)	0,0013 (2,02)
R&D	-0,003 (-1,14)	0,002 (1,35)	0,004 (2,20)
R&D - TSD	0,005 (1,57)		
EU support	0,013 (5,29)	0,007 (4,57)	0,014 (2,33)
EU support - TSD	-0,009 (-3,40)	-0,004 (-2,17)	-0,035 (-4,58)
D-Belgium		0,041 (4,33)	0,066 (4,27)
D-Germany		0,042 (4,65)	0,066 (4,22)
D-Greece		0,040 (5,07)	
D-Spain		0,049 (5,29)	
D-France		0,034 (3,78)	0,059 (3,87)
D-Italy		0,043 (4,69)	0,070 (4,35)
D-Portugal		0,050 (6,13)	
D-UK		0,045 (5,14)	0,071 (4,84)
D-Netherlands		0,034 (4,33)	0,052 (3,94)
Adjusted R^2	0,41	0,88	0,83
No. of regions	105	105	69

Note: OLS estimation on pooled data, with time slope dummy variables (TSD) as additional explanatory variables (t-values in brackets)

So, considering the different interpretations of the two models, after comparing columns the first two columns of Table 11.5, we observe the following. First, as it is usual in cross-country and cross-regions regressions, the overall explanatory power (measured by the adjusted R^2) increases significantly if we include country dummies[4].

Second, regarding the variable EU, its estimated coefficient is higher in the model without country dummies than in the one with country dummies, and this pattern does not seem to change over time. A possible interpretation of this result is that the EU structural funds contribute more to convergence "between regions in Europe as a whole", rather than "between regions within each country". Third we notice that some countries have dummies that are quite similar and we could clearly have used a common constant term for Belgium, Germany, Greece, Italy and the UK. Clearly the Iberian Peninsula has been growing faster than the other countries while France and the Netherlands are slow growers given their structural characteristics. The same pattern holds when the three poorest countries are excluded (column three). Thus the model does in fact go a long way in explaining also the variation in growth between regions of different countries in Europe.

Our econometric analysis has produced results that in general are quite supportive of the theoretical model referred to above. However, the results could perhaps also be interpreted according to a neoclassical model, but we have not attempted to link our choice of variables to such a model. What we would like to stress is our finding that the reform of regional policies in the EU has increased growth in the poor regions and contributed positively to sigma convergence. Thus the regional policy reform has reduced dispersion of regional incomes per capita. The question we now would like to address is what effect the reform has had on unemployment. The support has been allocated with reference to low income as well as high unemployment. We have shown that the support has had a positive impact on income as intended but has it also contributed to lower unemployment?

11.5. Effects of Regional Support on Regional Unemployment

We noticed in Section 11.2 a strong cyclical development in the dispersion of regional unemployment both within countries and for the EU as a whole. Our main focus here is how we can explain differences in unemployment performance between regions much in the same manner as we approach the modelling of economic growth in previous studies.

The level of unemployment in modern macroeconomic literature (Layard et al. 1991) is thought of as the outcome of wage bargaining between unions and firms and pricing behaviour of firms. The resulting level of unemployment or equilib-

[4] The presence of national dummies can also go a long way to eliminate the effects of residual autocorrelation, a point discussed in Chapter 1, although it may be useful to explore the impact on our conclusions of explicitly modelling spatial effects in future research.

rium unemployment is mainly dependant upon factors affecting pricing behaviour such as the competitive situation in the product market and factors affecting the outside options of employees (replacement ratio, probability of getting a job elsewhere and so forth). When trying to explain regional unemployment and in particular variation in unemployment between regions within a country, this framework has relatively little to offer. This is due to the fact that most variables included in this macroeconomic literature refer to the national level and will show little variation between regions. Although competitive pressure in goods markets may vary some in particular for non-traded goods even within a country, the replacement ratio will not as policies relating to unemployment benefits are similar within each country. However, since many of the relevant variables vary between countries, it is important to have country dummies in any regional unemployment equation because these will pick up the effect of various national policies not specified in the model.

Most recent theories of how wages are determined share one common feature namely "the wage curve", (Blanchflower and Oswald 1994). They conclude their extensive study by saying "Employees who work in areas of high unemployment earn less, other things constant, than those who are surrounded by low unemployment." (p. 360) This curve depicts an inverse relationship between the level of unemployment and the level of (real) wages and this relationship exists with either regional unemployment or industry unemployment. We do not have real wages by region for our whole sample only for a minor sub-sample, (Fagerberg, Verspagen and Caniëls 1997). However, we do have the level of regional income per capita that is a good proxy for real wages. Thus we would expect to find a negative relationship between the level of GDP per capita and unemployment. Notice the difference with the growth equation in the previous section that relates income growth with the level of unemployment. According to the wage-curve-hypothesis we should expect to find a negative relation between the level of income and unemployment. In our regional context and based on the role of central wage bargaining in many European countries, the causation may very well run in the opposite direction with wage-rates and in particular minimum wages, negotiated at the national level, with variations in regional unemployment being the outcome of this process.

The wage curve depends on factors that influence demand for labour as well as the supply of labour. The demand for labour obviously will be positively affected by factors that stimulate the productivity of business investment in the region. Thus economic growth may have a positive effect on growth in labour demand and thus a negative effect on unemployment. However, given a possible simultaneity between growth and unemployment, it is important to adjust for this when estimating an unemployment equation. This is the case even if traditional models of growth are expressed as growth per capita because population can be thought of as an exogenous variable.

It is far from obvious that growth in income per capita or productivity should have a negative effect on unemployment. In the long run it is reasonable to assume - in line with (Layard et al 1991) - that productivity growth is fully reflected in wages (or incomes) and thus picked up by the level of income per capital already included in the model with reference to the wage curve. Aghion and Howitt (1994) analyses the effects of growth on long-run unemployment. They use a search model of equilibrium unemployment where faster growth comes from faster increase in knowledge embodied in industrial innovations that is likely to increase job-destruction through creative destruction. More job-turnover will, according to search theories, increase unemployment. Accordingly more growth will increase unemployment. However, working in the opposite direction is the idea that higher growth would increase rate the returns from creating vacancies by establishing new firms and induce a faster exit rate from unemployment. This would lower unemployment. The two opposing effects may create a non-linear relationship between growth and unemployment. Low growth will imply that job destruction is low while high growth leads to the creation of many vacancies both leading to low unemployment. In between the job destruction effect out-weights vacancy creation and unemployment is high.

Infrastructure and R&D are expected to affect labour demand positively and are thus expected to reduce unemployment but these variables may turn out to be captured by the inclusion of the growth variable in the model.

The role of population density is less obvious. On the one hand economies of scale may have a positive effect on job creation and a negative effect on unemployment. However, population density can also be positively correlated with job search and unemployment among job seekers in labour markets with much job turnover as you typically can find in urban service businesses.

The industrial structure will normally affect unemployment. In agricultural communities with much self-employment, few will have incentives to register as unemployed. Even if our unemployment data are based on labour force surveys, those who say they are actively looking for jobs would tend to be fewer in regions where agriculture is relatively important either because if they were they would have migrated, or because labour markets are so transparent that they know they could not find any job and become "discouraged workers". For regions with a strong industrial base the effect on unemployment is more uncertain. Clearly a "rust belt" has developed in some regions that earlier were based on mining and heavy manufacturing. In this sense one would expect a positive effect on unemployment. On the other hand job-turnover is often lower in manufacturing than in service industries and this tends to work in the opposite direction.

Finally, we expect that regional policies that allocate resources to regions with certain characteristics, including high unemployment, will lower unemployment. It is also of interest to study whether or not the changes in regional policies introduced in 1989 has modified in any way the effect of these policies on unemployment in a similar way to its changed effect on economic growth, according to the results of our previous section.

The econometric model specification is very much the same as we have used when modelling regional growth in incomes, as described in Section 11.4. We test for the significance of country dummies and changes in the effects of our structural variables over time by introducing time period dummies (TSD), being set equal to one in the 1980s and zero thereafter. Note that the reported results in Table 11.6 include the growth of incomes per capita as an explanatory variable. Although there is no obvious argument in favour of including productivity growth, it is included to check that there is no simultaneity between the model of unemployment and that of economic growth. If the growth variable were to drop out, the growth equation and the unemployment equation would not even be recursively related, as only initial unemployment is assumed to affect growth. In the opposite case, we should use an instrument variables (IV) method to adjust for the simultaneity between the two equations and the growth equations in Table 11.4 are the obvious choice. It turned out that for the whole sample of 105 regions using instrument variables for the growth variable (as opposed to using OLS) makes a lot of difference for the estimates. However, when using IV the growth term is insignificant. For the small sample leaving out Greece, Portugal, and Spain the growth term remained significant even after using instrument variable estimation.

Table 11.6. Explaining regional unemployment using IV-estimation

	Large sample with dummies	Small sample with dummies
Initial unemployment	1.14 (6.85)	.96 (7.28)
Initial unemployment-TSD	-.97 (-5.89)	-1.06 (-9.02)
Agriculture	0.27 (.06)	-12.63 (-2.10)
Manufacturing	-6.84 (-1.35)	-9.35 (-2.90)
Infrastructure	0.21 (1,53)	0.23 (1.93)
Population density	-.71 (-1.32)	0.67 (1.94)
Population density - TSD	1.92 (2.86)	
R&D	1.35 (1.43)	1.41 (2.06)
R&D - TSD	-2.74 (-2.66)	-1.24 (-1.77)
Growth per capita	43.21 (0.29)	-231.7 (-4.23)
GDP-level per capita	-2.32 (-1.38)	-6.63 (-4.63)
GDP-level per capita -TSD	-.98 (-2.26)	
EU support	1.34 (1.62)	.74 (.45)
EU support - TSD	8.87 (4.34)	
D-Belgium	11.30 (1.51)	28.27 (6.00)
D-Germany	12.49 (1.62)	29.35 (6.18)
D-Greece	8.21 (1.16)	
D-Spain	14.96 (1.71)	
D-France	14.34 (2.20)	29.25 (6.41)

D-Italy	13.26 (1.70)	30.73 (6.26)
D-Portugal	8.42 (.96)	
D-UK	10.50 (1.32)	28.20 (5.11)
D-Netherlands	9.88 (1.54)	23.93 (5.85)
Adjusted R^2	0.93	0.96
No. of regions	95	69

The results of our IV-estimation are presented in Table 11.6. We note first of all that many of our hypotheses are not rejected. First of all high-income regions do usually have low unemployment. This is in line with the wage curve hypothesis. The effect is clearly much stronger in the small sample than in the large. This implies that real-wage flexibility is much larger in the small sample. We find this result intuitively appealing and one possible explanation is the well-known inflexibility of the Spanish labour market during most of the estimation period.

GDP-growth per capita has no effect on unemployment for the large sample but a clear negative effect for the smaller sample. One possible explanation for this is that the level of unemployment in Spain was much higher on average than in any other country included in the sample and that so the inclusion of Spain makes our assumption of linearity between growth and unemployment less reasonable. For the small sample higher growth reduces unemployment. Thus variables that will affect growth positively, such as R&D and perhaps EU-support, will have an effect on unemployment through the growth effects.

EU-support has a small positive effect on unemployment in the large sample including the three southern countries but for the smaller sample the direct effect is small and insignificant in the 1990s. However the effect is large and positive in the 1980s. One may speculate that before the reform of the regional support policies, these policies may have created a lock-in effect that led to higher unemployment. Again this direct positive effect is counteracted by a negative effect coming via the growth term. Results from a reduced form estimation (available on request) shows that in the 1980s EU-support led to an increase in unemployment while it had the opposite effect during the 1990s. Thus the reform of regional policies not only benefited growth, it also reduced unemployment for the small sample excluding the three southern countries. Note that EU-support policy to combat unemployment in general is not included in our definition of support because it cannot be allocated to specific regions.

R&D activities do not have a systematic effect on unemployment. In the 1980s R&D spending reduced unemployment both in the large sample but less clearly so in the small sample. However, taking the indirect effect via growth into account, there is a small negative effect from R&D on unemployment in the 1980s, but this disappears in the 1990s where the total effect is small and insignificant according to reduced form estimates.

Population density seems to increase unemployment in general supporting the view that job search and high labour turnover in urban areas leading to higher unemployment. This is in particular the case for the small sample of high-income countries. The infrastructure variable also has a positive impact on unemployment in both samples. This result is difficult to interpret but the effect is small in size and not very significant. This variable also has little systematic effect on economic growth, (Cappelen et. al., 2001).

There strong dependence between end of period unemployment with initial unemployment during the 1990s while that is not the case during the 1980s. This result is perhaps not surprising if one interprets it as being caused by "hysteresis" , meaning that path dependence increases over time. However, that the role of hysteresis should change so dramatically is harder to explain. Referring back to Figure 11.7 showing the level of unemployment in the EU on average we see that the choice of sample split may explain part of the result. The level of total unemployment in the EU did not change much between the early 1980s and the late 1980s, while it had increased by the end of the 1990s because the chosen years were at different points within the business cycle.

Finally economic structure matters for unemployment. For the small sample there is a negative direct impact of agriculture that survives also when we take into account the effects coming from through the growth variable. For manufacturing the negative direct effect is reduced when taking the indirect effect through growth into account. For the large sample there is on the other hand no effect of agriculture but a negative effect from manufacturing which again becomes small according to the reduced form estimates.

11.6 Conclusions

We have in previous work demonstrated that the process of regional convergence that characterized the member states of the European Union from the 1950s onwards came to an end around 1980 and that there has in general been little change since then. To the extent that there has been any tendency towards convergence, it has been on the country level, related to the catch up by the relatively poor Southern countries that joined the community during the 1980s. Within countries, however, there has at best been a standstill if not divergence. This Chapter, presenting new and more recent evidence, confirms these trends.

One important question is to what extent regional support from the EU, or national sources designed to help catch-up by relatively poor regions, has had a real impact on this situation. In recent years – following the 1988 reform - this support has increased in importance and it is thus natural to ask what the consequences of this reform are. Our most recent research has tried to investigate this question through the estimation of a technology gap model applied to the European regions in the period 1980-1997. We have found that the EU funds may have played a significantly positive role in fostering regional growth in all the models estimated. Nevertheless, this optimistic conclusion has to be refined in a number of ways.

First of all, the positive effect of EU regional funds turns out to increase over time, which indicates that the 1988 reform did have the expected effect of accelerating growth in the less advanced regions more than in the advanced ones. However, the complexity of this reform probably requires a longer period of estimation to better evaluate its effect on growth, so that further work is needed on this point in the future. Moreover, in the whole period 1980-1997 these funds have been more effective in determining convergence "between regions in EU as a whole" than "within each country". A possible interpretation of this trend is that the EU funds matter for regional growth much more if the region receiving the funds has some basic structural and institutional capabilities to use them in an efficient and productive way. We should also qualify our results by stating that although we are interested mainly in the effect of EU support on economic growth and convergence, there may be short-run demand effects from the increased regional support. This may lead us to overestimate the long-term effects of the support. Also, in order to assess the total economic effects of the reform, one should take into account how the reform is financed.

When it comes to explaining regional differences in unemployment, we notice that high- income regions have less unemployment. While income growth or productivity does not matter much for unemployment according to the results for our large sample of countries - a result that may be due to non-linearity effects - higher growth lowers unemployment in the richer countries. Factors closely linked to densely populated areas tend to increase unemployment. At best regional support plays a small positive role in reducing unemployment taking into account the indirect effects through the effect of growth on unemployment. Thus although the change in EU-support policy may have promoted growth, the effect on unemployment is less clear. If the purpose of regional policies is to promote social cohesion, there is still a job to do if one believes that high unemployment jeopardises cohesion.

References

Abramovitz, M. (1994), 'The Origins of the Postwar Catch-Up and Convergence Boom', in J. Fagerberg, B. Verspagen and N. von Tunzelmann (eds), *The Dynamics of Technology, Trade and Growth*, Aldershot, UK: Edward Elgar, pp. 21–52.

Aghion, P. and P. Howitt (1994), Growth and Unemployment, *Review of Economic Studies,* 61, 477-494.

Armstrong, H., J. Taylor and A. Williams (1994), 'Regional policy', in M. J. Artis and N. Lee (eds), *The Economics of the European Union, Policy and Analysis*, Oxford University Press, pp. 172-201.

Bachtler, J., and I. Turok, eds. 1997. *The Coherence of EU Regional Policy. Contrasting Perspectives on the Structural Funds.* London: Jessica Kingsley Publishers, Regional Studies Association.

Begg, I. and D. Mayes (1993), 'Cohesion in the European Community. A Key Imperative for the 1990s?', *Regional Science and Urban Economics*, 23, 427–48.

Begg, I. (1997) 'Reform of the Structural Funds after 1999', *European Planning Review*, 5, 675–89.

Blanchflower, D.G. and A.J.Oswald, (1994) *The Wage Curve*, MIT Press, Cambridge, MA.

Cappelen, A., F. Castellacci, J. Fagerberg and B Verspagen (2001), Regional Covergence, Clustering and European Policies, TIK Working Paper No.11 2001.

Cappelen, A., J. Fagerberg and B. Verspagen (1999), 'Lack of Regional Convergence', in J. Fagerberg, P. Guerrieri and B. Verspagen (eds), *The Economic Challenge for Europe. Adapting to Innovation Based Growth*, Edward Elgar, chapter 6.

Corvers, F. (1995). *Regional Policy within the European Union: The regional implications of a changing paradigm.* Coventry, Paper for the ESRC seminar on Regional and Local Responses to European Integration, 12 October, University of Warwick.

de la Fuente, A. (1996), On the Sources of Convergence: A Close Look at the Spanish Regions, CEPR Disc.Paper No. 1543.

European Commission. (1997). *The impact of structural policies on economic and social cohesion in the Union*, 1989-99. Luxembourg, Office for Official Publications of the European Communities.

European Commission. (1999). *Sixth Periodic Report on the social and economic situation and development of the regions of the European Union*, consulted from the www: http://europa.eu.int/comm/regional_policy/document/radi/radi_en.htm

Fagerberg, J. (1987), 'A Technology Gap Approach to Why Growth Rates Differ', *Research Policy*, 16, 87–99.

Fagerberg, J. (1988), 'Why Growth Rates Differ', in Giovanni Dosi et al. (eds), *Technical Change and Economic Theory*, London: Pinter Pubs, pp. 432–57.

Fagerberg, J. (1994), 'Technology and International Differences in Growth Rates', *Journal of Economic Literature*, 32, 1147–75.

Fagerberg, J. and B. Verspagen (1996), 'Heading for Divergence? Regional Growth in Europe Reconsidered', *Journal of Common Market Studies*, 34, 431–48.

Fagerberg, J., B. Verspagen and M. Caniëls (1997), 'Technology, Growth and Unemployment across European Regions', *Regional Studies*, 31, 457–66.

Heinelt, H., and R. Smith, eds. 1996. *Policy Networks and European Structural Funds.* Aldershot: Avebury.

Kaldor, N. (1967), *Strategic Factors in Economic Development*, Cornell University: Itacha.

Layard, R., S. Nickell and R. Jackman, (1991),*Unemployment*, Oxford Univ. Press, Oxford

Molle, W. (1980), *Regional Disparity and Economic Development in the European Community*, Westmead: Saxon House.

Molle, W. and R. Cappellin (eds) (1988), *Regional Impacts of Community Policies in Europe*, Aldershot, UK.

Neven, D. and C. Goyette (1995), 'Regional Convergence in the European Community', *Journal of Common Market Studies*, 33, 47–65.

Staeck, N. 1996. The European structural funds. Their history and impact. In *Policy Networks and European Structural Funds*, edited by H. Heinelt and R. Smith. Aldershot: Avebury

Verspagen, B. (1991), 'A New Empirical Approach to Catching Up or Falling Behind', *Structural Change and Economic Dynamics*, 2, 359–80.

Appendix. Data and Classification-codes of NUTS 2-Regions Included in our Sample in Section 2

Data

The data on regional GDP per capita are taken from the Eurostat REGIO-databank as of March 2000. We have throughout used data in purchasing power parities meaning that regional incomes have been converted to a common European purchasing standard. This conversion takes place at the national level only, so no regional price differences within a country are accounted for. A region within a country that has relatively low prices compared to other regions in that country will have its real income underestimated. The regional income data are in two accounting systems. The most recent data ie from 1995 and onwards, are based on ESA95 (European System of Accounts, Eurostat/EU-Commision, 1996). Data before 1995 are based on earlier standards and thus not fully comparable. We have solved this problem by using 1995 as a "chain - year". Since income data are available for 1995 according to both standards, we have simply used the relative difference in 1995 for all previous years (that is multiplied say the figures for 1983, with the factor given by the 1995 differences available for all regions).

Regional Classification

Belgium (11 regions)
BE1 Bruxelles
BE21 Antwerpen
BE22 Limburg (B)
BE23 Oost-Vlaanderen
BE24 Vlaams Brabant
BE25 West-Vlaanderen
BE31 Brabant Wallon
BE32 Hainaut
BE33 Liege
BE34 Luxembourg (B)
BE35 Namur

Germany (30 regions)
DE11 Stuttgart
DE12 Karlsruhe
DE13 Freiburg
DE14 Tûbingen
DE21 Oberbayern
DE22 Niederbayern
DE23 Oberpfalz
DE24 Oberfranken
DE25 Mittelfranken
DE26 Unterfranken
DE27 Schwaben
DE5 Bremen
DE6 Hamburg
DE71 Darmstadt
DE72 Giessen
DE73 Kassel
DE91 Braunschweig
DE92 Hannover
DE93 Lüneburg
DE94 Weser-Ems
DEA1 Düsseldorf
DEA2 Köln
DEA3 Münster
DEA4 Detmold
DEA5 Arnsberg
DEB1 Koblenz
DEB2 Trier
DEB3 Rheinhessen-Pfalz

DEC Saarland
DEF Schleswig-Holstein

Spain (18 regions)
ES11 Galicia
ES12 Principado de Asturias
ES13 Cantabria
ES21 Pais Vasco
ES22 Comunidad Foral de Navarra
ES23 La Rioja
ES24 Aragon
ES3 Comunidad de Madrid
ES41 Castilla y Leon
ES42 Castilla-la Mancha
ES43 Extremadura
ES51 Cataluna
ES52 Comunidad Valenciana
ES53 Islas Baleares
ES61 Andalucia
ES62 Region de Murcia
ES63 Ceuta y Melilla
ES7 Canarias

France (22 regions)
FR1 Ile de France
FR21 Champange-Ardenne
FR22 Picardie
FR23 Haute-Normandie
FR24 Centre
FR25 Basse-Normandie
FR26 Bourgogne
FR3 Nord-Pas-de-Calais
FR41 Lorraine
FR42 Alsace
FR43 Franche-Comte
FR51 Pays de la Loire
FR52 Bretagne
FR53 Poitou-Charentes
FR61 Aquitaine
FR62 Midi-Pyrenees
FR63 Limousin
FR71 Rhone-Alpes
FR72 Auvergne
FR81 Languedoc-Roussillon
FR82 Provence-Alpes-Cote d'Azur
FR83 Corse

Greece (13 regions)
GR11 Anatoliki Makedonia
GR12 Kentriki Makedonia
GR13 Dytiki Makedonia
GR14 Thessalia
GR21 Ipeiros
GR22 Ionia Nisia
GR23 Dytiki Ellada
GR24 Sterea Ellada
GR25 Peloponnisos
GR3 Attiki
GR41 Voreio Aigaio
GR42 Notio Aigaio
GR43 Kriti

Italy (20 regions)
IT11 Piemonte
IT12 Valle d'Aosta
IT13 Liguria
IT2 Lombardia
IT31 Trentino-Alto Adige
IT32 Veneto
IT33 Friuli-Venezia Giulia
IT4 Emilia-Romagna
IT51 Toscana
IT52 Umbria
IT53 Marche
IT6 Lazio
IT71 Abruzzo
IT72 Molise
IT8 Campania
IT91 Puglia
IT92 Basilicata
IT93 Calabria
ITA Sicilia
ITB Sardegna

Nederland (12 regions)
NL11 Groningen
NL12 Friesland
NL13 Drenthe
NL21 Overijssel
NL22 Gelderland
NL23 Flevoland
NL31 Utrecht

NL32 Noord-Holland
NL33 Zuid-Holland
NL34 Zeeland
NL41 Noord-Brabant
NL42 Limburg (NL)

Portugal (5 regions)
PT11 Norte
PT12 Centro
PT13 Lisboa
PT14 Alentejo
PT15 Algarve

UK (32 regions)
UKC1 Tees Valley and Durham
UKC2 Northumberland and Tyne and Wear
UKD1 Cumbria
UKD2 Cheshire
UKD3 Greater Manchester
UKD4 Lancashire
UKD5 Merseyside
UKE1 East Riding and North Lincolnshire
UKE2 North Yorkshire
UKE3 South Yorkshire
UKE4 West Yorkshire
UKF1 Derbyshire and Nottinghamshire
UKF2 Leicstershire, Rutland and Northamptonshire
UKF3 Lincolnshire
UKG1 Herefordshire, Woscerstershire and Warwickshire
UKG2 Shropshire and Staffordshire
UKG3 West Midlands
UKH1 East Anglia
UKH2 Bedfordshire and Hertfordshire
UKH3 Essex
UKI London (NUTS 1)
UKJ1 Berkshire, Bicks and Oxfordshire
UKJ2 Surrey, East and West Sussex
UKJ3 Hampshire and Isle of Wight
UKJ4 Kent
UKK1 Gloucerstershire, Wiltshire and North Somerset
UKK2 Dorset and Somerset
UKK3 Cornwall and Isles of Scilly
UKK4 Devon
UKL Wales (NUTS 1)
UKM Scotland (NUTS 1)
UKN Northern Ireland

12 Regional Growth in Western Europe: An Empirical Exploration of Interactions with Agriculture and Agricultural Policy

Roger S. Bivand and Rolf J. Brunstad

Norwegian School of Economics and Business Administration

12.1 Introduction

Much of the analysis of regional growth in Europe in recent years has concentrated on the concept of convergence, whether in the neo-classical model, or in alternative formulations. Empirical work has shifted from more confirmatory non-spatial estimation to the acknowledgement of the importance of spatial factors, including spillovers, and to exploratory spatial data analysis. This has occurred in conjunction with similar work in North America, and has led to a better understanding of the difficulties involved in relying simply on the convergence model without augmentations.

The work discussed in this chapter is an early attempt to throw light on apparent variability in regional convergence in relation to agriculture as a sector subject to powerful political measures. We would like to explore the possibility that some of the observed specification issues in current results are rooted in neglecting agricultural policy interventions, within the limitations imposed by data available at this stage. We would also like to use this as a case setting for evaluating the appropriateness of geographically weighted regression as a technique for assessing coefficient variability, over and above for instance country dummies, but possibly reflecting missing variables or other specification problems.

12.2 Convergence, Agriculture and Agricultural Policy

Rather than review the convergence literature broadly, we prefer to focus on suggestions pointing up issues to be explored here. We will therefore be taking some positions as given, and will only mention them briefly for clarity. We will be concerned with β-convergence as represented in empirical studies in the following way:

$$\frac{1}{T}\log\left(\frac{y_{i,T}}{y_{i,0}}\right)=\alpha+\beta\log(y_{i,0})+u_i \tag{12.1}$$

where α and β are coefficients and u_i is a disturbance term (Paci, 1997, p. 617). Given an estimate of β, the speed of convergence may be represented as: $\theta = -\log(1+T\beta)/T$, with 100θ expressing this speed in percentage points.

The underlying regularity in this representation is that the rate of growth $y_{i,T}/y_{i,0}$ of a regional economy i in the period up to T is related to its initial condition in period 0 for some measure $y_{i,0}$. The measure to be used here is gross value added per capita measured in EUR 1000 at constant 1990 prices. Figure 12.1 shows the regional distribution of this variable for 1989, the initial period to be used here except where stated. Details of the choice of period and regions will be given below in Section 12.2.2. The contrast between lower values in most of the Iberian peninsula and southern Italy and the rest of the study area is familiar, as are the effects of regional definition artifacts[1] in Benelux and western Germany, with high urban values contrasting with apparently lower surrounding rural values despite commuting.

[1] For example the underbounding of Hamburg is discussed by Fingleton, where the functional region of the agglomeration is clearly larger than the administrative area (2001, p. 147).

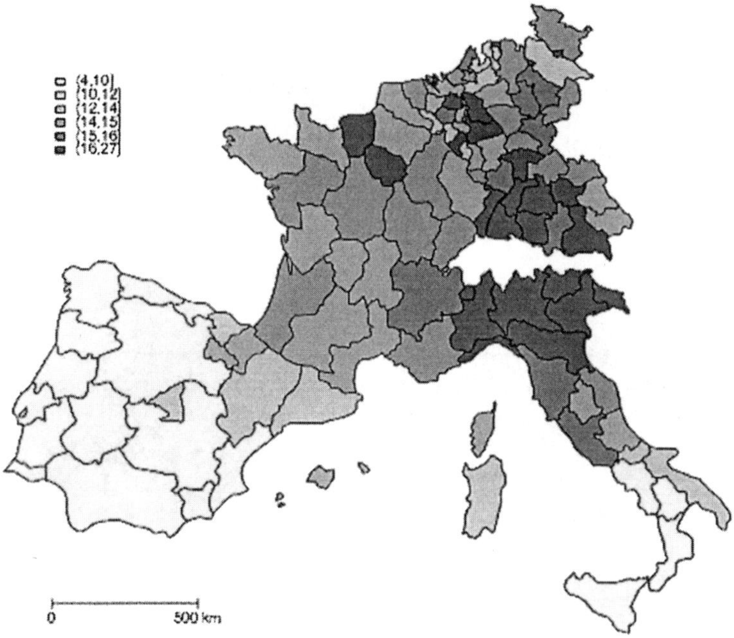

Fig. 12.1. Gross value added per capita, EUR 1000, 1989 ($y_{i,0}$).

Figure 12.2 appears to fit well to the initial conditions: regions such as those in Spain and Portugal with low initial condition values have high growth rates, while French regions have low growth rates and medium initial conditions. However, closer inspection suggests that the stories of particular regions, or clubs of regions, are more complex than our simple convergence model indicates. This is strengthened by an examination of Table 12.1, and by the insignificance of a χ^2 test (9.969, df = 9, p-value = 0.353) on the relationship.

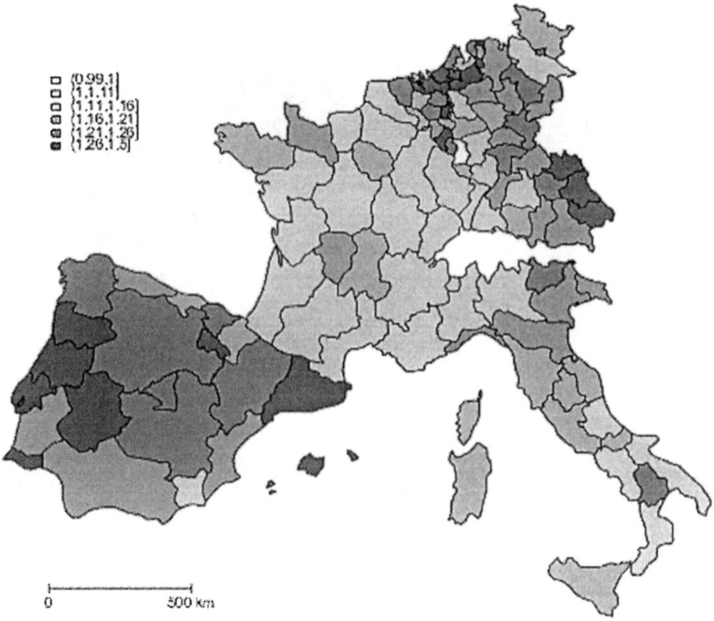

Fig. 12.2. Regional economic change: GVA per capita 1999 as proportion of GVA per capita 1989 $\left(y_{i,1999} / y_{i,1989}\right)$

Table 12.1. Contingency table of initial conditions (1989) by growth rates (1989-1999), both variables cut at quartiles

			$y_{i,1999}/y_{i,1989}$		
$y_{i,1989}$	0.999-1.11	1.11-1.17	1.17-1.23	1.23-1.45	sum
4.03-11.7	5	6	8	10	29
11.7-13.8	6	10	4	9	29
13.8-15.7	8	8	6	6	28
15.7-16.8	10	5	10	4	29
sum	29	29	28	29	115

To check whether this lack of a clear relationship is restricted to this choice of 1989 as base year, and the period to 1999 to measure growth, similar χ^2 tests for base years from 1989 to 1998, and growth rates from those base years to 1999, in total 55 tests, were conducted. They confirm that this value is not exceptional, and that it is difficult to sustain a general relationship between observed initial conditions and growth rates in European regions during this period. Only 14 of the 55 relationships pass as blunt a test as this. Returning to earlier empirical studies, it is clear that many have already pointed out not just the theoretical difficulties of the simple convergence model, but here more importantly the interesting structural features that it chooses to neglect.

Among others, Paci notes that "the observed process of aggregate convergence can hide important structural change phenomena" (1997, p. 617). In analyzing sectoral labour productivity, Paci finds that without using a "Southern" dummy or national dummies, the convergence relationship for agriculture is not significant (1997, p. 627). Others, including Fagerberg et al. (1997), Pons-Novell and Viladecans-Marsal (1999), Paci and Pigliaru (1999), and López-Bazo et al. (1999), also draw attention to the specific structural role of agriculture in empirical analyses of convergence in European regions. There are of course also other structural phenomena of interest, but here we will concentrate on agriculture.

12.2.1 The Impact of Agricultural Policy

The potential impacts of agricultural policy in the European Union on cohesion have been central in changes in the measures and component parts of the Common Agricultural Policy over the past decade. Cohesion is understood as accelerating regional economic growth in those parts of the EU with region GVA per capita markedly below that of developed regions, and thus most regions receiving cohesion support are agricultural regions, although not all agricultural regions are cohesion regions. The basic features of these measures, and changes taking place, including the Mac Sharry reforms and Agenda 2000, have recently been surveyed by Colman (2001), and links to regional policy are covered by Tondl (2001).

The second EU report on economic and social cohesion stresses the role of convergence, and concludes that specific measures will be needed to eliminate regional disparities (DG REGIO, 2001b). These will address differences in underlying conditions and factor endowments, among which labour force skills are seen as central. In addition, a preliminary study was devoted to the impact of community agricultural policies on cohesion (DG REGIO, 2001a). In this study, and more generally in the economics of tariff system structures, attention is drawn to difficulties in adequately measuring economic assistance.

Agricultural policy could be expected to interact with convergence in two ways. Firstly, and for practical reasons the only relationships to be explored empirically, one could expect the proportion of subsidised agriculture in a regional economy, and the intensity of the support, to influence the region's growth rate negatively. For present purposes we assume that agriculture may be treated as though it were uniformly subsidised. The reasons for these negative relationships are that subsidies attenuate the movement of labour and capital to other sectors (and/or regions) with higher productivity, conserving structures of factor allocation at the cost of those paying for the subsidies. The subsidies may also be expected to reduce or to distort incentives to farmers to change their mixes of products and/or methods of production.

In this sense the subsidies are counterproductive as they hamper the growth in GVA. However the recent discussion about the so-called multifunctionality of agriculture may indicate that agricultural activities produce benefits over and above the market value of agricultural production. In economic terms agricultural production may have positive external effects on perceived public goods like the amenity value of the cultural landscape. If this is the case, and if agricultural subsidies are used as a means to internalise these externalities, growth is reduced only because we are measuring the wrong thing, traditional GVA instead of an extended GVA including the willingness to pay for such amenities. Whether or not this is the case is of course of vital importance for the policy implications of a negative relation between agricultural support and regional growth. For a recent paper addressing this question see Brunstad et al. (1999).

The second approach not followed up here, would be to consider the impact on speed of convergence of changes in agricultural policy regimes; for such an approach to be considered, regionalised agricultural accounts would be needed. Since they are not available in a systematic form at scales and levels of detail needed for Europe-wide analysis, micro-level studies would probably be required, such as panel studies, to cast light on the detailed relationships between different subsidy regimes and the embedding of agriculture in regional economies.

While governments granting subsidies to producers can account for them from public expenditure, other forms of subsidy cannot be as readily measured. In particular this applies to subsidies based on tariffs, quota systems, import bans, etc., where the transfer is carried out with consumers within the trade barrier system subsidising producers by an amount equivalent to the difference between the local price and the price of the same good delivered to that market from an external source at world market prices. Measurement complications and production distortions here also affect markets in intermediate goods.

The most commonly adopted approach at the national scale is to estimate agricultural assistance using producer subsidy equivalents, an approach used in a number of international organisations, and described in detail by Cahill and Legg (1989). The data requirements are however substantial, not just for price and quantity series for the chosen commodities, but also insight into the intermediate agricultural goods involved in regionally varying production processes. Some of the work required to estimate regional PSE series has been carried out in the DG REGIO study (2001a), yet more in Heckelei and Britz (2000). Since the present study is only intended to flag the importance of the agricultural sector, it was found more appropriate to use less adequate but more accessible data, rather than further reduce the number of regions under consideration or increase uncertainties associated with estimating or interpolating variables.

12.2.2 Sources of Data in Agriculture

European agricultural accounts are available in two versions, EAA 89/92 and EAA 97 REV.1.1, and regional versions at the level of NUTS level 2 are available from Eurostat[2]. The accounting data used here is for agriculture gross value added at market prices, subsidies, taxes linked to production (including VAT balance), and gross value added at factor costs, for the period 1988-1998, in million ECU. From the middle of the period, there are very many missing values, and some countries either only report at NUTS level 1, or level 0. For this reason, the study area does not include Ireland, Denmark, Greece, UK, Berlin and former East Germany, and also drops overseas dependencies and Atlantic islands. It was further found necessary to aggregate three regions in Belgium (BE1 Brussels, BE24 Vlaams Brabant, and BE31 Brabant Wallon) in order to maintain the spatial series. The number of regions included in the analysis is consequently 115; not covering Ireland and Greece is unfortunate, because— had data been available – they would have cast additional light on the issues under study. In order to smooth the agricultural accounts data somewhat, and to accommodate further missing data problems, the variables to be used below are averages of values reported during 1988-1990, in most cases but not all, the averages of three values[3].

[2]The data used here are taken from Eurostat theme: theme1, domain: regio, collect: agri-r, table: a2acct. We are grateful to Lucy McKeever of RCADE, University of Durham, for help in accessing this and other Eurostat and GISCO datasets.

[3]It is unfortunate that the data are not more complete, because the choice of initial conditions means that some older members will appear to have higher support than newer members, and also because support to the older members declined during the chosen period while support to newer members probably rose.

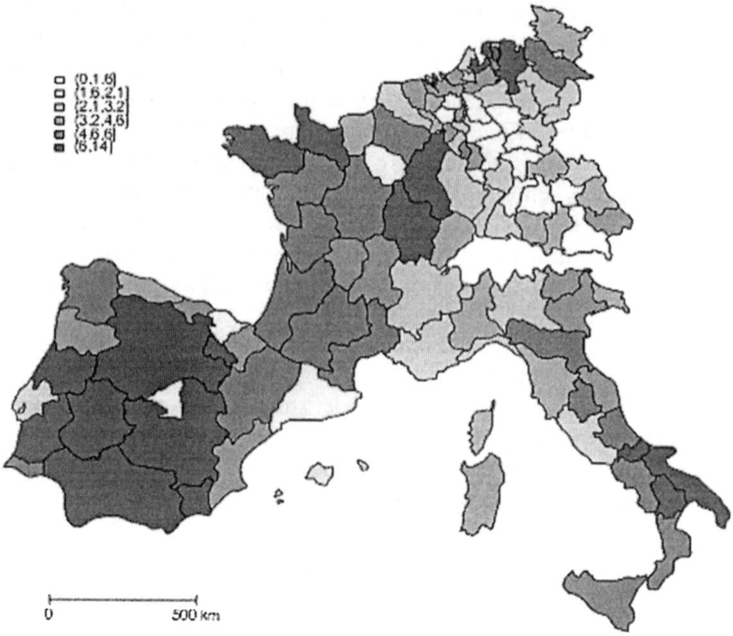

Fig. 12.3. GVA in agriculture at factor costs (average 1988-90) as percentage of total GVA 1989.

Figure 12.3 shows average agricultural GVA measured at factor costs 1988-90 (in EAA 89/92 nomenclature, GVA at factor costs is GVA at market prices plus subsidies minus taxes linked to production) as a percentage of total GVA for 1989. The underlying total GVA and population datasets have been taken from Cambridge Econometrics' European Regional Databank, and are measured in 1990m ECU and 1000 persons[4]. The two datasets have been merged after dropping NUTS level 2 regions that could not be used because of missing agricultural accounts data. There is a potential problem of double-counting involved in using regional GVA series, because they can and most often do include subsidies as a component of gross value added[5].

---

[4]We would like to thank Cambridge Econometrics and Sasha Thomas for their help.

[5]In addition, changes introduced in revisions of the System of National Accounts, knocked on to EAA 97, moving some subsidies into the definition of basic prices (OECD 2000); consequently the regional total GVA series may include varying amounts of subsidies depending on whether they are based on market prices or basic prices.

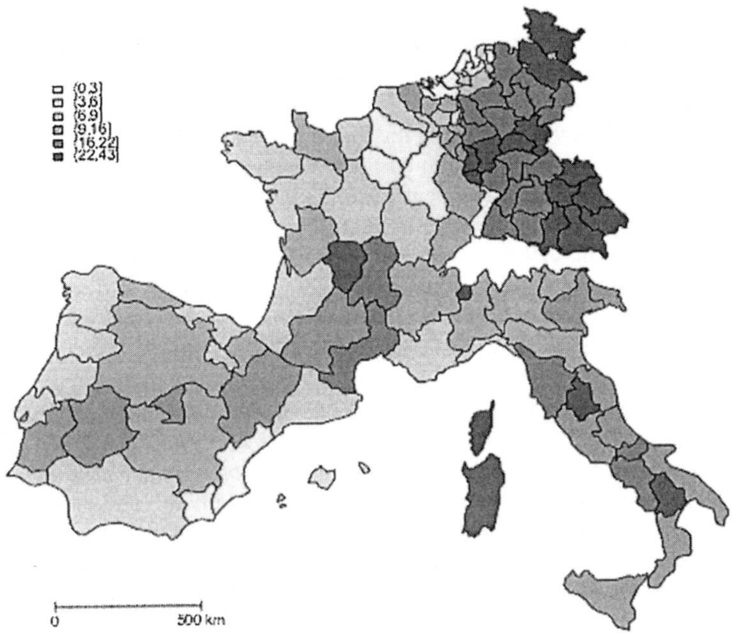

Fig. 12.4. Relative subsidy levels 1988-90: agricultural subsidies (average 1988-90) as proportion of GVA in agriculture at market prices (average 1988-90)

Figure 12.4 displays the other variable to be used to represent the influence of agricultural policy, agricultural subsidies as a proportion of agricultural GVA at market prices. This is necessarily a very inadequate substitute for a properly constructed measure of producer subsidy equivalents, because it does not encompass the effects of tariff structures, which are not simply proportional to subsidy payments, but vary regionally with produced quantities of commodities. It has been chosen to restrict the agricultural variables to the 1988-90 period, partly because of missing data in the mid-1990's in agricultural accounts, but also because the base year date for the convergence model is 1989 - also chosen to match the years with least missing agricultural data.

12.3 Estimating Convergence: Specification Issues

Attention has been drawn in a series of studies to specification problems found in estimating the standard convergence model using OLS. The roots of these problems are partly related to substantive spatial relationships, such as spillovers (Vayá et al., 1998), but may also involve missing variables, structural differences across the chosen study area, and functional form. Fingleton and McCombie (1998) and Fingleton (1999a, 2001) draw attention to the clear need to pay attention to specification, over and above the introduction of spatial econometric techniques also made by Vayá et al. (1998), in several Chapters in this volume, and in the North American context by Rey and Montouri (1999) and Rey (2001). Details of the estimation methods and tests are not repeated here for brevity, and may be found in the cited articles and in the earlier Chapters of this volume.

Results from the estimation of the standard convergence model, with the annual log GVA per capita growth rate (1989-1999) as the dependent variable and log GVA per capita in 1989 as the independent variable, are presented in Table 12.2. The table also shows the results of estimating the same model including the spatially lagged dependent variable[6] using maximum likelihood, and of estimating an augmented model including the two agriculture variables defined above in Section 12.2.2 for both estimation methods. The percentage speed of convergence for the OLS standard model, 0.75%, is comparable with that reported in Chapter 3 of 0.84% for 1980-1995 and 138 regions.

[6]The spatially lagged dependent variable can be understood as the average value of neighbours of the region examinied, for some definition of neighbour and row-standardised neighbourhood weights.

Table 12.2. Modelling convergence 1989-1999 (t-values or z-values in parentheses)

	OLS	ML lag	OLS	ML lag
Intercept	0.0346	0.0147	0.0400	0.0203
	(6.56)	(2.86)	(5.52)	(2.91)
log GVA pc 1989	-0.00725	-0.00288	-0.00987	-0.00498
	(-3.55)	(-1.61)	(-3.99)	(-2.24)
% speed of convergence	0.753	0.292	1.039	0.511
log % agriculture 1988-90			-0.00200	-0.00133
			(-2.19)	(-1.68)
log subsidy/GVA 1988-90			-0.00138	-0.000699
			(-1.90)	(-1.11)
σ	0.00704	0.00597	0.00687	0.00591
log likelihood	407.8	421.1	411.7	422.8
AIC	-809.6	-836.1	-813.3	-835.7
Chow F	3.772		1.284	
p-value	0.026		0.281	
RESET F	3.921		1.573	
p-value	0.0226		0.187	
Breusch-Pagan	0.0626		2.195	
p-value	0.969		0.699	
Moran's I	0.332		0.308	
Z	5.596		5.377	
p-value	<0.00001		<0.00001	
Robust LM (err)	0.444		0.092	
p-value	0.505		0.762	
Robust LM (lag)	3.860		2.414	
p-value	0.049		0.120	
Spatially lagged dependent variable		0.546		0.512
		(5.99)		(5.41)
LM (err)		2.00		0.485
p-value		0.157		0.486

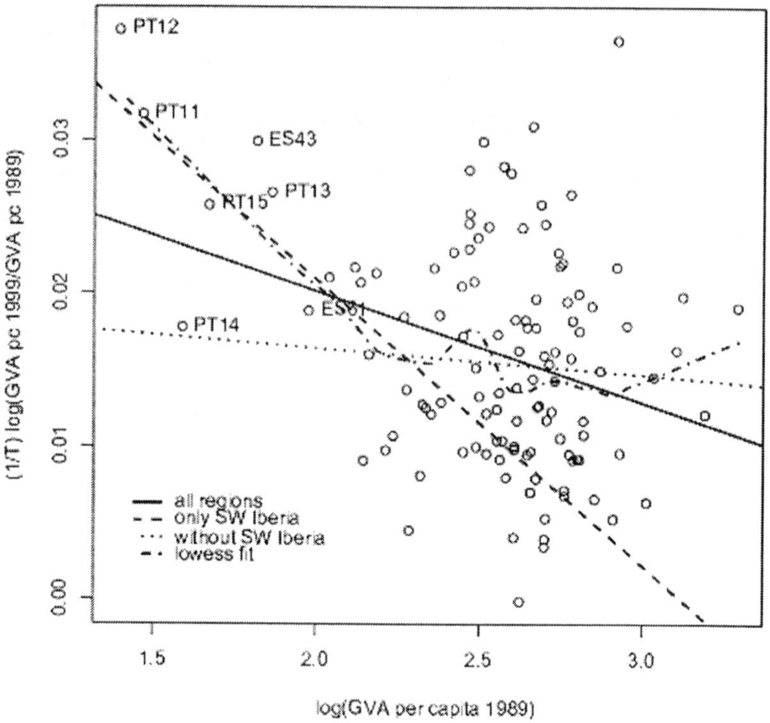

Fig. 12.5. Plot of standard OLS convergence model (1989-1999) with regression fitted lines for all regions, only SW Iberia, all without SW Iberia, and a lowess fit

Concentrating first on the OLS estimates of the standard model, it is worth noting that while the Breusch-Pagan test does not reject homoskedasticity[7], the RESET test (Johnson and DiNardo, 1997, p. 121) using the second, third, and fourth powers of the fitted values as extra variables indicates the presence of some specification error (the same result is found for all T=1990-1999 for base year 1989, and often for other base years). Figure 12.5 gives us a view of the relationship: it seems that the regions belonging to South West Iberia (the Portuguese regions and the Spanish regions of Extramadura and Andalucia) differ structurally from the remainder. They have much lower initial condition values and much higher growth rates, and the slopes of regression lines for all regions, just SW

[7]Tests of residual terms in all estimated models for non-normality were not significant; the outcomes of Spatial Breusch-Pagan tests on the ML lag models were not significant either.

Iberia, and all without SW Iberia are quite different. The plotted lowess fit (a robust local regression fit passing a moving window across the data set, see Cleveland, 1979) confirms that there are structural differences in the data set, which could be modeled by including the square of the independent variable, but may better be considered in a Chow test context (Johnson and DiNardo, 1997, p. 113-116). Subsetting the data set into SW Iberia and not SW Iberia, we can conduct a Chow test for structural difference in both slope and intercept coefficients, differences found to be significant as shown in Table 12.2. Again, this result applies to all T=1990-1999 for base year 1989, and for some other base years (1990, 1991, 1993, 1994 and 1995, and some T.

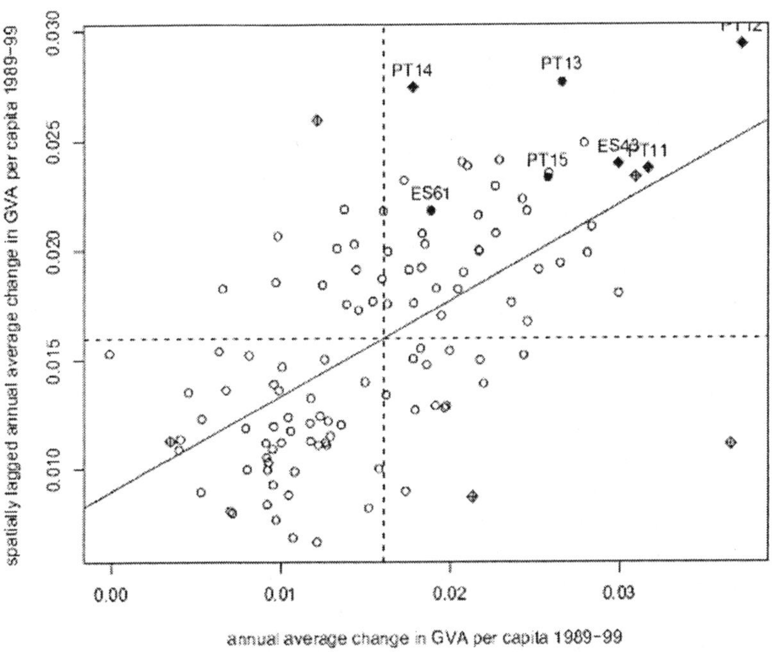

Fig. 12.6. Moran scatterplot of convergence model dependent variable $1/T \log(y_{i,T}/y_{i,0})$

The fact that the SW Iberia constitutes a block of outliers is also seen in Figure 12.6, a Moran scatterplot of the annual log growth rates 1989-1999 (see also Rey and Montouri, 1999, p.150). This plot of the values of the variable in question on the x-axis against its spatially lagged values on the y-axis lets us see how the inclusion of the spatially lagged variable might give more insight into the convergence process, where growth rates would depend on the average of growth rates across neighbours of regions, possibly representing spillover or perhaps underbounding of functional regions. Here neighbours have been defined using a sphere of influence graph[8] rather than boundary contiguity (Vayá et al. 1998) or distance (as in earlier Chapters of this book). Given a set of points representing the regions[9], a row standardised weights matrix was constructed and used in all analyses presented here.

The tests for spatial specification problems in the standard model, Moran's I (Cliff and Ord, 1981), and LM tests not reported here, are all highly significant. In addition, the robust LM test for a missing spatially lagged dependent variable in the presence of spatial error dependence is significant, but not the other way round (Anselin et al. 1996, Anselin and Bera 1998). It is not impossible that this lack of clarity in the results of the spatial specification tests is related to Fingleton's finding, that Moran's I may also detect spatial non-stationarity (1999b).

Following the augmentation of the standard model with the two agricultural variables, we can see that the non-spatial specification problems encountered in the standard model and reflected in the results of the Chow test and the RESET test are alleviated (column 3 in Table 12.2). Neither variable is strongly significant, but both have the expected signs, with both higher proportions of agriculture in regional GVA, and higher ratios of subsidies to agricultural regional GVA being associated with lower growth rates. The percentage speed of convergence is somewhat greater, just over 1%.

Stepping back to examine the ML lag estimates of the standard model, we can see that the inclusion of the spatially lagged growth rate reduces the percentage speed of convergence markedly, with the underlying coefficient value becoming much less significantly different from zero. The coefficient of the spatially lagged dependent variable is itself highly significant, and no residual spatial autocorrelation was found to remain in the model. Further details of estimation and testing methods may be found in Anselin (1988).

Introduction of the agricultural variables in the augmented ML lag model sees a further, but much less marked improvement in σ and log likelihood; however the value of the Akaike Information Criterion (AIC) rises a little compared to the standard ML lag model, suggesting possible interaction between the spatially lagged growth rate and the added variables. Adding the spatially lagged independent variables did not help, with AIC rising further to -831.2 (AIC for the standard ML lag model with spatially lagged initial conditions was -836.9).

12.4 Estimating Convergence: Exploring Non-Stationarity

As pointed out, Fingleton (1999b) has indicated that Moran's I may detect spatial non-stationarity in addition to residual autocorrelation. Because Moran's I continues to be significant in the augmented model, perhaps implying non-stationarity,

---

[8]Thanks to Nicholas Lewin-Koh for this suggestion - his implementation is released as soi.graph() in the R contributed package spdep, available from http://cran.r-project.org.

[9]GISCO NUTS2 label points projected to Lambert Azimuthal Equal Area using EU standard parameters.

and in the poor improvement of the augmented ML lag model over the standard ML lag model, it seems appropriate to explore the standard and augmented models using geographically weighted regression. Having originally been developed as a tool to exploratory spatial data analysis, its status is under revision at present, and it has been suggested that it may be used to indicate the presence of spatial non-stationarity in more formal ways. Exploratory methods have been used in the analysis of convergence models by Rey and Montouri (1999), Rey (2001) and Arbia (2001) and feature in earlier chapters of this volume..

12.4.1 Geographically Weighted Regression

Geographically weighted regression (GWR) is a technique for examining possible variability in coefficient estimates across study areas composed of regions represented by points, and was introduced by Brunsdon et al. (1996, a full description is found in Fotheringham et al. 2000). Instead of simply estimating global coefficient values over the whole data set, local parameter values are estimated for each region/point in the data set:

$$y_i = \beta_{i0} + \sum_{k=1}^{p} \beta_{ik} x_{ik} + \varepsilon_i$$

where y_i, $i = 1, n$, are observations of the dependent variable, x_{ik}, $i = 1, n$, $k = 1, p$ are observations of the p independent variables, ε_i are disturbance terms, and $\beta(\iota)$ unknown parameter vectors which are functions of location *i*.

The local estimates, one regression for each region, are made using weighted regression, with the weights assigned to observations being a function of the distance between the region for which coefficient estimates are required and all the other regions. The distance function is directly comparable to those used in kernel density methods, and here use has been made of a bisquare function:

$$w(i)_{jj} = (1 - (d_{ij}^2 / d^2))^2, \quad d_{ij} \le d$$

where $w(i)_{jj} = 0$ when $d_{ij} > d$; $w(i)_{jj}$ is the weight assigned to observation *j* in estimating the parameters for observation *i*. The parameter *d* is termed the bandwidth, and for the bisquare function is the maximum distance for non-zero weights; it may be found by cross-validation and the value used here is 563 km, found by cross-validation for the augmented model. The parameters are estimated by weighted least squares:

$$\hat{\beta}(i) = \left[\mathbf{X}^T \mathbf{W}(i) \mathbf{X}\right]^{-1} \mathbf{X}^T \mathbf{W}(i) \mathbf{y}$$

where $\mathbf{W}(i)$ is the diagonal weights matrix for observation *i*; $\mathbf{X}^T\mathbf{W}(i)\mathbf{X}$ is assumed to be invertible.

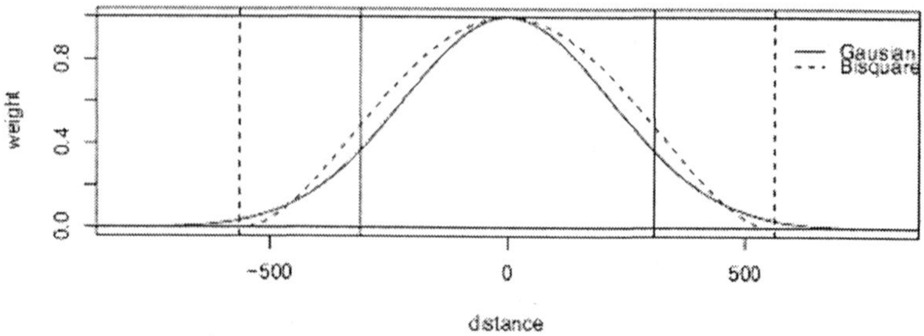

Fig. 12.7. Values of weighting functions by distance for bandwidths chosen by crossvalidation of the augmented model; bandwidth values shown by vertical lines. The very similar areas under the curves for the bisquare and Gaussian functions show that cross-validation has established adequate bandwidths in both cases.

Figure 12.7 shows the spatial range of this weighting scheme, and for comparison the Gaussian scheme and its equivalent cross-validated bandwidth (310 km). In current GWR techniques, bandwidth is first established globally, meaning that regions in denser parts of the study area have a larger number of weighted neighbours contributing to their estimates than in less dense parts. The bandwidth used here is about twice the greatest nearest neighbour distance in the data set (278 km), and between the lower quartile (420 km) and median (773 km) of all inter-region distances.

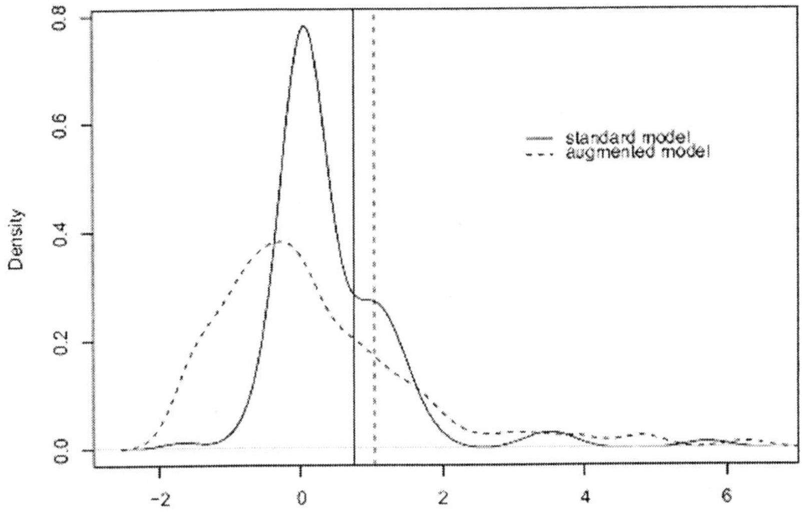

Fig. 12.8. Density plots (smoothed histograms) of percentage θ (speed of convergence) coefficient estimates from standard and augmented GWR models; OLS model point estimates are shown by vertical lines (see Table 12.2). Bandwidths in both cases 563 km.

Figure 12.8 shows the range of coefficient values estimated for the percentage speed of convergence by GWR for the standard and augmented models, plotted with vertical lines marking the global parameter estimates from the OLS models described above. As can be seen, in neither case do the GWR estimates group tightly or symmetrically around the global estimates. This is in itself not necessarily an issue, since some variation between multiples of standard errors is expected, but the extent and shape of the dispersion is worrying. For this weighting scheme (bisquare, bandwidth = 563km), it seems clear that in the standard model there is evidence of structural instability, with a peak just above zero, and a shelf centred at 1, with the global estimate at the root of the shelf. In the augmented model, the distribution is centred on zero, with many negative estimates, and a long right tail - the global estimate does not give a good description of the distribution at all. Of course here growth is also being driven (or rather restrained) by the values of the agricultural variables, but even so it is difficult to avoid the impression that non-stationarity may be present. Beyond non-stationarity, regions with negative percentage speeds of convergence would seem to imply not only that there is no unique steady state for all the regions, or that regions converge at different speeds, but that some regions are actually diverging, also when augmenting variables are taken into account.

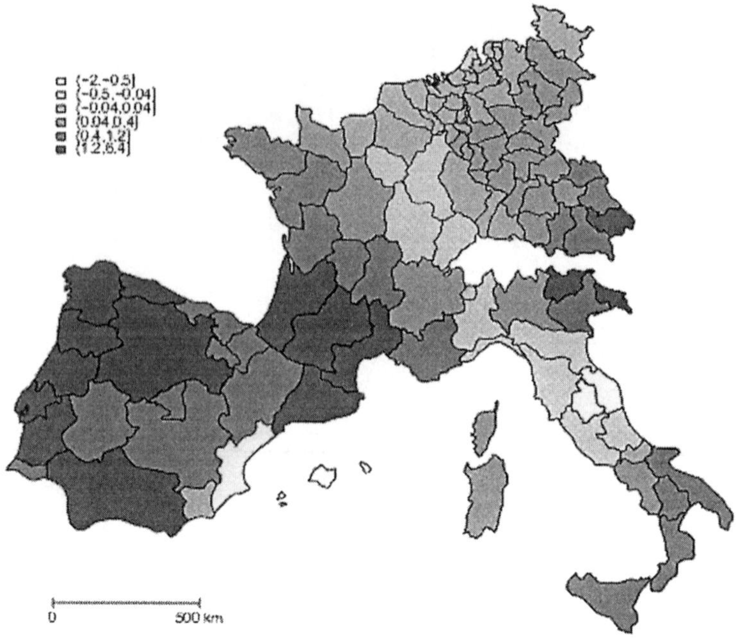

Fig. 12.9. Percentage θ (speed of convergence) coefficient estimates from standard GWR model. Bandwidth 563 km, bisquare weighting function.

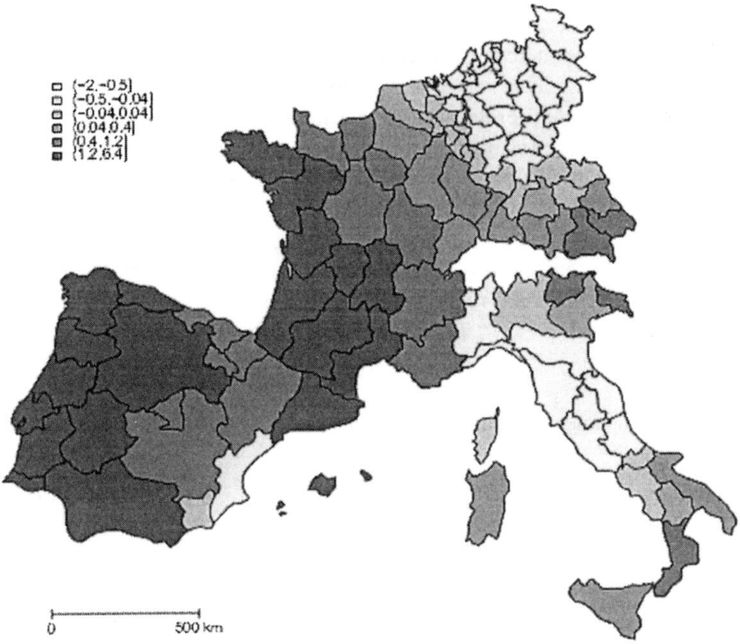

Fig. 12.10. Percentage θ (speed of convergence) coefficient estimates from augmented GWR model. Bandwidth 563 km, bisquare weighting function.

Mapping the standard GWR percentage speed of change estimates (Figure 12.9), we can see that our tentative conclusion of structural differences is sustained. The banded West-East pattern is even more pronounced in the mapped values of estimates from the augmented GWR model when agricultural variable have been taken into account (Figure 12.10). Introducing the agricultural variables has a major impact in Germany, for which values of the agricultural subsidies variable are relatively high (see Figure 12.4). In both cases, had the maps not displayed spatial patterning, we could have risked concluding that the variability seen in Figure 12.8 was of little importance. What remains is to try to establish whether the GWR estimates add substantially to the global results from Table 12.2.

12.4.2 Testing Stationarity Using Geographically Weighted Regression

So far, very few tests for GWR models have been published. Two, based on the use of analysis of variance, use generalised degrees of freedom (no longer integer because of the varying weights sums per estimate) to compare the improved sum of squares accounted for by the GWR estimates as compared with the global OLS estimates. Both Brunsdon et al. (1999) and Leung et al. (2000a) establish that a hat matrix may be constructed from rows from the GWR model, each row using its appropriate weights. The two F-tests they derive differ in that Brunsdon et al. (1999) compare the difference in the residual sums of squares of the OLS and GWR models with the residual sum of squares of the GWR model, while Leung et al. (2000a) compare the same difference with the residual sum of squares of the OLS model, choosing a different denominator for the F-ratio, as well as slightly different degrees of freedom. Results of these tests are shown in Table 12.3, from which we can see that the GWR estimates for both standard and augmented models significantly reduce the residual sum of squares over and above the OLS estimates.

Table 12.3. Comparison of standard and augmented OLS and GWR convergence models

	standard		augmented	
	OLS	GWR	OLS	GWR
σ	0.00704	0.00562	0.00687	0.00501
AIC	-809.6	-853.9	-813.3	-884.6
Brunsdon et al. ANOVA		2.85		2.16
p-value		0.0000		0.0003
Leung et al. F(2) ANOVA		2.20		1.64
p-value		0.0002		0.0102
Moran's I	0.332	0.122	0.308	0.062
Leung et al. p-value	< 0.00001	< 0.00001	< 0.00001	0.00148
Z under randomisation p-value	< 0.00001	0.0365	< 0.00001	0.2624

Table 12.3 also contains estimates of σ and the Akaike Information Criterion, based on work in progress reported in Brunsdon et al. (2000). Here again the properties of the GWR hat matrix are used to derive the residual sum of squares of the GWR model, and the appropriate degrees of freedom for the AIC measure. The reported values suggest that the level of non-stationarity present in the data under analysis is such that GWR does provide a more adequate representation than global parameter estimates, even when structurally-varying agricultural variables are included. Finally, following Brunsdon et al. (1998), the spatial autocorrelation of the GWR residuals is reported.

The results of the test procedure described by Leung et al. (2000b), a three-moment χ^2 approximation, is shown, together with the probability values of Moran's I under randomisation ignoring the fact that we are testing regression residuals. Both tests, and the estimated values of Moran's I statistic, suggest that the GWR models display less global spatial autocorrelation in their residuals than the OLS models, and that the GWR model augmented with agricultural variables performs best.

12.5 Conclusions and Directions for Further Research

Following up the impacts of agricultural policy on the estimation of empirical models of regional convergence does seem worthwhile, because on balance more of the variance of the average annual regional growth rates is accounted for in the augmented model than the standard model estimated by OLS, ML lag, and GWR. Since it is possible that the non-stationarity is related to further missing variables, including the inadequacy of the way in which agricultural subsidies are represented, it will be important to try to replace the agriculture variables with better estimates of producer subsidy equivalents for the base year. An issue that deserves further attention is how to handle changes in agricultural policy regime occurring between years 0 and T.

On balance however, we would argue that this exploratory analysis does give support to the role of agricultural subsidies in accounting for variations in regional growth. As argued in Section 12.2.1 these differences could stem from subsidies either preventing or slowing down transfer of factors of production to more productive sectors or regions, or distorting farmers' incentives to choose the most productive methods of production or mixes of products. Coefficient estimates for the agricultural variables in the OLS and ML lag global models had the expected negative signs, and were either significant or almost significant at conventional levels. We would also argue that the results of estimation using geographically weighted regression show that assuming the absence of non-stationarity may hide potentially interesting and useful differences in patterns of regional growth and change. We feel that while GWR is as yet not a mature analytical technique, the indications our results provide should encourage its use by others as a supplement and corrective to the spatial econometrics toolbox.

On the technical side, some further tests for specification problems in the spatial context are becoming available, but have not been applied to convergence models yet (Baltagi and Li, 2001, de Graff et al., 2001). The tests on GWR estimates also need to be more firmly established, although some Monte Carlo studies on the convergence models reported above did not give cause for concern that the conclusions drawn were inappropriate. The GWR results also need to be tested for spatial autocorrelation, and if necessary to accommodate a spatially lagged dependent variable in some form, although GWR does already involve a spatial weighting of the observations themselves.

It is clear that addressing concerns about non-stationarity is one of the major tasks which has to be tackled in order to promote spatial econometrics; it may also be the case that panel estimation techniques or multi-level modelling will provide further tools that may be employed to test for specification adequacy. It does seem to be clear, though, that the full spatial econometrics toolbox, including both exploratory and confirmatory techniques, needs to be applied to the unravelling of the stories involved in European regional growth.

We would like to acknowledge helpful comments from an anonymous referee and from participants in seminars at the departments of economics at the Norwe-

gian School of Economics and Business Administration and the Norwegian University of Science and Technology.

References

Anselin L (1988) *Spatial econometrics: methods and models.* Kluwer, Dordrecht

Anselin L, Bera AK (1998) Spatial dependence in linear regression models with an introduction to spatial econometrics. In: Ullah A, Giles DEA (eds) *Handbook of applied economic statistics.* Marcel Dekker, New York, pp. 237-289

Anselin L, Bera AK, Florax R, Yoon MJ (1996) Simple diagnostic tests for spatialdependence. *Regional Science and Urban Economics* 26: 77-104

Arbia G (2001) The role of spatial effects in the empirical analysis of regional concentration. *Journal of Geographical Systems* 3: 271-281

Baltagi BH, Li D (2001) LM tests for functional form and spatial error correlation. *International Regional Science Review* 24: 194-225

Brunsdon C, Fotheringham AS, Charlton ME (1996) Geographically weighted regression: a method for exploring spatial nonstationarity. *Geographical Analysis 28*: 281-298

Brunsdon C, Fotheringham AS, and Charlton, ME (1998) Spatial nonstationarity and autoregressive models. *Environment and Planning A* 30: 957-973

Brunsdon C, Fotheringham AS, and Charlton, ME (1999) Some notes on parametric signficance tests for geographically weighted regression. *Journal of Regional Science* 39: 497-524

Brunsdon C, Fotheringham AS, Charlton ME (2000) *Geographically weighted regression as a statistical model.* (Unpublished working paper, Department of Geography, University of Newcastle)

Brunstad RJ, Gaasland I, Vårdal E (1999) Agricultural production and the optimal level of landscape preservation. *Land Economics* 75: 538-546

Cahill C, Legg W (1989) Estimation of agricultural assistance using producer and consumer subsidy equivalents: theory and practice. *OECD Economic Studies* 13: 13-43

Cleveland WS (1979) Robust locally weighted regression and smoothing scatterplots. *Journal of the American Statistical Association* 74, 829-836.

Cliff AD, Ord JK (1981) *Spatial processes.* Pion, London

Colman D (2001) The Common Agricultural Policy. In: Artis M, Nixson F (eds) *The economics of the European Union.* Oxford University Press, Oxford, pp.97-124

DG REGIO (2001a) *Study on the impact of community agricultural policies on economic and social cohesion.* (http://www.inforegio.cec.eu.int/wbdoc/docgener/studies/pac_en.htm, accessed 2 Jan 2002)

DG REGIO (2001b) *Unity, solidarity, diversity for Europe, its people and its territory: Second report on economic and social cohesion.* (http://www.-inforegio.cec.eu.int/wbdoc/docoffic/official/reports/contentpdf_en.htm, accessed 2 Jan 2002)

Fagerberg J, Verspagen B, Caniels M (1997) Technology, growth and unemployment across European regions. *Regional Studies* 31: 457-466

Fingleton B (1999a) Estimates of time to convergence: an analysis of the regions of the European Union. *International Regional Science Review* 22: 5-34

Fingleton B (1999b) Spurious spatial regression: some Monte Carlo results with a spatial unit root and spatial cointegration. *Journal of Regional Science* 39: 1-19

Fingleton B (2000) Spatial econometrics, economic geography, dynamics and equilibrium: a 'third way'? *Environment and Planning A* 32: 1481-1498

Fingleton B (2001) Equilibrium and economic growth: spatial econometric models and simulations. *Journal of Regional Science* 41: 117-147

Fingleton B, McCombie JSL (1998) Increasing returns and economic growth: some evidence for manufacturing from the European Union regions. *Oxford Economic Papers* 50: 89-105

Fotheringham AS, Brunsdon C, Charlton ME (2000) *Quantitative geography: perspectives on spatial data analysis.* Sage, London

de Graff T, Florax RJGM, Nijkamp P, Reggiani A (2001) A general misspecification test for spatial regression models: dependence, heterogeneity, and nonlinearity. *Journal of Regional Science* 41: 255-276

Heckelei T, Britz W (2000) *Concept and explorative application of an EU-wide, regional agricultural sector model* (*CAPRI project*). (http://www.agp.uni-bonn.de/agpo/rsrch/capri/eaaecapri.pdf, accessed 2 Jan 2002)

Johnston J, DiNardo J (1997) *Econometric methods.* McGraw-Hill, New York

Leung Y, Mei CG, Zhang WX (2000a) Statistical tests for spatial non-stationarity based on the geographically weighted regression model. *Environment and Planning A* 32: 9-32

Leung Y, Mei CG, Zhang WX (2000b) Testing for spatial autocorrelation among the residuals of the geographically weighted regression. *Environment and Planning A* 32: 871-890

López-Bazo E, Vayá E, Mora AJ, Suriñach J (1999) Regional economic dynamics and convergence in the European Union. *Annals of Regional Science* 33: 343-370 OECD (2000) Eurostat's EAA97: *Main changes and the impact on the EAA-data.* OECD Statistics Directorate STD/NA/AGR(2000)10 (http://www1.oecd.org/std/AGRMeet2000/docs/naa00_10.pdf, accessed 2 Jan 2002)

Paci R (1997) More similar and less equal: economic growth in the European regions. *Weltwirtschaftliches Archiv* 133: 609-634

Paci R, Pigliaru F (1999) Is dualism still a source of convergence in Europe? *Applied Economics* 31: 1423-1436

Pons-Novell J, Viladecans-Marsal E (1999) Kaldor's Laws and spatial dependence: evidence for the European regions. *Regional Studies* 33: 443-451

Rey SJ (2001) Spatial empirics for economic growth and convergence. *Geographical Analysis* 33: 195-214

Rey SJ, Montouri BD (1999) US regional income convergence: a spatial econometric perspective. *Regional Studies* 33: 143-156

Tondl G (2001) Regional policy. In: Artis M, Nixson F (eds) *The economics of the European Union.* Oxford University Press, Oxford, pp. 180-212

Vayá E, López-Bazo E, Artís M (1998) *Growth, convergence and (why not?) regional externalities.* (Unpublished working paper E98/31, "Anàlisi Quantitativa Regional" Research Group, University of Barcelona)

13 Explaining the Distribution of Manufacturing Productivity in the EU Regions

Bernard Fingleton[1] and Enrique López-Bazo[2]

[1] *University of Cambridge, UK*

[2] *University of Barcelona, Spain*

13.1 Introduction

In this Chapter, simulations of manufacturing productivity levels across the EU provide detailed insights regarding possible long-run distributions under various alternative assumptions about the determinants of productivity growth. The explanation of manufacturing productivity growth by region is based on an econometric model embodying recent developments in urban economic theory and geographical economics, which includes both internal and external increasing returns and spatial externality (spillover) effects. The model implies market interdependence involving a competitive manufacturing sector and producer services under monopolistic competition. The assumption of technological externalities and the presence of cross-region spillovers in the model lead to a specification that is typical of recent approaches in spatial econometrics, which seeks to avoid bias by a specification involving spatial interdependence. The Chapter uses recent developments in the analysis of growth empirics, involving the application of the estimated density function and the stochastic kernel, in order to visualize the long-run stochastic distributions under various assumptions. We control for the effects of various ancillary factors assumed to influence the equilibrium distribution so as to isolate the impact of each individual factors of interest. The visualizations enabled by the stochastic kernel clearly identify the effect of the different model variables on the entire regional distribution of manufacturing productivity. We compare and evaluate the different equilibria and discuss implications for the future welfare and development of the regions of the EU.

13.2 Distribution Dynamics of EU Regional Productivity

13.2.1 Density Functions and Stochastic Kernels

Most contributions to the empirical growth literature have estimated growth equations in which growth rates are related to the rate of accumulation of factors of production, variables likely to affect the level of technology, and the initial level of output or income per capita (Barro, 1990; Mankiw et al, 1992; see also the survey of growth regressions in Durlauf and Quah, 1999). From estimates of the sign, magnitude and statistical significance of the coefficients, conclusions have been drawn about the causes of economic growth, and about the dispersion in the levels of development observed across economies. In this regard, attention has been focused in particular on the coefficient attached to the initial income level, which relates to the rate of convergence in a neoclassical growth model (Mankiw et al, 1992; Barro and Sala-i-Martin, 1992).

However, the so-called rate of β-convergence provides minimal information, by itself, about the tendency of the income per capita distribution in a set of economies to become concentrated at a common point or about the mechanisms associated with increased dispersion. It is evident that β-convergence analysis per se tells us little or nothing about other the processes associated with the emergence, stability and persistence of phenomena such as convergence clubs; i.e. the causes and dynamics associated with polarization (two groups) or stratification (more than two groups) in the distribution. These criticisms were originally formulated by Quah(1993, 1996a,b) in a flurry of papers outlining alternatives to conventional β-convergence analysis. In short, the basic idea behind this critique is that the richness of the distributional changes cannot be captured by single figures such as the first moments. Likewise the analysis of regression coefficients simply gives the change in growth per unit change in the factors determining growth (such as investment in physical and human capital, R&D, fiscal policy, etc), controlling for the other variables, so that the focus is on a single 'typical' region and the impact of, say, a unit increase in physical capital on a region's growth, rather than providing illumination on the distributional consequences from the perspective of an entire system.

To overcome these shortcomings, when what matters is the relative performance of a set of economies, Quah argues that we should be more concerned with the cross-section distribution dynamics of inter-regional or international income per capita, a form of analysis that had already become well established in the wages and personal income distribution literature, and in studies of firm and industry mobility, etc. This analysis focuses attention on the external shape of the distribution, and on movements within the distribution. With regard to the external shape, one scenario might be that the distribution is undergoing a process of collapsing to a single point, in which case the conclusion would be that the economies are involved in a convergence process in the sense that the poor regions are approaching the level of the rich regions. Another possibility is that the distribution is characterized by increasing dispersion over time. In both cases, the external distribution shape for different time periods allows us to visualize the ongoing process. Interestingly, the distribution shape also allows visualization of the presence or formation of convergence clubs, which are groups of economies (clubs) that show internal convergence to steady states specific to each club. It is important to recognize that the external shape can remain unaltered even when there is churning in the distribution. In other words, the shape of the distribution at two points in time might be the same, but the implications differ if poor economies remain poor and rich remain rich compared to the situation in which a significant degree of churning occurs so that those poor at the outset become rich, and vice versa. Thus, the focus on intra-distribution mobility reveals how economies transit from any point in the distribution to any other point at some future time, which at its simplest involves estimating the probability of a poor economy staying poor or becoming rich.

Since Quah's critique, a number of papers have applied methods that focus on mobility and dynamics in income per capita or related measures for systems of regions or sets of national economies (Bianchi, 1997; Fingleton, 1997; Magrini, 1999; López-Bazo et al, 1999; Johnson, 2000; Lamo, 2000). A general conclusion that can be drawn from this diverse set of papers is that the distribution of income per capita is characterized by notable persistence and, if anything, the movements that do occur are because of an ongoing process of polarization in which the poorest economies are being left behind, particularly in the case of a wide sample of countries.

An estimate of the external shape and mobility within the distribution is required to perform the type of analysis described above. This typically involves non-parametric estimation in which there is no a priori structure imposed on the data; on the contrary, non-parametric methods allow the data to speak for themselves. More specifically, the external shape of the distribution of income per capita for a sample of economies is measured by non-parametric estimation of the density function via the kernel method. It is useful to think of this kernel density estimator as a smooth version of a histogram describing the distribution, so that the "bars" in the histogram are replaced by smooth "bumps" (Silverman 1986). Smoothing is accomplished by putting less weight on observations that are further from the point being evaluated. More technically, the kernel density estimate of a series X at a point x is estimated by

$$f(x) = \frac{1}{Nh}\sum_{i=1}^{n} K\left(\frac{x - X_i}{h}\right) \tag{13.1}$$

where N is the number of observations, h is the bandwidth (or smoothing parameter) and $K(\ )$ is a kernel function that integrates to one. The kernel function is a weighting function that determines the shape of the bumps. We have used the Gaussian kernel in our estimates:

$$\frac{1}{\sqrt{2\pi}}\exp\left(-\frac{1}{2}u^2\right) \tag{13.2}$$

where u is the argument of the kernel function. The bandwidth h controls the smoothness of the density estimate; the larger the bandwidth, the smoother the estimate. Bandwidth selection is of crucial importance in density estimation, and various methods have been suggested in the literature. In this Chapter we have used the data-based automatic bandwidth suggested by Silverman (1986, equation 3.31):

$$h = 0.9N^{-1/5}\min\left\{s, Q^R/134\right\} \tag{13.3}$$

where s is the standard deviation, and Q^R is the interquartile range of the series.

Intra-distribution dynamics can be analyzed through the estimation of a stochastic kernel (Stokey and Lucas, 1989) for the distribution of income per capita over the period under analysis. This is merely the counterpart of a first order Markov probability of transitions matrix where the number of states tends to infinity, that is to say, where the length of the range of income defining each state tends to zero. Thus, the stochastic kernel provides the likelihood of transiting from one place in the range of values of income to the others[1].

Following Johnson (2000), let R (the ratio of productivity in each economy to the leading economy) be the variable under analysis, and $f_t(R = x)$ and $f_{t+k}(R = x)$ the probability density of $R = x$ in period t and $t+k$ respectively. Assuming a first-order time-invariant process for the evolution of the distribution of R, and existence of marginal and conditional density functions for the R distribution, the relationship between both distributions can be summarized by:

$$f_{t+k}(R = y) = \int_0^1 g_k(R = y \mid R = x) f_t(R = x) \tag{13.4}$$

where $g_k(R = y|R = x)$ is the density of $R = y$ in period $t+k$ conditional on $R = x$, k periods before. Then, $g_k(R = y|R = x)$ summarizes information on movements within the distribution over time. It is computed by first estimating the joint density for the distributions at t and $t+k$ by the kernel method[2] and then dividing it by the marginal density of R at t, obtained by integrating the joint density over R at $t+k$.

The stochastic kernel, that is the conditional density estimated from the data on R, is depicted in a three-dimensional graph such as the one in Figure 13.2 below. For each value of R at time t it shows the probability density 5 periods ahead, conditional on the density of that value at t. That is, it provides the probability of an

[1] See Durlauf and Quah (1999) for a formal definition and some properties of stochastic kernels in the study of distribution dynamics.

[2] A Gaussian kernel and bandwidth as described in Silverman (1986, Sect 4.3.2) was applied to estimate the joint densities. The Gauss procedure use to estimate the joint densities is that created by G. Suettrim.

economy starting at any ratio R at t ending up at any of the values after 5 years. The z-axis in the three-dimensional plot measures the conditional density of each pair of points in the x-y space that defines the values of the variable at t and $t+5$. The lines that run parallel to the $t+5$ axis measure the probability of transiting from the corresponding point in the t axis to any other point 5 periods ahead. The two-dimensional graph in the top right corner is a contour plot of the three-dimensional plot. Lines in this graph connect points at the same height on the three-dimensional plot, that is, points with the same density.

When the mass of probability is located along the positive diagonal, then this points to low mobility, in other words there is strong persistence. On the other hand, the kernel can shift above or below the diagonal, indicating increases or decreases in values of R at time $t+5$. It can also twist clockwise or counterclockwise. The former case would indicate convergence in the distribution, with the poor region having a high probability of moving to higher levels while a rich region would tend to move to a level below the starting level. In the limit, when the mass of probability is parallel to the t-axis, all economies end up at similar values in $t+5$ regardless of their position at time t. Finally, peaks in the kernel appear when there are attractors that encourage the formation of convergence clubs.

The ergodic density of R implied by the dynamics summarized by $g_k(R = y|R = x)$ is an approximation to the long-run shape of the distribution of the productivity ratios. It is the solution to:

$$f_\infty(R = y) = \int_0^1 g_k(R = y \mid R = x) f_\infty(R = x) \tag{13.5}$$

It can be compared with the actual density to predict what one might expect for the R distribution in the near future (under the assumption that the causes of the dynamics we observed remain stable into the future).

Finally, it is important to point out that the stochastic kernel can also be used to describe movements between two distributions, rather than being restricted to the analysis of the same variable at different points in time (Quah, 1996a). Cross-distribution analysis is carried out in Section 13.5 where we estimate stochastic kernels to analyze movements between actual (unconditional) and simulated (conditional) distributions under different assumptions that are hypothesized to affect future EU regional manufacturing productivity growth rates. The motivation here is to illustrate how conditioning factors influence the long-run distribution. Following the description given above, the resulting kernel is interpreted as the probability associated with each of the values in the simulated distribution, conditional on any given value in the actual distribution. In this case, when the mass of probability lies along the main diagonal, this indicates that the conditioning factor is not responsible for the variation in the actual distribution. In contrast, when a factor is responsible for most of the dispersion, conditioning out its effect will result in a kernel parallel to the unconditional distribution (so that economies share similar values when the effect of the factor is removed, regardless of their position in the original distribution).

13.2.2 Actual Manufacturing Productivity Dynamics in the EU Regions for the Period 1975-1995

In this Section we show estimated density functions based on our measure of the level of manufacturing productivity (manufacturing Gross Value Added per worker) for 178 EU regions over the period 1975-95. The analysis is based on the variable $R_{it}=P_{it}/P_t^*$ in which P_{it} refers to the level of labour productivity in region i ($i = 1,\ldots,178$) at time t (t = 1975, 1980, 1985, 1990, 1995), and P_t^* refers to the productivity level of the leading region at time t. Hereafter we refer to this variable as the "ratio" at time t. Of course R's upper bound is 1 (the leading region) and its lower bound approaches 0, depending on the productivity level for the least productive region. Hence the densities were estimated by using a truncated Gaussian kernel (see Silverman, 1986).

Figure 13.1 shows that the distribution in 1975 is centred on a ratio of approximately 0.5, with the majority of the distribution being somewhat distant from the leader. The two tails contain an important share of the overall distribution, and there appears to be a cluster of regions characterised by very low productivity (low R). Figure 13.1 also shows the distribution for 1985 and 1995, so we can trace the evolution of the shape of the distribution. It can be seen that there is clear persistence and development of an important "hump" at round about 0.2 and attenuation of the concentration of high productivity regions. We see that by 1985 the distribution has concentrated, although groups of regions with low and high ratios are even clearer in 1985 than in 1975. In the decade that follows, the distribution has become more dispersed. The reasons for this are complex, but seem to relate to the different regional responses to the deepening process of integration, EMU, etc. What is interesting is that the mass of probability at the right of the distribution seems to vanish, indicating that there is no obvious concentration of high productivity regions. In contrast, the cluster of low productivity regions remains a feature throughout the historical series. Generally, the distribution shows signs of polarisation.

The stochastic kernel in Figure 13.2 confirms what is revealed by visual inspection of densities over time: the distribution at the high and moderate values of R (high and medium productivity levels) is characterised by a comparative degree of fluidity in the "pecking order", while there is relative homogeneity in the very low productivity group, as indicated by the sharp peak on the vertical axis for low R. This suggests that there is minimal "churning" among the "poorer" regions, which are persistently near the bottom of the productivity ladder. However there is some change at the bottom because it is apparent that the very low productivity regions have become more homogenous, as illustrated by the clockwise turn in the kernel. Regions in the range of roughly 0 to 0.4 become more concentrated in the range 0.1 to 0.3.

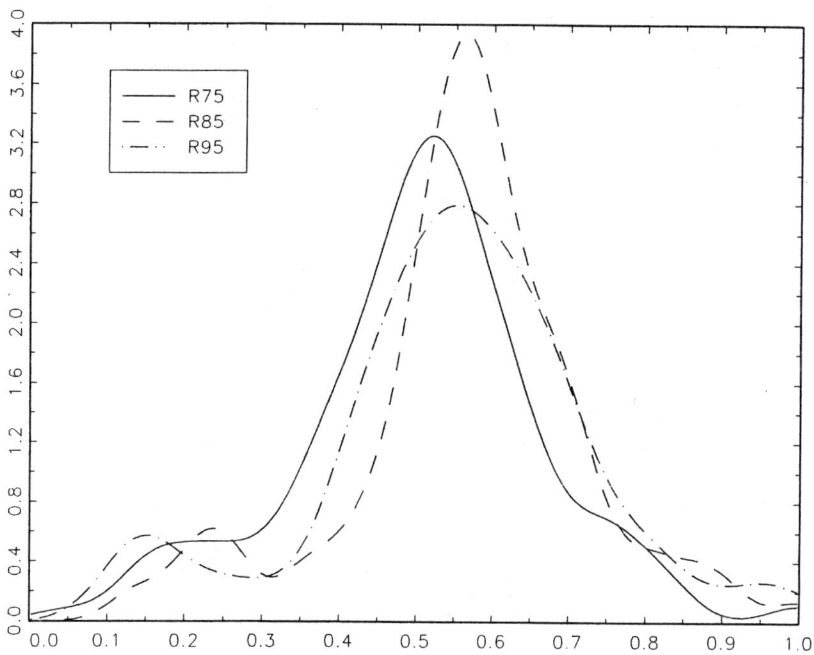

Fig. 13.1. Estimated density function for the productivity level ratios (R)

The ergodic distribution (Figure 13.3) is the long run equilibrium derived from the stochastic kernel based on the 5-yearly transitions described above. Some aspects of the historical data (Figures 13.1) are replicated in the equilibrium, for instance the mode is at roughly the same position, round about 0.4-0.5. Note that despite the presence of an important right tail, the group of high *R*'s is evidently not a long-run phenomenon. It appears that there is no persistence in the concentration of high productivity regions that was evident in the historical data. On the other hand, the cluster of low productivities (*R* around 0.2) does appear to be a persistent long-run phenomenon, as indicated by the contrasting shape of the distribution to the left of the mode in Figure 13.3.

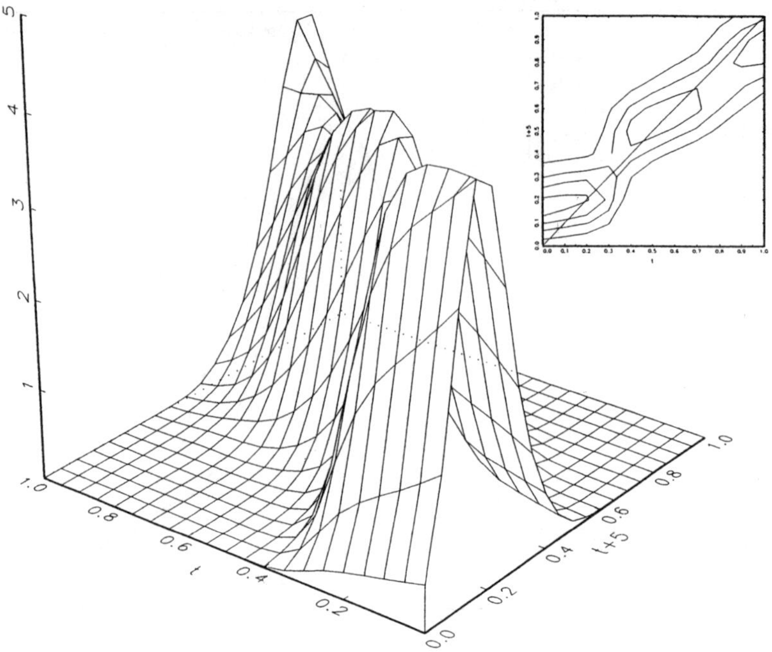

Fig. 13.2. Stochastic kernel for 5-yearly transitions for the productivity level ratios

13.3 A Productivity Growth Model – Theory

The subsequent empirical analysis of this Chapter is centred on a model of manufacturing productivity growth for the EU regions. At the core of this model is the concept of increasing returns, which has become popular in recent years within both urban economics (Rivera-Batiz, 1988; Abdel-Rahman and Fujita 1990; Quigley, 1998) and in geographical economics (Fujita, Krugman and Venables, 1999). The main technical innovation in this literature from the point of view of economic theory is that it allows increasing returns in city or region size while at the same time the decision problem for each actor is explicitly stated as one of profit or utility maximization. Increasing diversity or variety in producer inputs with increasing region size can yield external scale economies, even though firms are just breaking even (earning normal profits). So we see increasing returns to the economy as a whole in the context of competitive producers. This genre of model enables a general equilibrium solution by (almost invariably) utilizing the Dixit-Stiglitz theory of monopolistic competition. In the case of geographical economics, at its simplest, the market structure for industry is monopolistic competition while agriculture is competitive. In urban economics it is industry that is competitive, while non-traded producer "services" are treated as operating under monopolistic competition. The model in this Chapter, following Abdel-Rahman and Fujita (1990), likewise divides the economy into manufacturing and producer services, and follows the line of argument typified by the urban economics approach[3]. This ultimately leads to a simple and empirically tractable reduced form linking manufacturing productivity growth to the growth of manufacturing output, although because of the limitations of the basic theory, the model is extended by the inclusion of additional and necessary factors representing technological externalities (congestion, knowledge spillover) which go un-represented in the basic models outlined above[4].

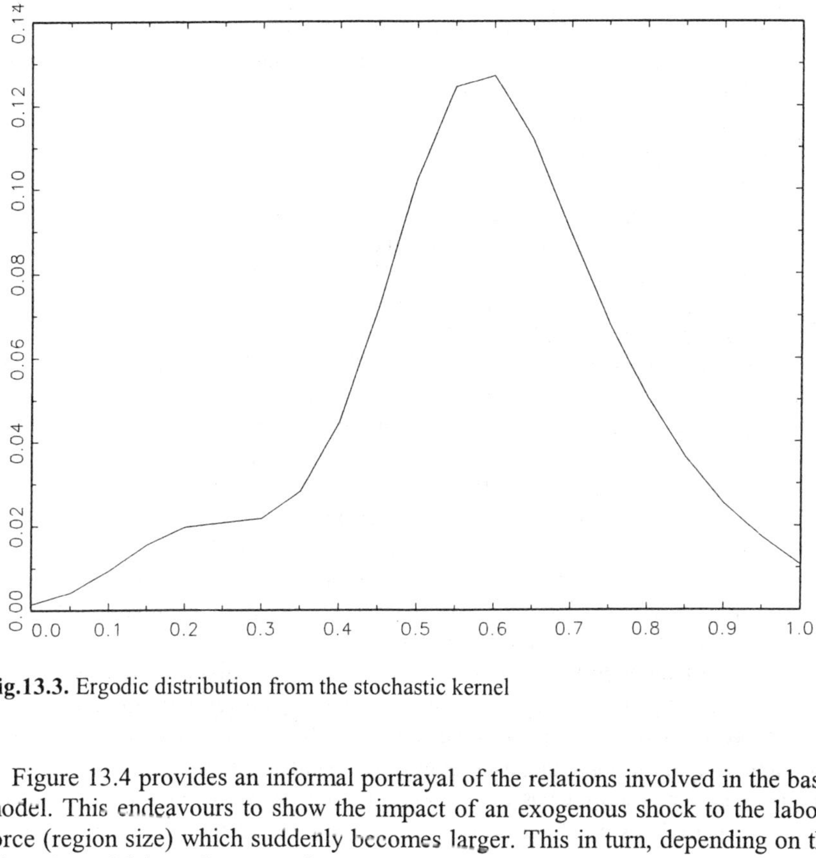

Fig.13.3. Ergodic distribution from the stochastic kernel

Figure 13.4 provides an informal portrayal of the relations involved in the basic model. This endeavours to show the impact of an exogenous shock to the labour force (region size) which suddenly becomes larger. This in turn, depending on the allocation of labour between the competitive manufacturing sector and the non-traded service sector, increases the level of composite services by increasing the

[3] The reduced form deriving from the model is identical to Verdoorn's law which is commonly used to represent increasing returns in manufacturing in the post-Keynesian literature (Fingleton ,2001b).

[4] Krugman(1991) argues that while technological externalities may be relevant, they leave no paper trail and prohibit formal analysis. Pecuniary externalities on the other hand are embodied within the theory.

number of service firms. The increase in manufacturing labour and the increase in the level of composite services increases the level of manufacturing output. The increased number of service firms is entirely equivalent to an increase in the number of varieties of services, and it is assumed that it is the variety of services available that is important to enhancing manufacturing production. A greater variety of intermediate services amounts to a finer division of labour. One has only to think of how much more efficient a manufacturer's production will be employing a range of specialized software writers each skilled in his or her own language (C++, Cobol, Fortran, Open GL etc) rather than a single 'jack of all trades' programmer, who has to get 'up to speed' on each different language as is required. Overall therefore, it is reasonable to assume that manufacturing output in a city or region will be related to the quantities of manufacturing labour, space and to the number of different producer inputs available in that city.

The theory of monopolistic competition provides the reason why an increase in service labour maps to an increase in service variety, rather than more of the same variety. Monopolistic competition envisages a large number of service firms producing differentiated services and firms freely entering the sector until profits go to zero (ie each firm earns normal profits). The existence of fixed costs means that firms prefer to concentrate on a single variety and reap internal economies of scale; there is no point in a variety's production being split. Without fixed costs, average costs do not fall with increasing output, and there are no internal scale economies. Since each firm is the producer of its own differentiated service, the ensuing monopoly power allows prices to be a mark up on marginal costs. There is an equilibrium level of output and therefore equilibrium labour requirement per service firm which is a constant, these equilibrium values depend on exogenous parameters and are therefore unaffected by an exogeous shock to labour. More service labour with a constant labour requirement per firm equals more firms and hence more varieties. The economics of the service firm are illustrated by the boxes below the dotted line in Figure 13.4 and the exogenous parameters are indicated by the *s, with the letter c denotes the constant equilibrium level of output and the labour requirement per service firm.

The level of monopoly power in the service sector is given by the exogenous parameter[5] μ. As μ increases, we see rising monopoly power and falling elasticity of substitution. More monopoly power enhances the level of composite services providing an input to manufacturing output. As μ falls back towards 1, the level of composite services approaches the number of firms times the equilibrium level of output per firm. This combines with β, the coefficient of the Cobb-Douglas production function for manufacturing. The set up is such that for $\mu > 1$ and $\beta < 1$, the relationship between the overall level of manufacturing output (Q) and the size of the total labour force (N) is nonlinear, in other words the initial shock to labour produces a greater than proportionate increase in the level of manufacturing output.

[5] This is the substitution parameter of the CES production function, which determines the elasticity of substitution, the price elasticity of demand and the internal returns to scale given by the average cost to marginal cost ratio for producer services in equilibrium.

Fig. 13.4. Relations in the basic model

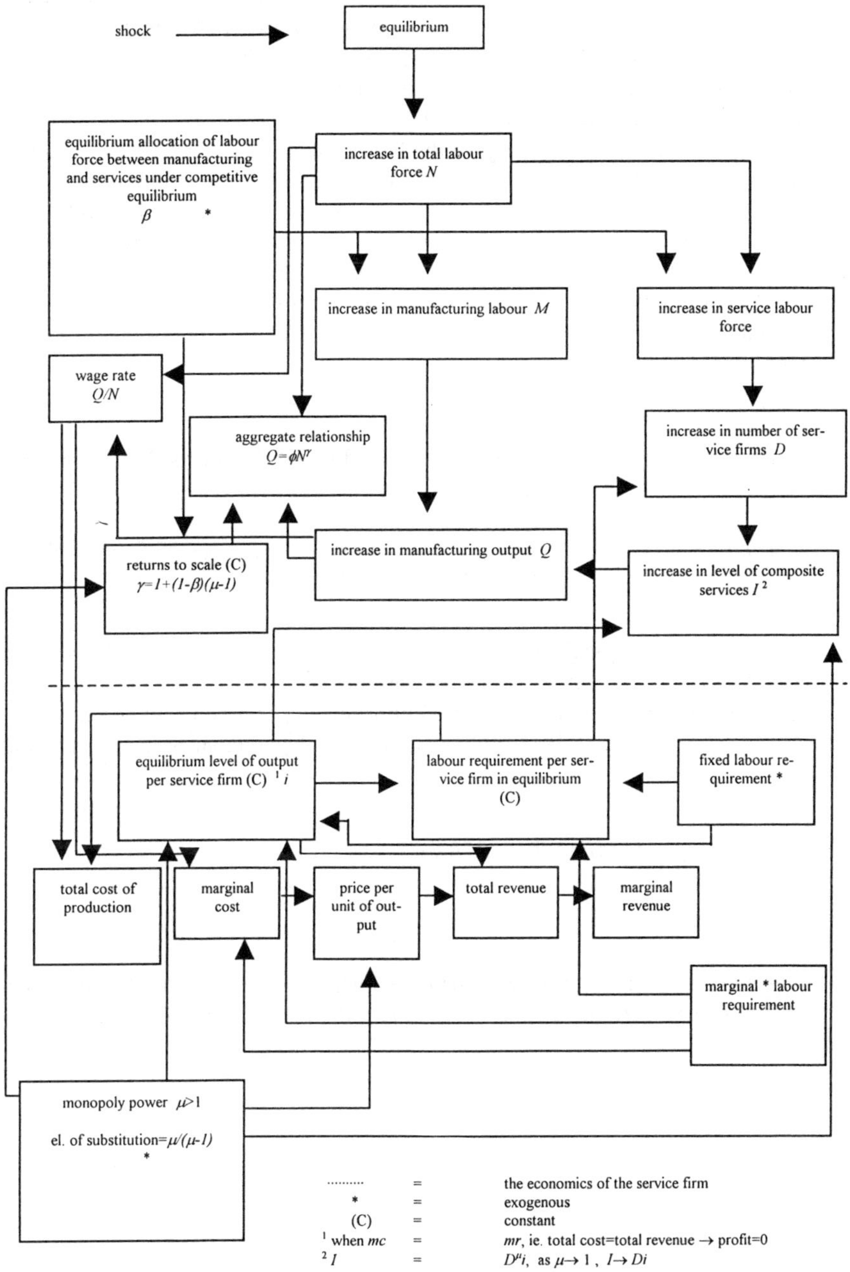

The elasticity γ depends on the exogenous parameters and is a measure of returns to manufacturing production with increasing N. For increasing returns, services have to be relevant ($\beta < 1$) to manufacturing output and each service firm has to exert a degree of monopoly power under monopolistic competition ($\mu > 1$). Since μ also equals the ratio of average to marginal costs at the equilibrium point for service firms, internal increasing returns in the service sector are translated into external increasing returns for manufacturing.

The equations in the Appendix give the formal model: these lead to the reduced form at the core of the empirical model of the Chapter linking manufacturing output (Q) to the intensity of activity in a unit area given by the total labour force (N).

We derive the reduced form by substituting the level of composite services I into the Cobb-Douglas production function giving the manufacturing production technology. In this the level of manufacturing output Q is a function of the input of manufacturing labour M and I. Note that manufacturing is competitive, with constant returns.

$$Q = M^{\beta} I^{1-\beta} \tag{13.6}$$

The level of composite services is determined by the CES production function. This is

$$I = \left[\int_{d=1}^{D} i(d)^{1/\mu} \partial d \right]^{\mu} \tag{13.7}$$

In which *i(d)* is the "typical" output of a service variety, and there are D varieties. Since we assume a very large number of varieties we approximate the continuous integral by the discrete summation. At equilibrium *i(d)* is a constant across all varieties (see Appendix) and therefore we can reduce the summation to a product as follows

$$I = [\sum_{d=1}^{D} i(d)^{1/\mu}]^{\mu} = [Di(d)^{1/\mu}]^{\mu} = D^{\mu} i(d) \tag{13.8}$$

Given this simplified form, we substitute for I and use the equilibrium values for the number of varieties D and M[6] to obtain the relationship between Q and N, hence

[6] This assumes that the economy is at a competitive equilibrium and workers are paid the value of their marginal product.

$$Q = M^{\beta} I^{1-\beta}$$
$$I = D^{\mu} i(d)$$
$$Q = M^{\beta} (D^{\mu} i(d))^{1-\beta}$$
$$Q = M^{\beta} D^{\mu-\mu\beta} i(d)^{1-\beta}$$
$$D = \frac{(1-\beta)N}{ai(d)+s} \tag{13.9}$$
$$M = \beta N$$
$$Q = N^{\beta+\mu-\mu\beta} \beta^{\beta} (ai(d)+s)^{\mu(\beta-1)} i(d)^{1-\beta} (1-\beta)^{-\mu(\beta-1)}$$
$$Q = N^{\beta+\mu-\mu\beta} \phi'$$
$$Q = \phi' N^{1+(1-\beta)(\mu-1)}$$

In which s is the fixed labour requirement and a is the marginal labour requirement of producer services (see Appendix).

In short,

$$Q = \phi' N^{\gamma'} \tag{13.10}$$

with

$$\gamma' = [1+(1-\beta)(\mu-1)] \tag{13.11}$$

The model outlined thus far excludes technological externalities, and yet it is necessary, in the interests of unbiased estimation and realistic simulation, not to exclude significant variables. The "two" externalities we are concerned with are the effects of congestion and the impact of knowledge flows on productivity growth. With regard to congestion, there are of courses many sources, for instance congestion involves interaction between firms, who "get in each others' way" or "step on each other's toes" (Cameron, 1996). Congestion arises when firms use common, but unpriced inputs[7] in short supply, such as roadspace or other communications networks or unpolluted air or water.

Congestion therefore comes from various sources which make production more costly on a restricted space. Measuring the impact of each is likely to be difficult, and as Gordon and McCann(2000) argue that we can only observe the net realized

---

[7] Since they are unpriced, the market has failed and therefore there is no market; we treat them as technological as opposed to pecuniary externalities. They are concerned with the technical relationship between inputs and outputs, with the structure of the production function rather than with prices in a market. If the congestion effect was due to outputs so that the firm's costs rise as the output of other firms rises, then we would have a pecuniary externality. However since congestion in a region is caused by the number of firms producing there, an input measure, it is a technological externality.

effects of diverse simultaneous externality mechanisms, rather than individual sources. We are more concerned with the effects of congestion on production rather than its diverse causes, and so simply model its impact in terms of a parameter in the manufacturing production function (see Ciccone and Hall, 1996). In fact it is easy to see the way we model congestion by introducing land (L) explicitly, hence

$$Q = \left[M^{\beta} I^{1-\beta} \right]^{\alpha} L^{1-\alpha} \tag{13.12}$$

Up to now we have also assumed that $\alpha = 1$, in other words the amount of land makes no difference to the level of production. By introducing $\alpha < 1$ we acknowledge that land is indeed a factor, and by assuming $L = 1$ and therefore working in terms of output per unit area, Q per unit of land is affected by congestion on that unit. Since $1^{1-\alpha} = 1$, we can leave L out of the production function), hence,

$$\begin{aligned} Q &= (M^{\beta} I^{1-\beta})^{\alpha} = (\phi' N^{\gamma'})^{\alpha} = \phi N^{\gamma} \\ \gamma &= \alpha[1 + (1-\beta)(\mu - 1)] \end{aligned} \tag{13.13}$$

Compared with what went before, this equation does not automatically imply external increasing returns. In order to move closer to a convenient reduced form, we linearize equation (13.13) by taking natural logarithms and rearranging to give an expression in terms of the level of manufacturing output per labour unit as a function of the level of manufacturing output, thus

$$\ln(Q/N) = \frac{\ln(\phi)}{\gamma} + \left[\frac{\gamma - 1}{\gamma} \right] \ln(Q) \tag{13.14}$$

This has the advantage of not requiring knowledge of the service sector per se, whereas information on manufacturing production is more accessible and reliable.

Also, assuming $M/N = \beta$, the preferred expression is in terms of the level of manufacturing output per unit of manufacturing labour, thus

$$\ln(Q/M) = \frac{\ln(\phi)}{\gamma} + \left[\frac{\gamma - 1}{\gamma} \right] \ln(Q) - \ln(\beta) \tag{13.15}$$

The second type of externality we wish to incorporate is knowledge spillovers. This we have said has been largely omitted from formal theory. However technological externalities involving information spillovers within and between regions and cities are increasingly recognized (Fujita and Thisse 1996, Quigley 1998) as an important contributor to the spatial concentration of economic activity, reflecting the earlier work of Jacobs (1969) among others[8]. In what follows, attention fo-

[8] We can distinguish between urbanization externalities or Jacobs externalities (after Jane Jacobs) and localization externalities or Marshall-Arrow-Romer (MAR) externalities.

cuses on the rate of technical progress as a manifestation of technological externalities. We see that technical progress is not confined by what exists within artificial region boundaries but influences and is influenced by technical progress in other regions, so that it varies by region rather than being an un-modeled constant.

In order to model the technical progress rate, consider the labour units to be labour efficiency units, hence

$$M_t = E_t A_t = E_t A_0 e^{\lambda t} \tag{13.16}$$

where E_t is the level of manufacturing employment and A_t is the efficiency level at time t that is determined by the initial level A_0 and the rate of technical progress λ. It then become possible to re-express the model in terms of the level of manufacturing (raw labour) productivity per unit area by substitution, hence

$$\ln(Q/E)_t = \frac{\ln(\phi)}{\gamma} + \left[\frac{\gamma - 1}{\gamma}\right] \ln(Q)_t - \ln(\beta) + \ln(A_0) + \lambda t \tag{13.17}$$

The assumption is that the technical progress rate λ depends on three groups of variables, namely human capital (H), the initial level of technology (G) and the spillover of knowledge across regional boundaries (S), plus an autonomous rate (ε) since we assume that technical progress will occur as a result of "learning by doing" irrespective of the other factors. Hence

$$\lambda = \nu H + \pi G + \rho S + \varepsilon \tag{13.18}$$

$$G = 1 - \frac{P}{P^*} = 1 - R$$

$$S = Wp$$

$$W_{ij}^* = Q_j^{\eta} d_{ij}^{-\delta}$$

$$W_{ij} = \frac{W_{ij}^*}{\sum_j W_{ij}^*}$$

Regions with substantial human capital assets are expected to make faster technical progress ($\nu > 0$) since human capital facilitates research, development and the spillover of knowledge on which technical progress depends. However there is a paucity of accurate pan-European data that we can use as a measure of human capital, but we do have access to data on educational attainment rates by EU region (see Chapter 1). These data are based on a labour force survey and give the share of the population aged 25-59 with higher educational attainment levels. There are undoubted problems of measurement error associated with these data.

While both involve knowledge spillovers between firms, MAR externalities restrict the spillover to firms in the same industry whereas Jacobs externalities refer to spillovers across different industries.

However we are able to nullify these effects in estimation via the use of instrumental variables and two-stage least squares.

The second factor is G which represents the start-of-period level of manufacturing technology gap, or one minus the ratio of the region's manufacturing technology level in 1975 to that of the leading region. The presence of G reflects the ubiquitous public good dimension of knowledge. There is no distance effect involved, reflecting the fact that some components of knowledge are free to diffuse to any region, irrespective of geography. However, the hypothesis is that the impact will be spatially variegated. The reason is differential acceptance of knowledge because of its varying usefulness. Some regions will gain great benefit from the diffusion of knowledge from the technological frontier region, but it seems reasonable to assume that those that are close to the technological frontier will see only minor gains. It is therefore envisaged that there will be a positive relationship ($\pi > 0$) between the technical progress rate and the initial technology gap, although this is a testable hypothesis and not an assumption of the model. A positive relationship indicates catching-up due to the faster growth of the initially lower level technology regions, but a negative relationship implies regional divergence.

Next we turn to spatially impeded information flows. The hypothesis is that regions with fast technical progress occurring in "neighbouring" regions will see faster than otherwise technical progress, and regions with slower technical progress neighbours will see a technical progress slowdown. "Who your neighbours are" matters in this context, because of the impedance to information flow across space. Working with the so-called NUTS 2 regional system of the EU means that we are likely to see flows across region boundaries. This is because this regionalizing does not define quasi-self-contained economic regions[9] but rather regions are largely defined on a formal or administrative basis. In addition, national barriers have been progressively reduced within the EU with the consequence that there are minimal barriers to remote cross-region spillover.

Two cross-region spill-over mechanisms are envisaged. One is based on the sharing of the same labour pool in a common local labour market area straddling regional boundaries, the thesis being that the rate of productivity growth occurring in one region will be transmitted to other nearby regions as workers embodying knowledge switch jobs within local labour market (i.e. journey to work) areas. It might be argued that this would be entirely a pecuniary externality operating via

---

[9] In a series of papers and books (eg Cheshire and Hay, 1989) Paul Cheshire and colleagues argue that functional urban areas (FURs) are more appropriate, but the problem with these is that they are dynamic rather than static entities, so that their definition is not fixed in time. Moreover, their precise definition is likely to be subjective and will not map to accessible data bases like the NUTs 2 system. The more or less self-contained nature of FURs is suggested as an advantage, but we argue that it is preferable to handle spatial interdependence at the modelling stage since this gives the data analyst the flexibility to estimate in the context of realised spatial interdependence, rather than accept a definition of spatial independence among FURs that may not be true.

the labour market[10]. Firms may spend money training labour but not appropriate all the benefits if labour migrates to other firms who gain by not having to pay the full cost of that training. However, some knowledge transfer is not mediated by market transactions. For example some aspects of knowledge will be of a learning-by-doing variety accumulated on the job, and carried to other firms as an incidental attribute (quality) of labour, without price or market, but which nevertheless contributes to technical progress. For example a programmer may incidentally learn how to install and service hardware, and this knowledge may be carried forward and benefit the next firm, even though this knowledge was hidden and played no part in the labour market transaction.

The second mechanism involves demonstration effects[11] of the efficacy of knowledge for productivity. This involves inter-firm interaction across region boundaries. Because of proximity, firms locally and in nearby regions may be competitors for the same local markets (a "creative destruction effect"), or collaborate as part of a localized production chain. In either case, fast technical progress in neighbouring regions will tend to induce technical progress and thus fast productivity growth locally.

The assumptions set out above results in Wp being specified as the weighted average of labour productivity in "surrounding" regions, with surrounding broadly defined so as to acknowledged that even the remotest regions interact and that the EU comprises a more or less integrated economy. Size also is considered relevant, and to a degree offsets remoteness, because of the extensive trade and labour market that a large diverse local economy naturally generates. This is apparent in the definition of the absolute (conditional) level of interaction between regions i and j, namely W^* in which Q_j is the economy size proxy, the 1975 level of output in region j, so that given the size of region i, the interaction with region j is likely to be stronger if region j possesses a larger economy. Proximity is represented by the great circle distance (d_{ij}) between the centres of regions i and j. Given a presumably negative parameter(δ), increasing distance reduces the absolute conditional interaction between i and j. It is assumed that $\delta = 2$ and $\eta = 1$ as a result of trials of different values reported in Fingleton (2001a). The standardized matrix W, so that cell entries sum to unity across rows with zeros on the main diagonal, means that the matrix product Wp is a vector of weighted averages.

The outcome of modelling spatial effects using the W matrix is

$$\lambda = \pi G + \upsilon H + \rho Wp + \varepsilon \tag{13.19}$$

Inserting (13.19) into (13.17), differentiating with respect to time, and inserting a well-behaved perturbance, gives

---

10 This is a version of the "surplus appropriability" problem, where an externality exists which is not a technological externality, due to the innovator not appropriating all the social gains from innovation because of imperfect price discrimination (Cameron, 1996).

11 Or as Cameron(1996) puts it, a "standing on shoulders effect" due to knowledge leaks, imperfect patenting, in addition to the movement of skilled labour to other firms reducing rivals' costs.

$$p = \varepsilon + \frac{\gamma - 1}{\gamma} q + \pi G + \upsilon H + \rho W p + \xi \tag{13.20}$$

$$\xi \sim N(0, \sigma^2 I)$$

where p is the logarithmic growth of manufacturing labour productivity (Q/E) and q is the logarithmic rate of growth of manufacturing output (Q).

The specification in (13.20), with the endogenous spatial lag, is a model that is familiar to spatial econometricians. It is interesting to note that it implies labour productivity growth in any region is affected by output growth, the technology gap and the endowment of human capital in all the other regions in the system, though the amount of influence is decreasing in the factors that define the contact matrix, W. The estimation of this model requires two stage least squares to attain consistency. In fact the Hausman test for exogeneity, jointly for all three variables q, H and Wp, gives a p-value equal to 0.146 in the $F_{3,170}$ distribution, which is sufficiently large to suggest that they are exogenous. The Hausman test of the exogeneity of each single variable in the presence of other endogenous variables[12] gives test statistics for Wp, q and H equal to 2.128, 3.959 and 0.437 respectively, with p-values equal to 0.1446, 0.0466 and 0.5085 in the appropriate χ^2 distribution with one degree of freedom. Hence there is some marginal evidence to suggest that q is endogenous, but the other two variables appear to be exogenous. With regard to Wp, which is by definition endogenous, it appears that the tests are insufficiently sensitive. In order to guard against any potential inconsistency not detected by these tests, it is safer to use instrumental variables.

The results given in Table 13.1 relate to 2sls estimation using instrumental variables in place of those regressors that are possibly correlated with the errors. Assuming that the technology gap variable G is exogenous, and therefore an instrument, leads to a second instrument, the spatial lag WG. There are two other instruments that we define, q_I and its lag Wq_I. The instrument q_I is defined by the 3 group method (described in Leser, 1966; Koutsoyiannis, 1977; Kennedy1992; Johnston 1984; see also Chapter 9) and takes values 1, 0 or -1 according to whether q is in the top, middle or bottom third of its ranking, which ranged from 1 to 178. The assumption is that this mapping eliminates any correlation between q_I and ξ induced by simultaneity. With four instruments, and given three presumed right hand side 'endogenous' variables the necessary order condition for identification is satisfied, hence the equation is identified exactly. There are additional or alternative additional instruments that could be introduced leading to over-identification, and the results of using different instruments or estimation techniques are given in the cited references. Further details of alternative instruments and alternative estimation methods, and diagnostic tests of model assumptions, are given in Fingleton (2000, 2001a, b). Here we simply focus on the new parameter

[12] Maddala, 2001, for details.

estimates and standard errors which are presented here for the first time[13]. These, together with the results of the simulation exercise, are the new research being reported in this Chapter. It is sufficient to note that all the variables are highly significant, for instance there is strong evidence that $\rho > 0$ suggesting that the spillover of knowledge across region boundaries has a significant impact on the growth of manufacturing productivity. This is supplemented by the effect of human capital within each region, which also indicates that technical progress is enhanced by higher levels of human capital. Likewise, the significant link between q and p points to the presence of increasing returns. The estimated value of the coefficient $\gamma = \alpha[1 + (1 - \beta)(\mu - 1)] \approx 1.5$, so that the effects of congestion (α) have not offset the effects of increasing returns in the service sector ($\mu > 1$) which are evidently relevant as an input to manufacturing production ($\beta < 1$). Finally, there is a strong positive link between productivity growth and the level of technology gap. This has implications for the dynamics associated with this model which are outlined below.

Table 13.1. Two-stage least squares groupwise heteroscedasticity estimates.

Variable	Coefficient estimate	Standard error	t-ratio
Constant	-0.0240	0.0084	-2.84
q	0.3265	0.0820	3.98
G	0.0476	0.0071	6.66
H	0.0716	0.0274	2.61
Wp	0.4327	0.1640	2.64
σ^2 (periphery)	0.00024		
σ^2 (core)	0.00011		

13.4 Simulation of the Long-Run Distribution of Manufacturing Productivity in the EU Regions.

A pivotal feature of the dynamics is the coefficient on G, which determines whether the regions converge to a stable equilibrium or not. Define equilibrium as the state in which the expected rate of productivity growth p is exactly the same across regions. A critical assumption of the dynamics is the relationship between productivity growth p and the level of technology gap $G = 1 - R$, which embodies the level of productivity P which we assume depends on p, in other words,

$$P_{t+1} = P_t \exp(E(p_t)) \tag{13.21}$$

This indicates that an inexorable rise in productivity levels occurs with positive productivity growth rates. Nonetheless, in equilibrium the ratio of productivity

[13] The estimates of Table 1 also include separate estimates of error variance for the two groups ($i = 1,2$) of regions defined as core (within 500 km of Luxembourg) and peripheral regions, which were shown to be necessary as a result of the diagnostic tests of error heterogeneity previously reported.

levels $R = P_t/P^*_t$ is non-increasing since both P_t and P^*_t grow at the same rate. This means that the change in R, $R^0 = R_t - R_{t-1} = 0$, hence the proportional rate of growth $R^0/R = 0$.

In equilibrium, all regions grow at the same rate and therefore $E(p - p^*) = 0$, where p^* is the productivity growth rate of the level-of-productivity leader. It is convenient to represent the above model in matrix form, hence

$$p = \rho Wp + Xb + \xi \tag{13.22}$$

$$\xi \sim N(0, \sigma_i^2 I)$$

in which X is the n by k matrix of regressors (q,G,H) and b is the k by 1 vector of coefficients. It then is the case that

$$E(p) = (I - \rho W)^{-1} Xb \tag{13.23}$$

In order to obtain the equilibrium level of technology ratio $R^e = 1 - G^e$ we use the equivalent expression

$$E(p - p^*) = (I - \rho W)^{-1} Xb - (I - \rho W)^{-1} X^* b = 0 \tag{13.24}$$

$$(I - \rho W)^{-1} X^{\otimes} b^{\otimes} + (I - \rho W)^{-1} X^G b^G - (I - \rho W)^{-1} X^* b = 0$$

in which X^* is a matrix with the same dimensions as X but containing identical rows which are each equal to the productivity leader's row of X. We split matrix X into two matrices, $X^{\otimes}$ which is identical to X apart from the variable G, and X^G which contains the single variable G. We denote the corresponding vectors of coefficients by $b^{\otimes}$ and b^G. From this it follows that

$$R^e = 1 - X^G = 1 - (X^* b - X^{\otimes} b^{\otimes})(b^G)^{-1} \tag{13.25}$$

While the above gives the equilibrium vector, it gives no information about the paths to equilibrium. It is worth noting that transitional paths leading to the same steady state are given by iteration. The following equations give one round of iteration; in order to obtain the dynamics, this sequence of calculations is repeated for $t = 1.....T$ and we observe how R evolves as t increases. In the dynamics, the vector of productivity growth rates at time t, $E(p_t)$, depends on the matrix X as at time t, but X evolves because one column of the matrix, denoted by $X_{t+1,G}$ contains variable G_{t+1} which depends on the level-of-productivity at time t, P_{t+1}, and the latter is determined by the rate of growth of productivity $E(p_t)$.

$$
\begin{aligned}
E(p_t) &= (I - \rho W)^{-1} X_t b \\
P_{t+1} &= P_t \exp(E(p_t)) \\
P^*_{t+1} &= P^*_t \exp(E(p^*_t)) \\
G_{t+1} &= 1 - \frac{P_{t+1}}{P^*_{t+1}} \\
X_{t+1,G} &= G_{t+1}
\end{aligned}
\tag{13.26}
$$

The results of this section provide the input into our conditioning analysis. That is, using the above expressions we obtain the long-run distribution from the model estimates, and obtain virtual distributions as a result of changing the model coefficient values so as to accommodate assumptions of constant returns to scale and equal endowments of human capital. The same procedure is also used to obtain distributions which allow us to assess sensitivity to the rate of catch-up and different amounts of increasing returns.

13.5 Effects of Increasing Returns, Human Capital and Catch-Up on the Long-Run Distribution of Manufacturing Productivity

In this section we first compare the actual distribution of productivity in the EU regions with long-run distributions derived from the empirical model described in the previous section. In essence, the observed 1995 distribution is compared to the steady-state R^e distribution. It allows us to illustrate the characteristic features of the equilibrium distribution and assess how it differs from the actual distribution. In addition, we compare the R^e distribution, based on an assumption that the actual observed values of factors influencing the equilibrium (i.e. amount of returns to scale, human capital, rate of catch-up) are maintained in the long-run, with simulated distributions based on an assumption that the currently observed differences in factors across the regions are eliminated (we term these conditional equilibrium distributions).

Stochastic kernels are used to summarize the relationship between the actual distribution in 1995 and the distribution for R^e, and the relationship between the unconditional and conditional equilibrium distributions. In the latter case, the stochastic kernel summarizes movements in the equilibrium distribution that are due to the conditioning factor.

13.5.1 Equilibrium Distribution: Full Effects Model

Figure 13.5 shows the stochastic kernel relating the distribution of *R* in 1995 to that of R^e. It is clear how the kernel twists clockwise for the whole range of *R*, though the twist is more intense for the medium and high values. For values around and below 0.2, it is evident that there is a high probability that there will be an increased *R*. Nevertheless, the equilibrium distribution is characterized by considerable dispersion (regions with a productivity level ratio in 1995 of around 0.2 can end up with similar values in the equilibrium distribution or with values around 0.3-0.5). For regions with medium and large *R*, the distribution is less concentrated and there is a high probability of R^e values similar to those for low *R* regions. This suggests that convergence at equilibrium is to a level which is only slightly higher on average to that for low *R* regions, but for the high *R* regions the dispersion is much greater and there is a much higher probability of attaining a value close to the leading region.

13.5.2 Conditioning to Increasing Returns

Increasing returns to scale makes the spatial concentration of manufacturing activities more attractive, which translates into increasing inequality in manufacturing productivity across regions. Therefore, increasing returns should be responsible for some of the dispersion in the equilibrium distribution and, is likely to affect (to some extent) the fortunes of the less productive regions. In this spirit, we have simulated the equilibrium R distribution based on the assumption that γ (the coefficient that captures the extent of returns to scale in our model) is equal to one, which in turn means that the coefficient for q in (13.20) equals zero. This amounts to an assumption that manufacturing output growth rates are equal across regions and therefore do not play a part in determining the rate of growth of manufacturing productivity. The determinants of manufacturing productivity growth under this scenario are therefore human capital, cross-region spillovers and catch-up. In the simulation, the coefficients on these variables are as estimated in the empirical model (see Section 13.3). The resulting movement between the unconditional and conditional equilibrium distributions is summarized by the stochastic kernel in Figure 13.6. The kernel shifts above the diagonal, indicating that the conditional distribution shift to the right of the unconditional one. That is, R^e for most of the EU regions would be higher in the case of constant rather than increasing returns. Additionally, we can see that increasing returns contribute to the poor situation of the less advanced regions. The clockwise turn in the kernel for low R^e values indicates clearly that were it not for the existence of increasing returns, the low productivity regions would have a much higher R^e. In contrast, the lack of increasing returns has relatively little impact on high R^e regions' productivity relative to the leading region.

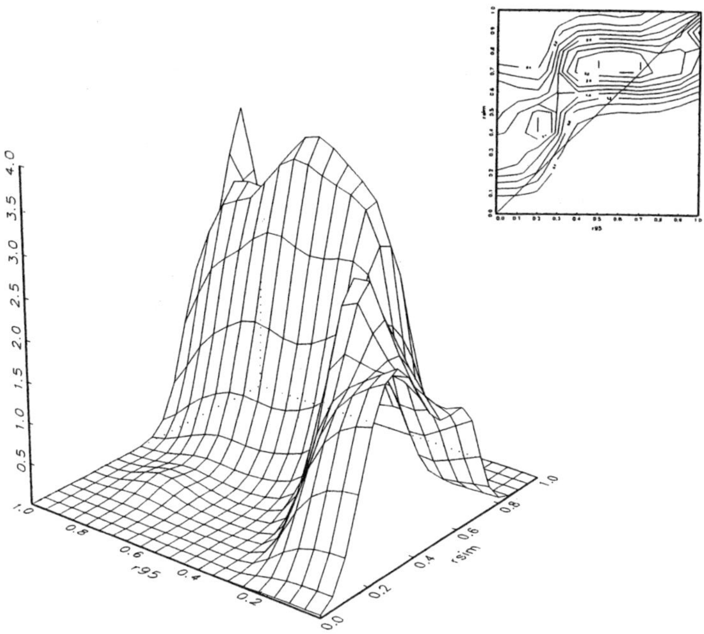

Fig. 13.5. Stochastic kernel for the distribution in 1995 and the steady-state distribution from model estimates

13.5.3 Conditioning on Human Capital

As mentioned in Section 13.3, human capital is an important determinant of technical progress, thus being a key factor for economic development. As such, we anticipate that human capital will also be a factor conditioning the equilibrium regional distribution of manufacturing productivity ratios in the EU. Deficits in human capital in some regions could be condemning them to a low productivity trap. The kernel in Figure 13.7 adds weight to this hypothesis. The equilibrium distribution, under the assumption of the same levels of human capital across regions, is more concentrated than the distribution assuming human capital differentials by region being maintained at their current levels. It is evident that low human capital is a cause of the low equilibrium level for the lowest technology regions in particular. Once human capital differentials are removed, they show a marked improvement. In contrast, regions with R^e above 0.5 show minimal change.

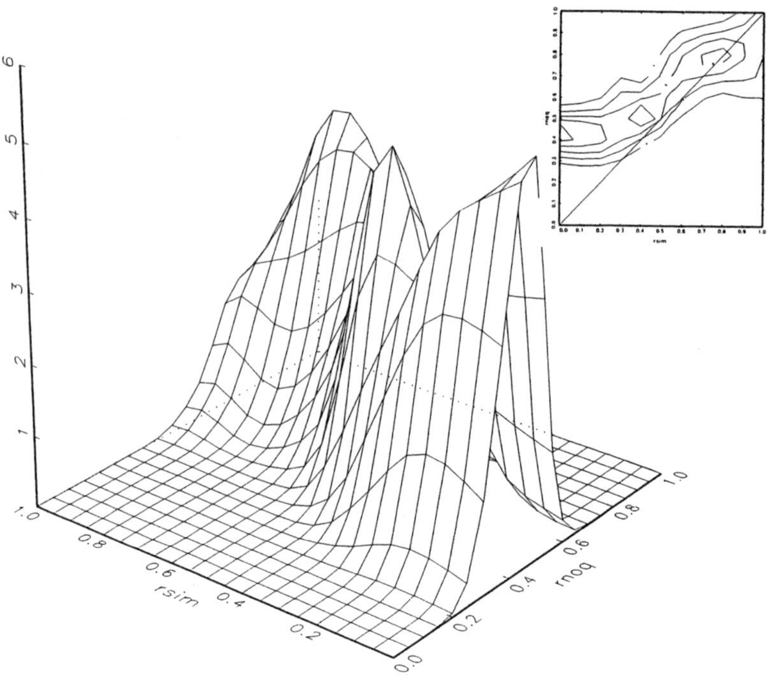

Fig. 13.6. Stochastic kernel: IRTS conditioning

13.5.4 Sensitivity to the Rate of Catch-Up

We cannot carry out an equivalent exercise for the catching-up effect because there is no equilibrium solution when catching-up is nullified. Under our model, with no catching-up, regions do not converge to a stable equilibrium. Instead, we explore how the equilibrium to which regions converge differs according to different rates of catch-up. To show this, we assume somewhat arbitrarily two different rates, setting the coefficient for G (the level of technology gap) equal to 80% and to 120% its estimated value. All the other coefficients are set to their estimated values.

The results of assuming faster catching-up are shown by Figure 13.8. There is a clockwise shift in the kernel, with a slightly less heterogeneous distribution of R^e than would be the case if the assumed parameters were simply the empirical estimates (this effect would be exaggerated if we had chosen 150% rather than 120%). With faster catching up, it is the regions with the lowest levels of technology that benefit most, as one would expect.

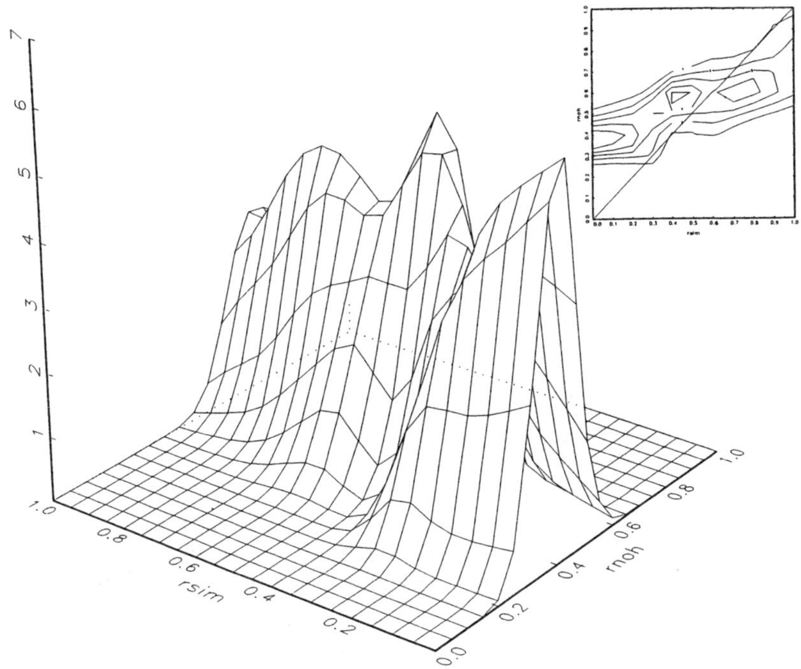

Fig. 13.7. Stochastic kernel: human capital conditioning

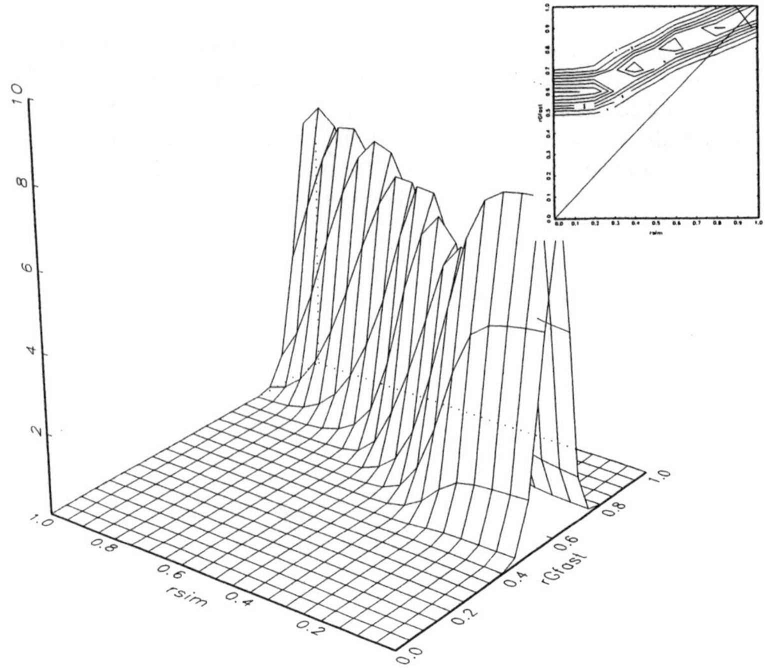

Fig. 13.8. Stochastic kernel: fast catch-up conditioning

Figure 13.9 gives the outcome of the alternative simulation based on slower than estimated catching-up. Here the outcome is not quite the opposite of what we have just seen. Slower catching up means R^e somewhat below what one would anticipate from the empirical estimate and much lower than what one would anticipate from fast catch-up. However the impact of slow catch-up is not confined to the regions with low R^e. The reason is that with slower catch-up, all regions lag behind the leading region which forges ahead and reinforces its leading position.

13.5.5 Sensitivity to the Amount of IRTS

We showed above that if returns to scale were completely eliminated then the lower technology regions would be much closer to the higher technology regions. However, this is not a very realistic scenario, so here we adjust rather than eliminate the returns to scale to gain more insight. Figure 13.10 illustrates the impact of smaller returns to scale, with the coefficient on q (manufacturing output growth) assumed to equal 75% of the estimated value. Figure 13.11 shows the result of assuming a coefficient equal to 125% of the estimated value. All other coefficients are set equal to their estimated values.

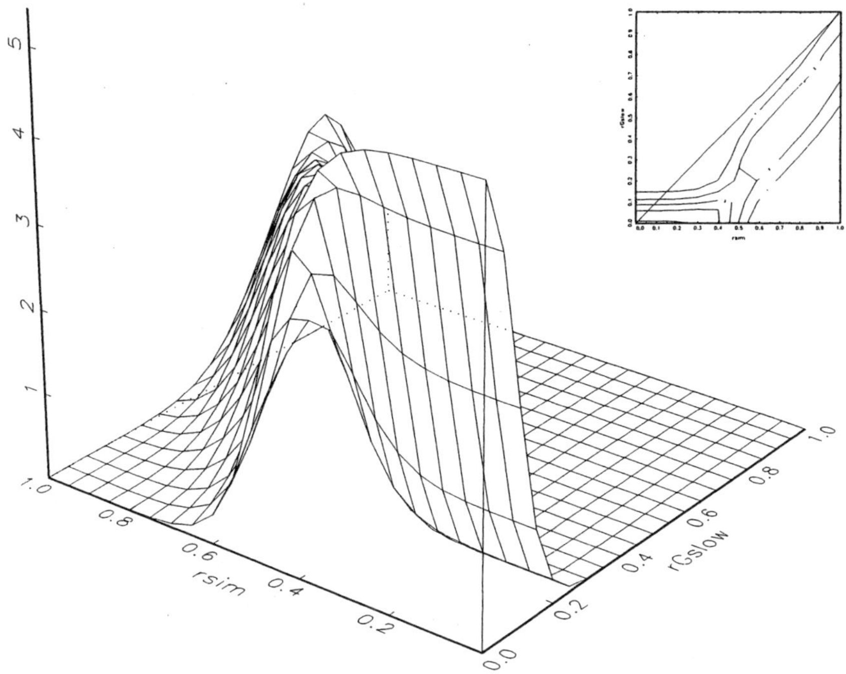

Fig. 13.9. Stochastic kernel: slow catch-up conditioning

With large returns to scale, regions tend to a lower R^e as the distribution becomes stretched, and one would expect that those with faster manufacturing output growth rates would see the greatest productivity boost and thus the greatest impact in terms of R^e. However, this impact is evidently confined to the very top end of the R^e range; under greater returns to scale the vast majority of regions are seen to be relatively worse off vis-à-vis the technological leadership. In contrast, Figure 13.10 shows that lower than estimated returns to scale has the effect of pulling the lowest regions upward, with the upward shift the greatest for the very lowest regions. The shift is in the same direction as in Figure 13.6 but of course less so. We could imagine that returns to scale might fall by this extent in the real world if congestion worsened significantly, as it is projected to for some regions (eg the South-East of England) unless there is a radical change of policy.

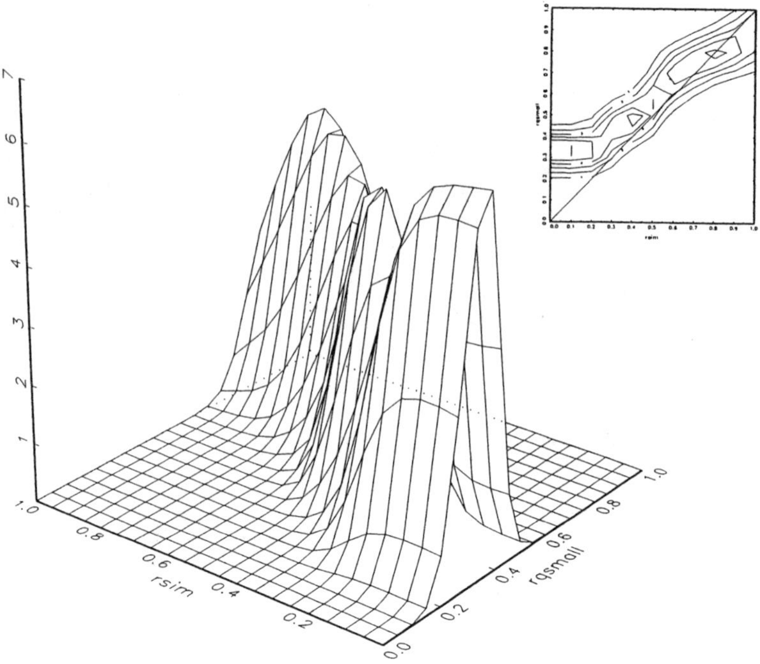

Fig. 13.10. Stochastic kernel: small RTS conditioning

13.6 Conclusions

This Chapter has combined recent developments in urban and geographical economics theory with an empirical spatial econometric model and with the method of stochastic kernels to produce visualizations of different long-run development scenarios under alternative assumptions. We have seen that if the impact and levels of output growth, the technology gap and human capital differences across EU regions remains the same as evident in the recent past, then the long-run distribution is a more concentrated one than "currently" exists (Figure 13.5), although it remains the case that in the long run equilibrium will be characterized by differences in the level of productivity. There is evidently no absolute converge on the distant horizon. The picture provided by the stochastic kernel approach gives more detail on the various nuances of the difference between the two distributions. We have chosen to simulate the separate impact of our variables by altering the coefficient so that the variable has a different to effect on productivity growth to that of the recent past. By setting the coefficient to zero, as in the case of returns to scale and human capital, we have in effect simulated the outcomes that could occur if by some magic wand (or as a result of enhanced economic integration so that many regions were in fact one) the variable assumes the same values in every region.

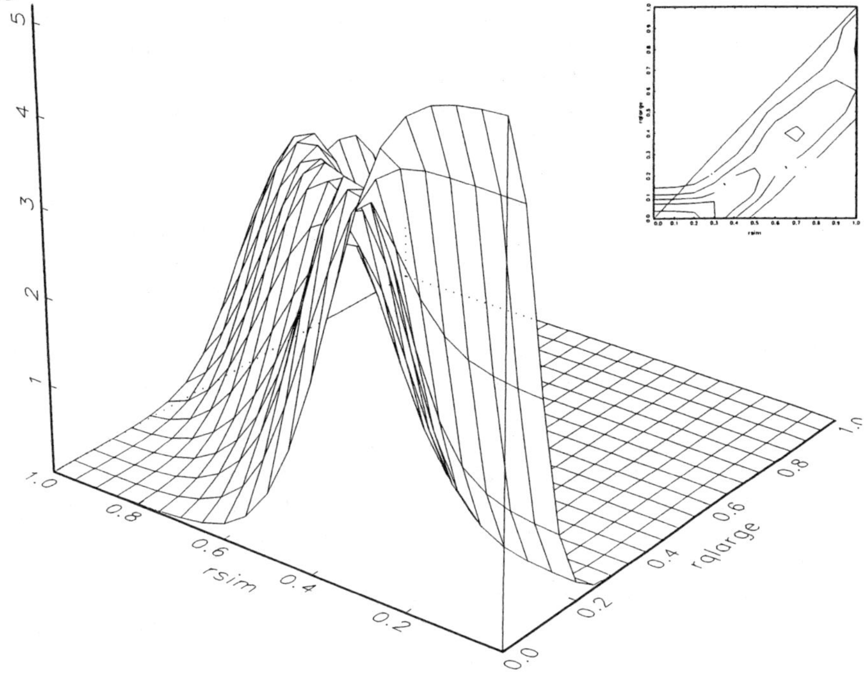

Fig. 13.11. Stochastic kernel: large RTS conditioning

The effect of assuming that returns to scale are constant across regions is to enhance the comparative status of the poorer regions, although again the impact is a subtly differentiated one as is apparent from Figure 13.6. Likewise equalized human capital brings up the lower level regions relative to the higher ones, thus pointing to the importance of this factor for regional development. The estimates given in Table 13.1 also highlight the significance for growth over the historical period of the initial level of technology, with low technology level regions with a large technology gap growing faster than regions already at a high level, since diffusing innovations in the latter case will have little or no impact for regions already at or near the technology frontier. However, unlike human capital and output growth, technology gap equalization across regions has the effect of destroying the tendency for regions to converge to a steady state. This may be a possibility, but the empirical evidence is to the contrary and so we discount it. As an alternative, we therefore set up scenarios in which different amounts of catching up are envisaged (Figures 13.8 and 13.9), and this facilitates illustration via the stochastic kernel approach. Likewise in Figure 13.10 we do the same for different amounts of increasing returns to scale. This could occur as a result of increased congestion, since as assumed in equation (13.13) the net effect could be a lowering of the coefficient on returns to scale. One variable which does not enter directly into our simulations is the impact of spatial external economies as represented by the endogenous lag in our model, which Table 13.1 shows has been a significant factor affecting productivity growth and is therefore necessary to avoid biased estimates. Nevertheless, it is apparent from the equilibrium equation derived in Section 13.4 that the steady state, to which regions are tending, is the same regardless of the strength of these spillovers, although the paths leading to steady state do depend on these spatial interactions.

There are other simulations that are possible that would be based on altering the assumed levels of the causal variables in the different regions rather than altering the coefficients. For example, the convergence of the EU regional economies makes it reasonable to envisage a situation in which variance of manufacturing output growth rates across regions is smaller than it has been in the recent past, due to the lowering of barriers to interregional trade and the exposure of industry to similar demand and supply conditions. While this might not produce the exact equality of output growth, which we might anticipate if equilibrium output growth were itself a function of productivity growth, that we have in effect simulated in this Chapter, there is scope for looking at different growth assumptions other than exact equality or the assumption that current output growth differences will prevail ad infinitum. Similarly, while ideally equalized educational attainment across regions as we have supposed would be of maximum benefit for regional economic and social cohesion, it is likely that this will remain a goal rather than be a real outcome. It would however be quite feasible to explore the ramifications of different assumptions about educational attainment disparities across Europe and show the consequences for relative productivity levels.

Appendix: The Urban Economics Model Underlying the Reduced Form

Endogenous variables

1. Manufacturing labour (workers):

$$M = N\beta \tag{13.A1}$$

   Manufacturing labour (workers) (M) equals total labour (N) times β which is the equilibrium allocation of labour to manufacturing under competitive conditions.

2. Manufacturing output:

$$Q = M^{\beta} I^{1-\beta} \tag{13.A2}$$

This is a Cobb-Douglas production function. Output (Q) equals workers (M) raised to the power β, multiplied by the level of composite services (I) to the power ($1-\beta$).

3. Composite services:

$$I = \left[\int_{d=1}^{d=D} i(d)^{1/\mu} dt \right]^{\mu} \tag{13.A3}$$

$$I = [Di(d)^{1/\mu}]^{\mu} = D^{\mu} i(d) \tag{13.A4}$$

This is the CES (constant elasticity of substitution) (sub) production function for I, which is a function of the output of the typical services firm ($i(d)$), the number of services firms(D) and the elasticity of substitution, which diminishes with increasing μ. As μ approaches 1, then the services level approaches the number of firms times their output, as $\mu >> 1$ it is more than this due to the effect of the number of varieties (D), so that increasing firms results in a proportionately larger I.

4. Equilibrium output level of typical service firm:

$$i(d) = \frac{s}{a(\mu - 1)} \tag{13.A5}$$

When firms are at equilibrium, so that (marginal) costs equal (marginal) revenues and profits are driven to zero, the output per firm can be shown to equal the fixed labour requirement (s) divided by the marginal labour requirement (a) times μ-1.

5. Cost :

$$c = w(ai(d) + s) \tag{13.A6}$$

Cost of production equals wage rate (w) times amount of labour ($ai(d) + s$)
Marginal cost equals wage rate (w) times marginal labour requirement (a):

$$mc = wa \tag{13.A7}$$

6. Revenue equals wage rate (w) times marginal labour requirement (a) times markup on costs (μ) [$wa\mu = p =$ price] times equilibrium output ($i(d)$):

$$r = wa\mu i(d) \tag{13.A8}$$

Marginal revenue equals price ($p = wa\mu$) times (1-1/E) where E is the constant (subjective) price elasticity of demand [which can be shown to equal $1/(1-1/\mu)$], thus $(1-1/E) = 1/\mu$.:

$$mr = \frac{wa\mu}{\mu} = wa \tag{13.A9}$$

Hence $mr = p$ times $1/\mu = p/\mu$. Note, here we are talking about imperfect competition so that price is unequal to marginal revenue. In fact price (p) = wage rate (w) times marginal labour requirement (a) times markup (μ)

$$p = wa\mu \tag{13.A10}$$

.If $\mu = 1$ we have perfect competition so then $mr = p$

7. The number of service firms (varieties):

$$D = \frac{(1-\beta)N}{ai(d)+s} \tag{13.A11}$$

The number of firms (D) equals the total services labour force $(1-\beta)N$ divided by the labour force per firm ($L = ai(d)+s$) at equilibrium

8. Labour requirement:

$$L = s + ai(d) \tag{13.A12}$$

The labour requirement equal to fixed labour requirement(s) plus marginal labour requirement (a) times firm's output ($i(d)$)

Exogenous Variables

1. Marginal labour requirement (a): this is the exogenously determined increase in labour needed by the firm per unit increment of output (note that since output can be measured in any units, this can be left as 1).
2. Fixed labour requirement ($s > 0$): this is the fixed cost in terms of service labour that must be incurred to produce any variety. It implies that increasing returns to scale exist in the service sector.
3. Monopoly power/elasticity of substitution (μ): as μ increases, the elasticity of substitution diminishes, as μ approaches 1, the services approach being perfect substitutes and variety diminishes in importance as a determinant of I. Note that the elasticity of substitution is

$$\frac{\mu}{\mu-1} \tag{13.A13}$$

4. Total labour force (N): note how total manufacturing output (Q) is a nonlinear function of N, showing increasing returns with city size. However the latter is not modeled here and we treat N as exogenously determined.
5. The relative importance of workers versus services (β)

Equilibrium

Occurs when the level of output is such that marginal revenue(mr) equals marginal cost (mc), firms have entered shifting the demand curve to the left, driving down profits to zero, at which point entry stops. This is the equilibrium, when total revenue equals total costs and there are zero profits. This determines the equilibrium output level *i(d)*.

Hence, at equilibrium, profits are zero and costs equal revenues,

$$c = w(ai(d) + s) = r = wa\mu i(d) \qquad (13.A14)$$

Hence,

$$ai(d) + s = a\mu i(d) \qquad (13.A15)$$

$$i(d) = \frac{s}{a(\mu - 1)}$$

We can choose units of output to be anything we want, which means we can choose them so that the marginal labour requirement $a = 1$. This gives the simplified version

$$i(d) = \frac{s}{\mu - 1} \qquad (13.A16)$$

References

Abdel-Rahman H, Fujita M (1990) Product variety, Marshallian externalities, and city sizes. *Journal of Regional Science,* 30: 165-183

Barro R (1991) Economic growth in a cross section of countries. *Quarterly Journal of Economics,* 106: 407-443

Barro R, Sala-i-Martin X (1992) Convergence. *Journal of Political Economy,* 100: 223-251

Bianchi M (1997) Testing for convergence: Evidence from non-parametric multimodality tests. *Journal of Applied Econometrics,* 12: 393-409

Cameron G (1996) *Innovation and Economic Growth.* DPhil thesis, Ch 2, University of Oxford

Ciccone A, Hall RE (1996) Productivity and the density of economic activity. *American Economic Review,* 86: 54-70

Cheshire PC, Hay DG (1989) *Urban problems in Western Europe: an Economic Analysis.* Unwin Hyman, London

Durlauf SN, Quah D (1999) The New Empirics of Economic Growth. In: Taylor J, Woodford, M (eds) *Handbook of Macroeconomics.* Elsevier Science, North-Holland, New York and Oxford, pp 235-308

Fingleton B (1997) Specification and testing of Markov chain models: An application to convergence in the European Union. *Oxford Bulletin of Economics and Statistics,* 59: 385-403

Fingleton B (2000) Spatial econometrics, economic geography, dynamics and equilibrium : a third way? *Environment & Planning A*, 32: 1481-1498

Fingleton B (2001a) Theoretical economic geography and spatial econometrics: dynamic perspectives *Journal of Economic Geography*, 1: 201-225

Fingleton B (2001b) Equilibrium and economic growth: spatial econometric models and simulations. *Journal of Regional Science*, 41: 117-148

Fujita M, Krugman P, Venables AJ (1999) *The Spatial Economy: Cities, Regions, and International Trade.* MIT Press, Cambridge and London

Fujita M, Thisse JF (1996) Economics of agglomeration. *Journal of the Japanese and International Economies,* 10: 339-78

Glaeser EL (1999) Learning in cities. *Journal of Urban Economics,* 46: 254-277

Gordon I, McCann P (2000) *Industrial clusters: Complexes, agglomeration and/or social networks?* Urban Studies, 37: 513-32

Jacobs J (1969) *The Economy of Cities.* Random House, New York

Johnson, PA (2000) A nonparametric analysis of income convergence across the US States. *Economics Letters,* 69: 219-223

Johnston J (1984) *Econometric Methods.* McGraw Hill, New York

Kennedy P (1992) *A Guide to Econometrics.* Blackwell, Oxford

Koutsoyiannis A (1977) *Theory of Econometrics.* Macmillan, London

Krugman P (1991) *Geography and Trade.* Leuven University Press, Leuven

Lamo A (2000) On convergence empirics: some evidence for Spanish regions. *Investigaciones Economicas* 24: 681-707

Leser C (1966) *Econometric Techniques and Problems.* Griffin, London

López-Bazo E, Vayá E, Mora AJ, Suriñach J (1999) Regional economic dynamics and convergence In The European Union. *The Annals of Regional Science*, 33: 343-370

Maddala G S (2001) *An Introduction to Econometrics.* Wiley New York

Magrini S (1999) The evolution of income disparities among the regions of the European Union. *Regional Science and Urban Economics*, 29: 257-81

Mankiw NG, Romer D, Weil DN (1992) A contribution to the empirics of economic growth. *Quarterly Journal of Economics*, 107: 407-437

Puga D (1999) The rise and fall of regional inequalities. *European Economic Review*, 43: 303-34

Quah D (1993) Empirical cross-section dynamics in economic growth. *European Economic Review*, 37: 426-434

Quah D (1996a) Convergence empirics across economies with (some) capital mobility. *Journal of Economic Growth,* 1: 95-124

Quah D (1996b) Regional convergence clusters across Europe. *European Economic Review,* 40: 951-958

Quigley J (1998) Urban diversity and economic growth. *Journal of Economic Perspectives,* 12: 127-138

Rivera-Batiz F (1988) Increasing returns, monopolistic competition, and agglomeration economies in consumption and production. *Regional Science and Urban Economics,* 18: 125-153

Silverman BW (1986) *Density Estimation for Statistics and Data Analysis.* Chapman and Hall, New York

Stokey N, Lucas R E Jr (1989) *Recursive Methods in Economic Dynamics.* Harvard University Press, Cambridge, Mass. and London

Author Index

Subject Index

Figures

Tables

Contributors

Harald Badinger	Research Institute for European Affairs, Wirtschaftsuniversitaet Wien, Althanstrasse 39-45, A-1090 Wien, Austria Harald.Badinger@wu-wien.ac.at
Catherine Baumont	Université de Bourgogne, Laboratoire d'Analyse et de Techniques Economiques (LATEC - UMR CNRS 5118) Pôle d'Economie et de Gestion B.P. 26611 21066 Dijon Cedex catherine.baumont@u-bourgogne.fr
Roger Bivand	Norwegian School of Economics and Business Administration, Breiviksveien 40, N-5045 Bergen, Norway Roger.Bivand@nhh.no
Rolf Brunstad	Norwegian School of Economics and Business Administration, Breiviksveien 40, N-5045 Bergen, Norway rolf.brunstad@nhh.no
Marjolein Caniëls	Universitair Docent Economie, Bestuurskunde en Bedrijfswetenschappen (EBB) Open Universiteit Nederland Postbus 2960 6401 DL Marjolein.Caniels@ou.nl
Ådne Cappelen	Research Department, Statistics Norway, Postboks 8131 Dep, N-0033 Oslo aadne.cappelen@ssb.no
Fulvio Castellacci	Centre for Technology, Innovation and Culture, PO box 1108 Blindern N-0317 Oslo Norway fulvio.castellacci@tik.uio.no

Cem Ertur	Université de Bourgogne, Laboratoire d'Analyse et de Techniques Economiques (LATEC - UMR CNRS 5118) Pôle d'Economie et de Gestion B.P. 26611 21066 Dijon Cedex cem.ertur@u-bourgogne.fr
Jan Fagerberg	Centre for Technology, Innovation and Culture, PO box 1108 Blindern N-0317 Oslo Norway jan.fagerberg@tik.uio.no
Bernard Fingleton	Department of Land Economy, University of Cambridge, 19 Silver Street Cambridge CB3 9EP UK bf100@cam.ac.uk
Julie Le Gallo	Université de Bourgogne, Laboratoire d'Analyse et de Techniques Economiques (LATEC - UMR CNRS 5118) Pôle d'Economie et de Gestion B.P. 26611 21066 Dijon Cedex JLeGallo@aol.com
Lydia Greunz	Université Libre de Bruxelles, DULBEA CP-140 Av. F.D. Roosevelt, 50 B-1050 Bruxelles Belgique lgreunz@ulb.ac.be
Danilo Camargo Igliori	Department of Land Economy, University of Cambridge, 19 Silver Street Cambridge CB3 9EP UK dci21@cam.ac.uk
Enrique López-Bazo	Regional Quantitative Analysis Research Group (AQR), Department of Econometrics, Statistics and the Spanish Economy, Universitat de Barcelona Av. Diagonal, 690, 08034 Barcelona, Spain elopez@eco.ub.es
Barry Moore	Department of Land Economy, University of Cambridge, 19 Silver Street Cambridge CB3 9EP UK bcm1@cam.ac.uk
Raffaele Paci	CRENoS, Facoltà di Scienze Politiche, via Sant'Ignazio 78, 09123 Cagliari, Italy paci@unica.it

Jordi Suriñach	Regional Quantitative Analysis Research Group (AQR), Department of Econometrics, Statistics and the Spanish Economy, Universitat de Barcelona Av. Diagonal, 690, 08034 Barcelona, Spain surinach@eco.ub.es
Gabriele Tondl	Research Institute for European Affairs, Wirtschaftsuniversitaet Wien Althanstrasse 39-45, A-1090 Wien, Austria gabriele.tondl@wu-wien.ac.at
Stefano Usai	CRENoS and Faculty of Political Science, University of Sassari, Italy, DEIS, Viale Margherita di Savoia, 15 I-07100 Sassari, Italy usai@uniss.it
Esther Vayá	Regional Quantitative Analysis Research Group (AQR), Department of Econometrics, Statistics and the Spanish Economy, Universitat de Barcelona Av. Diagonal, 690, 08034 Barcelona, Spain evaya@eco.ub.es
Bart Verspagen	Eindhoven Center for Innovation Studies (ECIS), Eindhoven University of Technology, PO Box 513 (DG 1.21), 5600 MB Eindhoven The Netherlands b.verspagen@tm.tue.nl

Titles in the Series (continued)

M. M. Fischer and *J. Fröhlich* (Eds.)
Knowledge, Complexity and Innovation Systems
XII, 477 pages. 2001. ISBN 3-540-41969-1

M. M. Fischer, J. Revilla Diez and *F. Snickars*
Metropolitan Innovation Systems
VIII, 270 pages. 2001. ISBN 3-540-41967-5

L. Lundqvist and *L.-G. Mattsson* (Eds.)
National Transport Models
VIII, 202 pages. 2002. ISBN 3-540-42426-1

J. R. Cuadrado-Roura and *M. Parellada* (Eds.)
Regional Convergence in the European Union
VIII, 368 pages. 2002. ISBN 3-540-43242-6

G. J. D. Hewings, M. Sonis and *D. Boyce* (Eds.)
Trade, Networks and Hierarchies
XI, 467 pages. 2002. ISBN 3-540-43087-3

G. Atalik and *M. M. Fischer* (Eds.)
Regional Development Reconsidered
X, 220 pages. 2002. ISBN 3-540-43610-3

Z. J. Acs, H. L. F. de Groot and *P. Nijkamp* (Eds.)
The Emergence of the Knowledge Economy
VII, 388 pages. 2002. ISBN 3-540-43722-3

R. J. Stimson, R. R. Stough and *B. H. Roberts*
Regional Economic Development
X, 397 pages. 2002. ISBN 3-540-43731-2

S. Geertman and *J. Stillwell* (Eds.)
Planning Support Systems in Practice
XI, 578 pages. 2003. ISBN 3-540-43719-3

Advances in Spatial Science

G. Clarke, M. Madden (Eds.)

Regional Science in Business

2001. VIII, 363 pp. 79 figs., 61 tabs. Hardcover **€ 74,95;**
£ 52,50; sFr 124,- ISBN 3-540-41780-X

This book provides a new collection of papers on applied regional science. It takes each of the main methodologies of the discipline and provides a case study of their use for policy making. These case studies show how the discipline has matured and how it is now capable of providing answers to a wide range of what-if scenarios.

B. Johansson, C. Karlsson, R.R. Stough (Eds.)

Theories of Endogenous Regional Growth

Lessons for Regional Policies

2001. IX, 428 pp. 85 figs., 48 tabs. Hardcover **€ 79,95;**
£ 56,-; sFr 132,50 ISBN 3-540-67988-X

The contributions in the book develop these these advances into a theoretical framework for endogenous regional economic growth and explain the implications for regional economic policies in the perspective of the new century.

M.M. Fischer, J. Fröhlich (Eds.)

Knowledge, Complexity and Innovation Systems

2001. XII, 477 pp. 44 figs., 68 tabs. Hardcover **€ 84,95;**
£ 59,50; sFr 141,- ISBN 3-540-41969-1

The book addresses the relationship between knowledge, complexity and innovation systems. In particular, it offers insights about knowledge creation and spillovers, innovation and learning systems, innovation diffusion processes and innovation policies.

M.M. Fischer, J. Revilla Diez, F. Snickars

Innovation Systems

Theory and Evidence from Three Metropolitan Regions in Europe

2001. VIII, 270 pp. 40 figs., 159 tabs. Hardcover **€ 74,95;**
£ 52,50; sFr 124,- ISBN 3-540-41967-5

This book makes an attempt to provide evidence of the innovation processes across metropolitan regions studies based on the use of a common more rigid methodology. The comparison is based on three metropolitan regions such as Barcelona, Vienna and Stockholm. These regions are sufficiently different in terms of history, culture and economy. The book offers a compact review of current conceptual and theoretical developments and valuable insights deriving from a cross-national interregional comparative empirical study.

M.M. Fischer, Y. Leung (Eds.)

GeoComputational Modelling

Techniques and Applications

2001. XII, 275 pp. 72 figs., 44 tabs. Hardcover **€ 74,95;**
£ 52,50; sFr 124,- ISBN 3-540-41968-3

Please order from
Springer · Customer Service
Haberstr. 7 · 69126 Heidelberg, Germany
Tel.: +49(0)6221-345-0 · Fax: +49(0)6221-345-4229
e-mail: orders@springer.de
or through your bookseller

All prices are net-prices subject to local VAT, e.g. in Germany 7% VAT for books.
Prices and other details are subject to change without notice. d&p · BA 43087/1

Springer

Advances in Spatial Science

Furthermore titles...

L. Lundqvist, L.-G. Mattsson, Royal Institute of Technology, Stockholm, Sweden (Eds.)

National Transport Models

Recent Developments and Prospects

The increase of interregional and international mobility requires forecasting tools that go beyond the urban or regional the appropriate spatial level. This volume presents the state of the art and prospects of a sample of the most advanced national and European transport models within a comparative framework.

2002. VIII, 202 pp. 35 figs., 23 tabs. Hardcover **€ 64,95**; £ 45,50; sFr 107,50 ISBN 3-540-42426-1

J.R. Cuadrado-Roura, University of Alcala, Madrid, Spain; **M. Parellada,** University of Barcelona, Spain (Eds.)

Regional Convergence in the European Union

Facts, Prospects and Policies

This book explores these important questions and examines the relationship between economic integration, regional convergence, and divergence. Central and peripheral regions are analysed as well as those in a re-conversion economic process. The authors address the question of whether regions tend to converge or not from different points of view. No doubt, this is an excellent book for anybody interested in learning more about regional trends and prospects in an increasingly integrated world.

2002. VIII, 368 pp. 61 figs., 57 tabs. Hardcover **€ 69,95**; £ 49,-; sFr 116,- ISBN 3-540-43242-6

Please order from
Springer · Customer Service
Haberstr. 7 · 69126 Heidelberg, Germany
Tel.: +49 (0) 6221 - 345 - 217/8
Fax: +49 (0) 6221 - 345 - 229
e-mail: orders@springer.de

All prices are net-prices subject to local VAT, e.g. in Germany 7% VAT for books.
Prices and other details are subject to change without notice. d&p · BA 43722/2

G.J.D. Hewings, University of Illinois, Urbana, IL, USA; **M. Sonis,** Bar Ilan University, Ramat-Gan, Israel; **D. Boyce,** University of Illinois at Chicago, Chicago, IL, USA (Eds.)

Trade, Networks and Hierarchies

Modeling Regional and Interregional Economies

The book provides an overview of some of the recent techniques that have been applied to an understanding of the structure of regional and interregional exchange within national economies. The issues range from an evaluation of NAFTA, comparisons of regional economies, structral change over time and issues related to measurement and interpretation.

2002. XI, 467 pp. 100 figs., 111 tabs. Hardcover **€ 94,95**; £ 66,50; sFr 157,50 ISBN 3-540-43087-3

G. Atalik, Istanbul Technical University, Istanbul, Turkey; **M.M. Fischer,** University of Vienna, Austria (Eds.)

Regional Development Reconsidered

The book integrates research findings from a broad area including: economics, transportation, geography and regional science. The contributions provide an excellent coverage of current conceptual and theoretical developments, and valuable insights from both empirical and conceptual work. Written for graduate students of economic and social sciences; broad research coverage makes it invaluable reading for researchers and professionals in the subject area.

2002. X, 220 pp. 35 figs., 32 tabs. Hardcover **€ 69,95**; £ 49,-; sFr 116,- ISBN 3-540-43610-3

Z.J. Acs, University of Baltimore, MD, USA; **H.L.F. de Groot, P. Nijkamp,** Free University, Amsterdam, The Netherlands (Eds.)

The Emergence of the Knowledge Economy

A Regional Perspective

This volume first deals with the theoretical background for understanding the knowledge economy, with knowledge spillovers and development externalities. It then examines aspects of the relationship between knowledge inputs and innovative outputs in the information, computer and telecommunications sector (ICT) of the economy at the regional level. Case studies focusing on a wide variety of sectors, countries and regions finally illustrate important regional innovation issues.

2002. VIII, 388 pp. 52 figs., 86 tabs. Hardcover **€ 79,95**; £ 56,-; sFr 132,50 ISBN 3-540-43722-3

Springer

L. Schätzl, J. Revilla Diez (Eds.)

Technological Change and Regional Development in Europe

This volume presents the main findings of the research programme Technological Change and Regional Development in Europe, backed by the German Research Council. It examines the relationship between technological change and regional development by means of empirical surveys. The empirical surveys focused on Europe and covered regions in Germany, Great Britain, Netherlands, Spain, Slovenia, and Sweden.

2002. IX, 421 pp. 88 figs., 81 tabs. (Contributions to Economics)
Softcover **€ 60,95**; £ 42,50; sFr 101,- ISBN 3-7908-1460-1

L. Hedegaard, B. Lindström (Eds.)

The NEBI YEARBOOK 2001/2002

North European and Baltic Sea Integration

The NEBI Yearbook 2001/2002 provides a balanced picture of integrationist developments in the North European and Baltic Sea area. The special focus of this volume is the impact of EU enlargement on developments in the North European and Baltic Sea area.

2002. XIII, 486 pp. (JG 2001/2002) Hardcover **€ 84,95**; £ 59,50;
sFr 141,- ISBN 3-540-43004-0

N. Konsolas, A. Papadaskalopoulos, I. Plaskovitis

Regional Development in Greece

This unique text has as its main themes the development of Greek regions and the impact of the structural policies of the European Union on their progress and prosperity. Emphasis is given to the geographic and socio-economic characteristics of the 13 regions of Greece and the objectives of the Community Support Framework for the period 2000 - 2006.

2002. XIV, 216 pp. 76 figs., 35 tabs. Hardcover **€ 74,95**; £ 52,50;
sFr 124,- ISBN 3-540-42395-8

P. Meusburger, H. Jöns (Eds.)

Transformations in Hungary

Essays in Economy and Society

The book analyses related political, legal, institutional and socio-economic structures and processes in order to contribute to a further understanding of Hungary's ongoing transformation processes and its current situation as one of the leading candidates for EU membership. Drawing together comprehensive empirical data and a geat variety of viewpoints, the book offers innovative examples of the application of different theoretical approaches to transformation studies and studies of economy and society in general.

2001. XIV, 382 pp. 97 figs., 92 tabs. (Contributions to Economics)
Softcover **€ 56,95**; £ 40,-; sFr 94,50 ISBN 3-7908-1412-1

Please order from
Springer · Customer Service
Haberstr. 7
69126 Heidelberg, Germany
Tel.: +49 (0) 6221 - 345 - 0
Fax: +49 (0) 6221 - 345 - 4229
e-mail: orders@springer.de
or through your bookseller

Visit our homepage:
http://www.springer.de/economics

Springer

All prices are net-prices subject to local VAT, e.g. in Germany 7% VAT for books.
Prices and other details are subject to change without notice. d&p · BA 00366/2

Journal of Geographical Systems

Geographical Information, Analysis, Theory, and Decision

Editors:
M.M. Fischer, University of Economics and Business Administration, Vienna, Austria
A. Getis, San Diego State University, San Diego, CA, USA

Editorial Board:
L. Anselin, R.G.V. Baker, R.S. Bivand, B. Boots, P.A. Burrough, A.U. Frank, M.F. Goodchild, G. Haag, R.P. Haining, K.E. Haynes, W. Kuhn, Y. Leung, P. Longley, B. Macmillan, P. Nijkamp, A. Okabe, S. Openshaw, Y.Y. Papageorgiou, D. Pumain, A. Reggiani, G. Rushton, F. Snickars, V. Tikunov, D. Unwin, G.G. Wilkinson

Please order from
Springer · Customer Service
Haberstr. 7 · 69126 Heidelberg, Germany
Tel: +49 6221 345-0
Fax: +49 6221 345-4229
e-mail: subscriptions@springer.de
or through your bookseller

Plus carriage charges. Price subject to change without notice.
In EU countries the local VAT is effective. d&p · BA 00366/1

The Journal of Geographical Systems, a journal dedicated to geographical information, analysis, theory, and decision, aims to encourage and promote high-quality scholarship on important theoretical and practical issues in regional science, geography, the environmental sciences, and planning. One of the distinctive features of the journal is its concern for the interface between mathematical modelling, the geographical information sciences, and regional issues. An important goal of the journal is to encourage interdisciplinary communication and research, especially when spatial analysis, spatial theory and spatial decision systems are the themes. In particular, the journal seeks to promote interaction between the theorists and users of the geographical information sciences and practitioners in the fields of regional science, geography, and planning.

Subscription information 2003:
Volume 5, 4 issues
€ 210,-
ISSN 1435-5930 (print)
ISSN 1435-5949 (electronic)
Title No. 10109

Springer